# Educational Communications and Technology Yearbook

**Series Editor**

Will W. K. Ma, Technological and Higher Education Institute of Hong Kong, Hong Kong, Hong Kong

The Hong Kong Association for Educational Communications and Technology (HKAECT) was established in 1989. Its first conference was organized in 1990, addressing "The Role of Educational Communications and Technology in Year 2000," with speakers coming from the United States, China, and Taiwan to discuss the outlook on educational communication and technology. Throughout these years, the HKAECT has held a number of international conferences, symposiums, workshops, and talks with various themes to provide a platform to enable rich exchanges for academicians, practitioners, and professionals in the fields of communication and education to discourse about the shaping and changing issues on education, communication, and technology. This Yearbook series collect presentations from the annual international conferences held by the HKAECT. Chapters would come from the annual global call for submission, and be selected based on blind review from international review board. Subject areas include but not limited to communication, new media, news media, broadcast journalism, democracy and the media, entertainment and education, learning analytics, AI in education, game-based learning, ubiquitous learning, MOOCs, open education, instructional design, social context and learning environment, social media, risk and ethics in new media, etc.

More information about this series at http://www.springer.com/series/16077

Will W. K. Ma · Kar-wai Tong ·
Wing Bo Anna Tso
Editors

# Learning Environment and Design

## Current and Future Impacts

Springer

*Editors*
Will W. K. Ma
Hong Kong Association for Educational Communications and Technology (HKAECT)
Hong Kong, Hong Kong

Kar-wai Tong
Hong Kong Association for Educational Communications and Technology (HKAECT)
Hong Kong, Hong Kong

Wing Bo Anna Tso
Hong Kong Association for Educational Communications and Technology (HKAECT)
Hong Kong, Hong Kong

ISSN 2524-4078 ISSN 2524-4086 (electronic)
Educational Communications and Technology Yearbook
ISBN 978-981-15-8166-3 ISBN 978-981-15-8167-0 (eBook)
https://doi.org/10.1007/978-981-15-8167-0



This Springer imprint is published by the registered company Springer Nature Singapore Pte Ltd.
The registered company address is: 152 Beach Road, #21-01/04 Gateway East, Singapore 189721, Singapore

# HKAECT

The Hong Kong Association for Educational Communications and Technology (HKAECT; http://www.hkaect.org/) was established in 1989. It organized its first conference in 1990 with the theme "The Role of Educational Communications and Technology in Year 2000" and invited speakers from the United States, China, and Taiwan to address and discuss the outlook on educational communications and technology. Throughout the years, the HKAECT has held a number of international conferences, symposiums, workshops, and talks with various themes to provide a rich exchange platform for academics, practitioners, and professionals in the fields to discourse on the shaping and changing issues on education, communications, and technology.

## *List of HKAECT Presidents*

| | |
|---|---|
| 1989–2010 | Leo P. K. Yam |
| 2010–2016 | Allan H. K. Yuen |
| Current President | Will W. K. Ma |

## *Themes of the HKAECT Conferences since 1990*

| | |
|---|---|
| 1990 | The Role of Educational Communications and Educational Technology in Year 2000 |
| 1992 | Instructional Technology: Design, Utilization and Evaluation |
| 1994 | Telecommunications in Education |
| 1996 | Innovations and Quality in Teaching and Learning |
| 1997 | Tertiary Teaching in the Use of Technology: Vision and Practice |
| 1998 | New Challenges and Innovations in Teaching and Training into the twenty-first Century |

| | |
|---|---|
| 2001 | Education Reform: Integrating Information Technology, Communication, and Curriculum |
| 2004 | Media Innovations in Education: Input and Outcome in New Society |
| 2007 | Educational Communications and Technology as Learning Experiences |
| 2010 | Multiliteracies for the 21st Century: Education, Communication, and Technology |
| 2014 | New Media, Knowledge Practices, and Multiliteracies |
| 2017 | New Ecology for Education: Communication X Learning |
| 2018 | New Media for Educational Change |
| 2019 | Shaping the Future of Education, Communication and Technology |
| 2020 | Learning Environment and Design: Current and Future Impact |

# HKAECT 2020 Organization

## International Advisory Board

Hsin-Tzu (Tommy) Chen, TAECT/Chinese Culture University
Michael M. Grant, AECT
Phillip Harris, AECT
Brad Hokanson, AECT
Siu Cheung Kong, The Education University of Hong Kong
Eugene Kowch, AECT/University of Calgary
Feng-Qi Lai, Indiana State University
Alice Y. L. Lee, Hong Kong Baptist University
In-sook Lee, KSET/Sejong University
Kedong Li, South China Normal University
Lin Lin, SICET/University of North Texas
Juhong Liu, SICET/James Madison University
Meifeng Liu, Beijing Normal University
Trey Martindale, AECT/Mississippi State University
Kay A. Persichitte, AECT
Michael J. Spector, AECT
Chiu-Kay Tang, Chu Hai College of Higher Education
Timothy Teo, University of Macau
Allan H. K. Yuen, The University of Hong Kong
Jianhua Zhao, Southern University of Science and Technology

## *Organizing Committee*

### Conference Chair

Will W. K. Ma, HKAECT/Technological and Higher Education Institute of Hong Kong (THEi), Hong Kong SAR

### Program Co-Chairs

Kar-wai Tong, HKAECT/City University of Hong Kong, Hong Kong SAR
Anna Tso, HKAECT/Hang Seng University of Hong Kong, Hong Kong SAR

### Members

Shurui Bai, The University of Hong Kong, Hong Kong SAR
Wendy Chan, Chu Hai College of Higher Education, Hong Kong SAR
Cat Miaoting Cheng, The University of Hong Kong, Hong Kong SAR
Chenggui Duan, Chu Hai College of Higher Education, Hong Kong SAR
Heidi Fung, HKAECT, Hong Kong SAR
Timothy K. F. Hew, The University of Hong Kong, Hong Kong SAR
Rose Fong, QAHE/Northumbria University London, UK
Cora Junru Yang, Harbour School, Hong Kong SAR
Carol Yiu, Chu Hai College of Higher Education, Hong Kong SAR

## *International Program Committee*

Haruna Abdulkadir, Hassan Usman Katsina Polytechnic, Nigeria
Beth Teagarden Bair, University of Miami
Rik Bair, University of Miami
Chi-Keung (Alex) Chan, Hong Kong Shue Yan University
Shih Chien Chang, Chu Hai College of Higher Education
Yang-Hsueh Chen, National Chengchi University
Michele Della-Ventura, Music Academy "Studio Musica"
Liping Deng, Hong Kong Baptist University
Angeline Guo, Macquarie University
Tracy Lee, Chu Hai College of Higher Education
Jiutong Luo, The University of Hong Kong
Scott Mcleod, University of Colorado Denver

Steven K. K. Ng, University of Sunderland
Shaun Nykvist, Norwegian University of Science and Technology
Aynur Sarisakaloglu, Turkish-German University
Hing Wa (Helena) Sit, The University of Newcastle (Australia)
Ying Tang, The University of Hong Kong
Jessica To, The University of Hong Kong
Chi Hung Wong, The Open University of Hong Kong
Meimei Xu, The University of Georgia
Eunice Pui-yu Yim, The Open University of Hong Kong

## *Organizers*

Hong Kong Association for Educational, Communications & Technology (HKAECT)

The University of Hong Kong Libraries

# Preface

The Hong Kong Association for Educational Communications and Technology (HKAECT) has been organizing for decades international conferences in Hong Kong to provide a platform for academic and professional exchanges. In the past few years, scholars, academics, and practitioners coming from different corners of the world joined the conferences. They shared empirical research results, novel applications, and case studies in the fields of education, communication, and technology. In due course, the HKAECT selected appropriate manuscripts and published them in independent volumes of the *Educational Communications and Technology Yearbook Series*, published by Springer. In September 2019, the HKAECT sent out a usual global call for papers and presentations on a theme of *Learning Environment and Design: Current and Future Impact* for the 2020 conference scheduled originally for June 2020. The subsequent development illustrated that the global call was not as usual as those in the previous years.

There was a social unrest in Hong Kong in 2019, spreading over local university campuses and influencing the normal teaching and learning. Seeing the developments, the HKAECT had no choice but made a difficult decision in November 2019 to cancel the conference. In lieu of a conference proceeding, the HKAECT determined to publish a special edition for the *Educational Communications and Technology Yearbook Series* ("Special Edition" hereafter). When the HKAECT made this decision, they never imagined that another global challenge had been on its way. Coronavirus (COVID-19) has given a hard hit to the whole world since January 2020. Many countries implemented travel bans and people commenced to work from home. Online teaching has become a new norm at universities. At the time of writing (late October 2020), the worldwide number of confirmed infected cases has been more than 40 million and over a million people died, as reported by the World Health Organization.

Under such a context, we were appointed co-editors of the Special Edition. It is by grace that the responses to the original global call for papers were overwhelming that over 35 submissions were received from territories across continents, including Australia, Ireland, Italy, Malaysia, New Zealand, Norway, Singapore, Taiwan, Turkey, the United States, Mainland China, and Hong Kong. Owing to the space constraint, we could only accept around half of these submissions after a series of blind peer reviews.

The Special Edition bears a title *Learning Environment and Design: Current and Future Impact*. We appreciate the tremendous changes of the environment in which individuals used to teach and learn in the past several decades. New media and technology have infiltrated and shaped the learning environment from mere physical spaces into multifaceted possibilities. This gives rise to several important issues, such as designs of learning environments to harness technology appropriately to engage learners better and the roles of learners and educators that play in this changing learning environment. This Special Edition provides a timely forum to link local and international academics, researchers, educators, and practitioners in the global village to share their theoretical and practical insights. We hope that it catalyzes the successful exchanges from a wide perspective, with inspirational insights into how emerging technologies can be adapted in the fields of education and communication to facilitate the current and future designs of learning environments to improve learners' performances.

We are pleased to present to you this Special Edition of the *Educational Communications and Technology Yearbook Series*. On behalf of the HKAECT, we take the opportunity to express our deepest gratitude to all the authors who submitted their manuscripts to us, in particular, those chapter contributors who stayed working hard for the accomplishment of this book project. Our heartfelt thanks and appreciation also go to all the reviewers whose excellent works and advice have made this monograph a success. Finally, yet importantly, we owe the publisher, Springer, a debt of thanks for their incessant support to the HKAECT throughout these years!

Hong Kong, Hong Kong
July 2020

Will W. K. Ma, Ph.D.
Kar-wai Tong, JSD
Wing Bo Anna Tso, Ph.D.

# Contents

# Redefining Quality Education

# Effective Learning Through Deep Learning, What Matters: Self, Others, Way of Thinking, and/or Design of Learning Environment?

**Will W. K. Ma**

**Abstract** Deep learning is important for achieving deep knowledge, but traditional teaching often falls short of providing a deep learning environment for learners. This study reviews existing studies about deep learning to examine whether there is any convergence in the explanations and definitions of deep learning based on evidence-based research. Content analysis of the 299 articles reviewed resulted in a set of keywords that were in turn grouped into four main themes corresponding to four models of learning: *social*, *personal*, *information processing,* and *behavioral*. The findings of this study provide practical guidance to academics and practitioners to design for deep learning, with reference to the learner, to the peers, to the way of thinking, and to the design of the learning environment. In the past, research studies would focus on one specific perspective to find its relationship to deep learning. This study suggests that all the four perspectives could be considered together, whenever it is appropriate. This study also finds that deep learning is a choice: as an approach of learning. It suggests that it is not the learner who could not think deep, but whether the learning design could encourage learners to choose more the deep learning approach.

**Keywords** Deep learning · Social learning · Personal learning · Information processing · Behavioral systems

## 1 Introduction

The traditional role of educators is to deliver curriculum content. However, this is not the same as teaching deep knowledge that leads to intelligent performance. In the 1980s, cognitive scientists discovered that if learners learn deep knowledge rather than surface knowledge, their performance improves: they retain material better and are able to generalize and apply knowledge to a broader range of contexts (Sawyer,

W. W. K. Ma (✉)
Learning Commons & Digital Innovation, Technological and Higher Education Institute of Hong Kong (THEi), 133 Shing Tai Road, Chai Wan, Hong Kong SAR, China
e-mail: will.wkma@gmail.com



W. W. K. Ma et al. (eds.), *Learning Environment and Design*,
Educational Communications and Technology Yearbook,
https://doi.org/10.1007/978-981-15-8167-0_1

2014, pp. 4–5). In the twenty-first century, the application of knowledge—that is, learners' ability to use knowledge in real-world social and practical settings—has come to be identified as a key competence (World Economic Forum, 2018, p. 12).

A comparison of the deep learning of knowledge versus traditional classroom practices (instructionism) based on findings from cognitive science is best summarized by Sawyer (2014, p. 5) as follows:

- Deep learning requires learners to relate new ideas and concepts to previous knowledge and experience, whereas in traditional classroom practices, learners treat course material as being unrelated to what they already know.
- Deep learning requires learners to integrate their knowledge into interrelated conceptual systems, whereas in traditional classroom practices, learners treat course material as disconnected bits of knowledge.
- Deep learning requires learners to look for patterns and underlying principles, whereas in traditional classroom practices, learners memorize facts and carry out procedures without understanding how or why.
- Deep learning requires learners to evaluate new ideas and relate them to conclusions, whereas in traditional classroom practices, learners have difficulty making sense of new ideas that are different from what they encountered in the textbook.
- Deep learning requires learners to understand the process of dialogue through which knowledge is created and to examine the logic of an argument critically, whereas in traditional classroom practices, learners treat facts and procedures as static knowledge handed down from an all-knowing authority.
- Deep learning requires learners to reflect on their own understanding and their own process of learning, whereas in traditional classroom practices, learners memorize without reflecting on the purpose of learning or on their own learning strategies.

In this study, a review was conducted of previous studies on deep learning. Its purpose was to explore patterns in these studies, to summarize and categorize the findings from these studies to devise a roadmap for future work on deep learning, and ultimately, to deepen our understanding of what constitutes effective education. The research questions were as follows:

RQ1: What previous studies have been undertaken about deep learning?

RQ2: Are there any patterns in these deep learning studies?

The next section reports the methods used to select the studies included in this review, and the methods of analysis used to extract the major themes from the studies. We then summarize and discuss our findings.

## 2 Methodology

### *2.1 Data and Sample*

One way to understand previous work that has been done on deep learning is to review previous research studies on the topic. Important sources include empirical studies that review the literature, collect human subjects' opinions, or observe human subjects' behavior in relation to deep learning. The papers reporting these studies typically include a review of the literature to identify important constructs, a discussion of the research designs used, a summary of the methods used to collect empirical data, and an analysis of the data for results and conclusions. Each of these studies, therefore, provides us with a better understanding of deep learning. Analysis of these empirical studies can help answer our research questions.

## 2.2 *Data Collection*

The data for this study were empirical studies on deep learning that aimed to identify effective teaching and learning. These studies were found using the online database *Academic Search Complete*, which provides the full text of more than 5,900 journals in multiple disciplines, including the full text of more than 5,030 peer-reviewed titles. The content in this database goes back as far as 1887. A summary of the search process is listed in the table below (see Table 1).

The search was limited to the abstracts of the articles in the database. The keyword "deep learning" was used as the initial search string, and the search was limited to research articles in English published in scholarly journals, resulting in 4,276 articles from 1995 to 2019. However, a browse through the first few articles returned revealed that many of the articles were related to machine learning development and application, irrelevant to the scope of this study. Therefore, the search was narrowed down by adding the subject term, "education," resulting in 342 articles. A few more articles were filtered out when the subject term "machine learning" was excluded, resulting in 339 articles. The term was excluded because this study focused on exploring effective teaching and learning of learners, rather than machines. All 339 articles were read. Articles that were not empirical studies—for example, conceptual framework papers, editorials, and book reviews—were discarded. If an article was related to

**Table 1** Search process and results of deep learning studies from the e-database *Academic Search Complete*

| Steps | Keywords | Duration | Articles |
|---|---|---|---|
| 1 | Deep learning | 1968–2019 | 7,690 |
| 2 | "Deep Learning" | 1995–2019 | 4,276 |
| 3 | AND Subject term, "education" | 1998–2019 | 342 |
| 4 | NOT Subject term, "machine learning" | 1998–2019 | 339 |
| 5 | Remove irrelevant studies | 1998–2019 | 299[a] |

*N.B.* The scope of the search was limited to the abstract field. Limiters included *Scholarly (Peer Reviewed) Journals*; *Research Article* document type; and *English language*

[a]*Due to the length of the list of all 229 references, it is not included in this article but can be presented on request*

either machine learning or artificial intelligence, it was also discarded. Finally, 299 articles were included for further analysis.

### *2.3 Data Analysis*

All of the 299 article abstracts were analyzed inductively using NVivo 12.0 to identify the key topics and concepts in line with qualitative textual and content analysis procedures (e.g., Neuman, 2006; Punch, 1998). The analysis produced lists and word clouds showing the most frequently occurring terms, concepts, and phrases. The initial stage of open coding focused on identifying the key terms; after this, further analysis tried to group terms and concepts into categories (e.g., Glaser, 1978; Glaser & Strauss, 1967). These were derived from a mixture of inductive and deductive analysis. For instance, terms were categorized inductively where possible based on grouping synonyms and derivative words in the abstracts, while a priori categories were also used to deductively categorize terms into specific fields of practice from the literature (Joyce, Weil, & Calhoun, 2015).

## 3 Findings

### *3.1 Year and Number of Studies*

The 299 articles appeared between January 1998 and June 2019 across 203 peer-reviewed scholarly journals. There was a general increasing trend of studies from 1998, when the first relevant article on deep learning appeared. Across the last 21 years, peak years in the number of relevant articles were 2000 (7), 2007 (18), 2013 (25), 2015 (25), and 2017 (25), which was also the year that registered the highest count. A chart summarizing the number of relevant studies per year is given in Fig. 1. The year 2019 was excluded from the chart because the figure did not show the number of studies for the entire year (Table 2).

### *3.2 Subjects/Disciplines for Deep Learning Studies*

The subjects/disciplines mentioned in the deep learning studies were analyzed, and 136 studies were identified. The three disciplines in which deep learning was studied most frequently were medicine and nursing, sciences, and engineering (see Table 3).

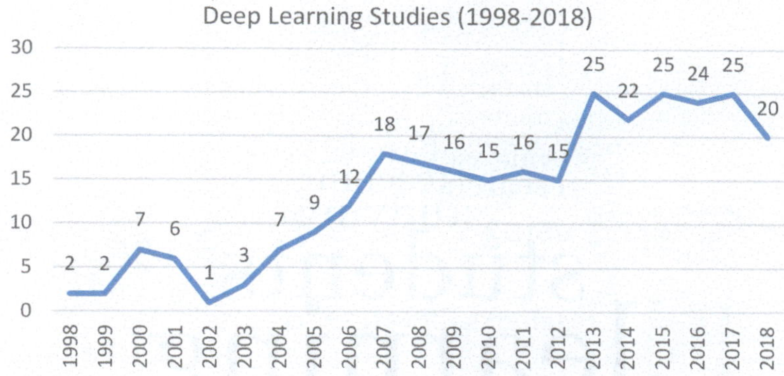


**Fig. 1** Year versus number of articles on deep learning

**Table 2** Year of publication and number of articles on deep learning (1998–2018)

| Year | 98 | 99 | 00 | 01 | 02 | 03 | 04 | 05 | 06 | 07 | 08 | 09 | 10 | 11 | 12 | 13 | 14 | 15 | 16 | 17 | 18 |
|---|---|---|---|---|---|---|---|---|---|---|---|---|---|---|---|---|---|---|---|---|---|
| No | 2 | 2 | 7 | 6 | 1 | 3 | 7 | 9 | 12 | 18 | 17 | 16 | 15 | 16 | 15 | 25 | 22 | 25 | 24 | 25 | 20 |

**Table 3** Appearance of subjects/disciplines in deep learning studies (1998–2019)

| Disciplines | No. of studies | Subtotal |
|---|---|---|
| Medicine/nursing | 43/7 | 50 |
| Science/physics/chemistry/biology/ | 37/2/6/4 | 49 |
| Engineering | 16 | 16 |
| Mathematics/statistics | 9/1 | 10 |
| Geography/business/music/architecture/economics/fashion/accounting | 8/6/3/1/1/1/1 | 21 |
| | Total | 136 |

## 3.3 *Instrument Development: Keywords and Themes*

*Step 1*

NVivo 12.0 was used to analyze the qualitative data (Jackson & Bazeley, 2019). First, it helped run a word frequency query with stemmed words to generate both a list of exact word counts and a word cloud. The query results listed the word frequency counts from highest to lowest. The most frequently appearing words included *learning* ($f = 1814$), *students* ($f = 1223$), *education* ($f = 580$), *deep* ($f = 537$), *surface* ($f = 104$), *approach* ($f = 478$), *engagement* ($f = 136$), and *curriculum* ($f = 119$). These results showed that the abstracts returned by the search process were relevant to this study; the studies were related to learning and education with a

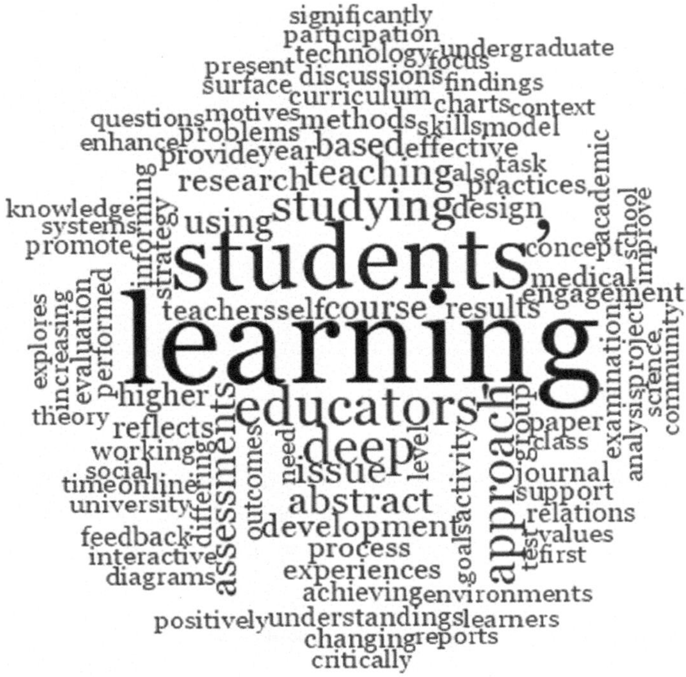


**Fig. 2** Word cloud generated from word frequency query results

focus on deep learning and learning approaches. The keyword frequency is presented graphically as a word cloud (see Fig. 2).

*Step 2*

The list was reviewed to remove irrelevant words. For example, the word *abstract* was removed as it appeared at the beginning of every abstract. Generic terms, such as *learning*, , *issue*, *study*, *teaching*, *using*, and *course*, were also removed. Other words related to the understanding of deep learning, factors in deep learning, and the consequences of deep learning were highlighted and analyzed.

*Step 3*

Finally, four main themes emerged from the relevant keywords, referring to four aspects of learning: social, personal, information processing, and behavioral. It was noted that for some keywords, there was an ambiguity as to which of the four themes best matched. For example, some mentions of "online learning community" referred to how learners build learning communities and help peer learners (i.e., social and behavioral). "Online" could also refer to any networked system that helped adjust behavior for learning, such as e-tutoring systems that were online but did not include any peer interaction activity (i.e., only behavioral). The multiple possible interpretations of such keywords required that the corresponding abstracts be read through to determine the precise meanings.

*Step 4*

Using the keywords from the query, the abstracts were reviewed again. A few additional conceptual phrases related to the keywords were identified. For example, *problem* in itself might not be relevant to deep learning but "*a problem-based learning approach*" (Westhues, Barsen, Freymond, & Train, 2014) would be a relevant pedagogy. These phrases were entered again into the frequency query. This exploration process was conducted to explore and identify additional important concepts.

*Step 5*

Table 4 below summarizes the emerging keywords and concepts. The successive step involved analyzing the studies, extracting the findings and evidence, and summarizing the explanations of deep learning.

## 3.4 Four Families of Learning

The identified keywords were categorized into four main themes, which are defined and discussed below.

## 3.5 The Social Family

The social family of learning is defined in this study as the learning theories that *cultivate the environment or structure for social interactionamong peer learners to achieve deep learning*. Identified keywords in the studies included *group, discussions, project, support, community, social, collaboration, peer, help, team, sharing, dialogue, connection, cooperative, constructivist, guidance, and partners.* Some of the empirical studies examined social models of learning and found evidence for deep learning. For example, Pack (2013) described a community of a learning model in which "students elect to work on local practice partners' projects throughout the semester (p. 1011)." Pack further elaborated, "This mix of academia working in partnership with frontline social and humanitarian workers brings immediacy to the learning. Opportunities for 'deep learning' are facilitated by this partnership approach (p. 1011)." In a study by Kong and Song (2015), it was found that "[the] learners perceived that the initiative could help achieving their learning outcomes. They also perceived that group interaction and experience sharing with peers, teachers and related experts in this context could help advance their knowledge. These results imply that the designed initiative can promote learners to be engaged in reflective

**Table 4** Frequency ($f$) table of keywords under the four main themes

| Social | Personal | Information processing | Behavioral |
|---|---|---|---|
| Group (151)<br>Discussions (111)<br>Project (106)<br>Support (106)<br>Community (82)<br>Social (81)<br>Collaboration (71)<br>Peer (68)<br>Help (52)<br>Team (29)<br>Sharing (24)<br>Dialogue (33)<br>Connection (19)<br>Cooperative (14)<br>Constructivist (11)<br>Guidance (22)<br>Partners (10) | Development (214)<br>Reflects (161)<br>Self (157)<br>Experience (173)<br>Goals (96)<br>Motivation (89)<br>Needs (88)<br>Values (87)<br>Cultures (71)<br>Styles (68)<br>Professional (65)<br>Personally (60)<br>Perceptions (59)<br>Attitudes (48)<br>Beliefs (28)<br>Efficacy (28)<br>Placement (27)<br>Playing (16)<br>Confidence (15)<br>Enjoyment (13)<br>Growth (22)<br>Identity (10)<br>Emotions (10) | Questioning (85)<br>Critical (77)<br>Cognitive (73)<br>Thinking (63)<br>Solving (48)<br>Construct (47)<br>Conceptual (31)<br>Competence (23)<br>Inquiry (23)<br>Scientific (12) | Assessment (314)<br>Practices (140)<br>Scores/grades/marks (123)<br>Test/exam (121)<br>Evaluation (104)<br>Online (117)<br>Technology (105)<br>Feedback (102)<br>Environment (95)<br>Systems (90)<br>Interactive (85)<br>Responses (62)<br>Tutoring (44)<br>Game (40)<br>Behavior (35)<br>Memory/recall/retention (34)<br>Adaptive (30)<br>Simulations (28)<br>Stimulate (21)<br>Assignments (19)<br>Intervention (18)<br>Flexible (16)<br>Tutorial (16)<br>Multimedia (11) |
| *Concept phrases* | | | |
| | Self-regulated learning (10) | Problem-solving (40)<br>Problem-based Learning (30)<br>Reasoning (32)<br>Critical thinking (28) | e-learning (30)<br>Online learning (13) |

[a]Keywords or conceptual phrases that had a frequency count lower than 10 are not included

inquiry for deep learning and personal growth (p. 227)." Luby (2014) studied interfaith dialogue and found that such dialogue provides opportunities for peer teaching, critical reflection, and deep learning. Another study (Chang & Chang, 2008) examined the effect of using an online concept mapping activity (CMA) featuring peer learning and found that students who participated in the CMA applied the concept with significantly higher performance and greater fidelity than those who did not. The authors concluded that CMAs motivate students to adopt deep learning approaches and develop effective cognitive information processing ability for better concept application.

### 3.6 The Personal Family

The personal family of learning is defined as the learning theories that *promote the pursuit of personal growth, with clear goals, self-awareness, and self-confidence so that learners are more motivated and engaged to achieve deep learning*. Identified keywords in the studies included *development; reflects, self, experience, goals,motivation, needs, values, cultures, styles, professional, personally, perceptions, attitudes, beliefs, efficacy, placement, playing, confidence, enjoyment, growth, identity, emotions*, and *self-regulated learning*. Studies suggested that reflective learning achieves deep learning. For example, the study by Gavin (2010) presented "details of attempts to move from a traditional instructor-centered model to a student-centered model of education in order to promote reflective (deep) learning (p. 175)." Moreover, empirical results in the study by Sommet and Elliot (2017) showed relationships among goals, autonomy, and deep learning: "First, when testing goals and reasons separately, mastery (-approach) goals and autonomous reasons explained variance in beneficial experiential (interest, satisfaction, positive emotion) and self-regulated learning (deep learning, help-seeking, challenging tasks, persistence) outcomes (p. 1141)." Similarly, another study (Soria & Weiner, 2013) "[encouraged] students to connect with their audience(s) and develop a sense of purpose for writing tasks (p. 181)" to achieve deep learning. Lefevre (2015) argued that "Qualifying social work education must provide students with a variety of experiential, personalized, participatory, didactic and critically reflective learning opportunities across both the taught curriculum and in practice placements if deep learning of the capabilities needed for effective communication with children and young people is to be ensured (p. 211)." Zhou and Urhahne (2017) explored the relationships between learning goals and intended learning strategies. It was found that learning goals could predict intended deep-learning strategies.

### 3.7 The Information-Processing Family

The information-processing family of learning is defined in this study as *ways to develop scientific inquiry and systematic thinking to solve problems in order to achieve deep learning*. Identified keywords included *questioning, critical, cognitive, thinking, solving, construct, conceptual, competence, inquiry, scientific, problem-solving, problem-based learning, reasoning*, and *critical thinking*. Studies found relationships between concept mapping and deep learning. Kinchin and Cabot (2007), for example, claimed that their findings "indicate that students understand the strategic value of bullet-pointed presentations for short-term assessment goals and the benefits of deep learning mediated by concept mapping that may support longer term professional development (p. 194)." Shield and Dole (2013) examined the relationship between proportion and proportional reasoning teaching in mathematics and deep learning. They found that "the five textbook series examined provided limited support for the development of multiplicative structures required for proportional reasoning, and hence would not serve well the development of deep learning of mathematics (p. 183)." Chapman (2001) studied an introductory college molecular cell biology class that explored the effect of emphasizing critical thinking and deep learning skills in addition to disciplinary content (concepts and reasoning skills), and found that the development of critical thinking skills could be integrated successfully with the study of the process of science, and that this approach was consistent with content learning. Problem-based learning is regarded as a method of deep learning, as described by Ong (2014): "deep learning methods such as flexible learning and problem-based learning can be used to encourage students to take greater responsibilities for their learning outcomes (p. 156)." To identify the problems in assessing reflective learning and to seek ways to tackle them, Bourner (2003) suggested that a questioning approach to reflective learning could provide a solution. Therefore, the nature of reflective learning could differentiate surface learning and deep learning generalized to the domain of reflective learning.

## 3.8 The Behavioral Systems Family

The behavioral systems family of learning is defined in this study as the learning theories that *promote a structured system or environment to provide feedback to adjust behavior to achieve deep learning*. Identified keywords in the studies included *assessment, practices, scores/grades/marks, test/exam, evaluation, online,technology, feedback, environment, systems, interactive, responses, tutoring, game, behavior, memory/recall/retention, adaptive, simulations, stimulate, assignments, intervention, flexible, tutorial, multimedia, e-learning, and online learning*. The empirical findings suggested that "adaptive eLearning method can facilitate deep learning as it provides improved engagements, interactions, real-time feedback and improved flexibility" (Kamardeen, 2014, p. 79). Carstensen and Bernhard (2009) analyzed and described a novel learning environment design, the problem-solving lab, for learning transient response, claiming that "the design merges problem-solving classes and labs, allowing students to engage in deep learning through the integrated use of tools (p. 393)." Evans and Gibbons (2007) examined the addition of interactivity to a multimedia computer-based learning package by the incorporation of control of pace, self-assessment questions, and an interactive simulation. They found that learners using the interactive system needed less time to complete both memory and problem-solving tests than the students who used the system without interactivity design. They concluded that interactivity facilitated deep learning by actively engaging the learner in the learning process.

## 4 Discussion

This study reviewed deep learning studies from the period 1998 to June 2019 using content analysis to identify keywords in the studies. Keywords that appeared more frequently in the empirical studies were grouped into categories. The categories converged into four main themes that define deep learning, explain deep learning, and help explain the effects of deep learning. The four themes correspond to four models of learning: *social*, *personal*, *information processing*, and *behavioral.*

### *4.1 The Social Family*

The social models of learning are those that include partners in learning, that develop interdependent strategies of social interaction, and that develop an understanding of self-other relationships and emotions (Johnson & Johnson, 2009). These include social inquiry, academic inquiry, social and personal development, and cooperative strategies for approaching academic study (e.g., Slavin, 1983). In the literature that was reviewed for this study, peer learning and teaching, group projects, collaborative group work, and online learning communities were frequently identified as activities that contribute to deep learning and that result in more engaging and effective learning.

### *4.2 The Personal Family*

The personal models of learning are those that view learning from the perspectives of the selfhood of the individual. When activities follow these models of learning, learners come to understand themselves better, take responsibility for their education, and learn to reach beyond their current development to become stronger, more sensitive, and more creative in their search for high-quality lives (Joyce et al., 2015, p. 13). For example, Rogers (1961, 1982) proposed nondirective teaching that aims to build capacity for personal development, self-understanding, autonomy, and self-esteem. In this study, personal models of learning were identified as one of the main factors that contribute to deep learning, because they lead to greater participation and are characterized by a stronger motivation to learn.

### *4.3 The Information-Processing Family*

Information-processing models have their origins in knowledge acquisition and learning. They aim to enhance human beings to make sense of the world by acquiring and organizing data, sensing problems and generating solutions, and developing concepts and the language for conveying them. The literature suggests various information-processing learning theories, such as inductive thinking (Taba, 1966), scientific inquiry (Schwab, 1982), and concept attainment (Bruner, 1961). Several of the reviewed empirical studies examined the effects of how information processing and related ways of thinking affect deep learning. Learning through cognitive thinking and reasoning in scientific inquiry helps learners understand better and leads to more effective learning.

## 4.4 The Behavioral Systems Family

Human beings are self-correcting communication systems that modify their behavior in response to information about how successfully tasks are navigated. As behavioral models of learning are used, information is acquired by understanding how to respond to task and feedback mechanisms and has been used in programs to alleviate phobies, to replace fear with relaxation, and to learn the nuances of essential intellectual, social and physical skills (Joyce et al. 2015, p. 18). The behavioral systems family also includes the notion of mastery learning developed by Bloom (1971). Several of the reviewed studies evaluated the effectiveness of newly created integrated learning systems leading to deep learning. While learning is increasingly driven by emergent technologies, some commonalities in the design of these learning systems are ultimately found to lead to effective deep learning, such as interactivity and instant feedback that helps adjust behavior of learners to achieve the expected learning outcomes.

## 4.5 Practical Contribution

The findings of this study suggest that there is evidence that all four learning families—social, personal, information-processing, and behavioral systems—help achieve deep learning. These are important references that can be used by academics and practitioners to design their curriculum to cultivate the necessary structures or environments for learners to achieve deep learning. Interestingly, because there is evidence that each of the learning families supports deep learning, practitioners should not design learning activities based on only one of these learning families. Rather, they should try to incorporate as many families as possible to the structure or environment; different models of learning may be needed at different stages of the course to encourage more participation, engagement, and deep learning. For example, a course could be created with a well-defined goal of personal growth, but could be taught through hands-on experiential learning to solve real-life problems where face-to-face teaching is supplemented by an out-of-class learning environment built with a peer learning community to encourage learners' interaction among peers. The implementation of appropriate learning strategies facilitates deep learning. Another important finding in the study alerts us to the fact that it is often the learners' choice that determines whether a deep learning approach or a surface learning approach is employed. The role of the educator is to provide and cultivate the environment to encourage learners to choose the deep learning approaches and strategies.

### *4.6 Limitations and Further Studies*

This study reviewed deep learning studies published in the past 20 years and included in a single multidisciplinary database. The review is not exhaustive. Further studies could conduct similar reviews using other databases to examine whether the same conclusions are found. This study analyzed the past 20 years' deep learning research altogether. Nevertheless, major improvements have been made to technology-enhanced learning. More analysis should be performed in order to assess the changes over time. Interesting insights may have emerged. An additional limitation, as in other qualitative studies, is that the instrument developed for analysis might have involved the subjective judgment of the researcher, leading to validity issues in different parts of the analysis process: the generation of keywords, the grouping into categories, and the extraction of the main themes. Further studies in the area could increase the generalizability of the results. This highlights the importance of evaluating the logic and reasons used to explain the relationship between the keywords and deep learning. This study has attempted to cite the findings and arguments from selected references to provide the rationale behind the analysis.

## 5 Conclusion

This study sampled deep learning studies in the last 20 years and came up with four learning families identified in these studies, namely, social, personal, information processing, and behavioral. The studies reviewed demonstrated empirical evidence that shows the relationship between models of learning and deep learning. This study contributes to a better understanding of deep learning: it defines the learning strategies that are useful for achieving deep learning, identifies the factors that explain deep learning, and summarizes the effects of deep learning.

## References

Bloom, B. S. (1971). Master learning. In J. H. Block (Ed.), *Mastery learning: Theory and practice.* New York, NY: Holt, Rinehart & Winston.

Bourner, T. (2003). Assessing reflective learning. *Education + Training, 45*(5), 267272.

Bruner, J. (1961). *The process of education*. Cambridge, MA: Harvard University Press.
Carstensen, A.-K., & Bernhard, J. (2009). Student learning in an electric circuit theory course: Critical aspects and task design. *European Journal of Engineering Education, 34*(4), 393–408.
Chang, S. L., & Chang, Y. (2008). Using online concept mapping with peer learning to enhance concept application. *Quarterly Review of Distance Education, 9*(1), 17–27.
Chapman, B. S. (2001). Emphasizing concepts and reasoning skills in introductory college molecular cell biology. *International Journal of Science Education, 23*(11), 1157–1176.
Evans, C., & Gibbons, N. J. (2007). The interactivity effect in multimedia learning. *Computers & Education, 49*(4), 1147–1160.
Gavin, K. G. (2010). Design of the curriculum for a second-cycle course in civil engineering in the context of the Bologna framework. *European Journal of Engineering Education, 35*(2), 175–185.
Glaser, B. (1978). *Theoretical sensitivity*. Mill Valley, CA: Sociology Press.
Glaser, B., & Strauss, A. (1967). *The discovery of grounded theory: Strategies for qualitative research*. Chicago, IL: Aldine.
Jackson, K., & Bazeley, P. (2019). *Qualitative data analysis with NVivo* (3rd ed.). London, UK: Sage.
Johnson, D. W., & Johnson, R. T. (2009). An educational psychology success story: Social interdependence theory and cooperative learning. *Educational Research, 38*(5), 365–379.
Joyce, B., Weil, M., & Calhoun, E. (2015). *Models of teaching* (9th ed.). Upper Saddle River, NJ: Pearson.
Kamardeen, I. (2014). Adaptive e-Tutorial for enhancing student learning in construction education. *International Journal of Construction Education & Research, 10*(2), 79–95.
Kinchin, I. M., & Cabot, L. B. (2007). Using concept mapping principles in PowerPoint. *European Journal of Dental Education, 11*(4), 194–199.
Kong, S. C., & Song, Y. (2015). An experience of personalized learning hub initiative embedding BYOD for reflective engagement in higher education. *Computers & Education, 88*, 227–240.
Lefevre, M. (2015). Integrating the teaching, learning and assessment of communication with children within the qualifying social work curriculum. *Child & Family Social Work, 20*(2), 211–222.
Luby, A. (2014). First footing inter-faith dialogue. *Educational Action Research, 22*(1), 57–71.
Neuman, W. (2006). *Social research methods: Qualitative and quantitative approaches* (6th ed.). Boston, MA: Pearson Education.
Ong, C. H. (2014). Goal orientation of adult students towards learning strategies: The Malaysian context. *Psychological Thought, 7*(2), 156–167.
Pack, M. J. (2013). A tale of two programmes: Developing communities of learning with practice partners in social work and humanitarian studies at Charles Darwin University, Northern Territory, Australia. *Social Work Education, 32*(8), 1011–1020.
Punch, K. (1998). *Introduction to social research: Quantitative and qualitative approaches*. London, UK: Sage.
Rogers, C. (1961). *On becoming a person*. Boston, MA: Houghton Mifflin.
Rogers, C. (1982). *Freedom to learn in the eighties*. Columbus, OH: Merrill.
Sawyer, R. K. (Ed.). (2014). *The Cambridge handbook of learning sciences* (2nd ed.). New York, NY: Cambridge University Press.
Schwab, J. (1982). *Science, curriculum, and liberal education: Selected essays*. Chicago, IL: University of Chicago Press.
Shield, M., & Dole, S. (2013). Assessing the potential of mathematics textbooks to promote deep learning. *Educational Studies in Mathematics, 82*(2), 183–199.
Slavin, R. E. (1983). *Cooperative learning*. New York, NY: Longman.
Sommet, N., & Elliot, A. J. (2017). Achievement goals, reasons for goal pursuit, and achievement goal complexes as predictors of beneficial outcomes: Is the influence of goals reducible to reasons? *Journal of Educational Psychology, 109*(8), 1141–1162.
Soria, K. M., & Weiner, B. (2013). A "virtual fieldtrip": Service learning in distance education technical writing courses. *Journal of Technical Writing & Communication, 43*(2), 181–200.

Taba, H. (1966). *Teaching strategies and cognitive functioning in elementary school children.* (Cooperative Research Project 2404.) San Francisco, CA: San Francisco State College.
Westhues, A., Barsen, C., Freymond, N., & Train, P. (2014). An outcome evaluation of a problem-based learning approach with MSW students. *Journal of Social Work Education, 50*(3), 472–489.
World Economic Forum. (2018). *The future of jobs report 2018.* Retrieved from https://www.weforum.org/reports/the-future-of-jobs-report-2018.
Zhou, J., & Urhahne, D. (2017). Self-regulated learning in the museum: Understanding the relationship of visitor's goals, learning strategies, and appraisals. *Scandinavian Journal of Educational Research, 61*(4), 394–410.

**Will W. K. Ma, Ph.D.** Dr. Ma's research focuses on the use of information systems for communication, knowledge sharing, and knowledge creation. He has papers in *Computers & Education, Computers in Human Behavior, Journal of Computer Assisted Learning, International Journal of Communications, Law, and Policy*, and *Information & Management*. He has served as the Editor of the Educational Communications and Technology Yearbook Series, published by Springer, since 2018. He has also been the President of the Hong Kong Association for Educational Communications and Technology (HKAECT) since 2016.

# The Brazil Project: A Continuing Education Partnership

**Rik Bair** and **Beth Teagarden Bair**

**Abstract** Partnerships are growing portals to connect businesses, industries, and communities to the field of continuing education. From individual courses to programs and initiatives, the University of Miami is developing a unique international program model with a micro-learning entry point to boost interest and enrollment. This chapter discusses the foundation model for forming the first university and international business partnership and how to prepare faculty to teach learners from other cultures. The authors also discuss how instructional designers should prepare continuing education courses and programs. The discussion concludes with the description of the credit and noncredit program offerings with the Brazil sports club partnerships.

**Keywords** Higher education · Postsecondary · Distance learning · Certificates · Micro-credentials · International partnership

## 1 The Brazil Project: A Continuing Education Partnership

Many countries have professional sports clubs that develop players from an early age for the purpose of entertainment, health, and potential careers. With millions of social media followers, the leaders of these sports clubs understand the importance of raising the skill level for coaches and heads to properly manage every aspect of a player's development. As they consider all possible ways, they realize that a partnership with a well-respected university could result in the best avenue to achieve their desires. With the growth of technology and digital access, an online

R. Bair (✉)
Allan Hall, University of Miami, 5050 Brunson Drive, Coral Gables, FL 33146-1610, USA
e-mail: rikbair@miami.edu

B. T. Bair
Ashe Building Suite 240, University of Miami, 1252 Memorial Drive, Coral Gables, FL 33146-1610, USA
e-mail: bethbair@miami.edu



W. W. K. Ma et al. (eds.), *Learning Environment and Design*, Educational Communications and Technology Yearbook,
https://doi.org/10.1007/978-981-15-8167-0_2

program would allow these leaders to equitably reach all levels of coaches to ensure a consistent and professional training.

The Brazil sports clubs have connected with the University of Miami (UM) to partner in the development of continuing education options to the club members in three possible phases. The first phase of this program is a 5-week course that provides five micro-learning units where the participants get an exposure to each topic, in either English or Portuguese. If learners are inspired to learn more about one or more of the units, they may then enroll in a noncredit certificate program for a specific topic that is 6 weeks in length. Upon completion of the 6-week program, they will receive a certificate and a digital badge. Finally, if the learners would like to advance their education to a Master level, they could enroll in one of the online UM Sport Master programs.

## *1.1 Foundation of Continuing Education Partnerships*

Online continuing education partnerships often begin with a conversation between two interested parties, but it requires far more than a conversation to establish a successful venture. Moving beyond a conversation, UM can leverage their knowledge and expertise of online course design to develop an engaging learning environment for the Brazil sports clubs. When considering effective design, UM could implements the triple helix model of economic development, which proposes that when the government, universities, and industries work together to solve problems, economic development occurs (Zheng & Harris, 2007). Although the desire of the university to research and develop knowledge for the creation of goods and services by industries is currently in place, the lack of funding by the government makes the triple helix model ineffective. A more conducive model is the Carnegie classification, where the university and the community engage and collaborate as partners, which enriches both entities and will potentially lead to economic development (Carnegie Foundation, 2009). Therefore, the University of Miami (UM) is pursuing a combination of credit and noncredit continuing education international partnerships through the implementation of the Carnegie classification model.

UM's location in the southern peninsula of Florida paired with the power of the university's brand generates ideal partnerships with Latin American businesses and organizations. UM has traditionally relied on the physical attendance of Latin American students for their various programs, including Intensive English; however, a global transformation has occurred over the past decade that includes advances in technology, global migration, and English as the global language of commerce and society (Olcott, 2009). A market that recognizes credentials and a growing inventory of skills that will expand employment opportunities drives continuing education for students. The continuing education element in the university acts as an entrepreneur that brings together curriculum, instructors, and instructional methods to provide solutions for the ever-changing workforce needs (Moroney & Boeck, 2012). Through a shift in the university's mindset and by leveraging the various available online technology instructional tools, UM is poised to make a powerful impact on their southern neighbors.

A vital key to the success of this venture will be the development of the relationships between the University of Miami and the Latin American organizations. A common vision and shared values will establish a solid foundation, but both sides must continue to nurture the relationship beyond the starting line. Clear and ongoing communication between the entities will result in responsiveness and flexibility to adapt to each other's needs. They both must agree upon what "success" looks like for both entities, which will sustain the focus of both parties in a common direction (Moroney & Boeck, 2012). By further integrating curriculum "that addresses language, culture, and geographic preferences", the University of Miami will develop a strong Latin American niche (Olcott, 2009, p. 7).

The Distance Learning Institute in the Division of Continuing and International Education (DCIE) is establishing partnerships within the University of Miami schools and colleges to provide distance learning options that include credit and noncredit options. DCIE is positioned as the hub for the University of Miami to support the development and ensure the spread of the academic constituents' offerings. DCIE must stay alert to the innovations and changes in the labor market and match the subject matter experts with the needed programs to fill the workforce gaps. Often, these short programs may lead to a longer degree program or a series of stackable certificates. For instance, a Coding Bootcamp may steer a learner to later enroll in a Cybersecurity program or into a series of various types of coding certifications.

## *1.2 Potential Obstacles*

Distance learning provides a flexible avenue for working professionals to access continuing and professional education courses. Programs and courses that are short in length, affordable, and that deliver a meaningful learning experience are important factors to consider during design and development. Automation of the workforce will continually shift employees in new positions that require a different set of skills. Industries and organizations will value employees who are willing and able to acquire multiple skills sets throughout their careers and adapt as change occurs to ensure the continued success of their organization. Although industries and universities may look at data and try to match the latest advancements, it is virtually impossible to predict each innovation and the needs that will arise.

Another obstacle is finding the right resources to develop a course or program and ensure an industry recognizable credibility. Depending on the topic, it may require a combination of both industry and university subject matter experts to develop content. The search for the right experts can delay the project from being completed in a timely manner. A close, communicative partnership between a university and a corporation could make a difference in the timely completion and offering of new programs.

The future use of blockchain technology will provide employers immediate and accurate verification of professional and educational achievements (Miranda, 2018). Blockchain is an incorruptible digital ledger that is duplicated on millions of computers at a time. Since blockchain is not stored in one centralized location, it is not hackable, but is available to everyone on the Internet. This ability creates an ideal legal document depository where all parties have access to the same document for verification (Tapscott & Tapscott, 2016).

Another factor that will result in timely development is the offering of noncredit pathways. Continuing education learners are not necessarily seeking new 2–4 year degrees that must progress through the various university approval steps before they can be developed. Continuing education learners are pursuing avenues to rapidly build a skills portfolio to further enhance their career advancement. DCIE can quickly develop noncredit options to fill an identified workforce skills gap, but the learners must ascertain that the content is relevant and will have an immediate impact on their growth and development.

## *1.3 Transitioning to Instructional Technology*

When postsecondary teachers find instructional practices comfortable, they will more readily embrace the transition to teaching fully online and the addition of technology components (Buus & Georgsen, 2018). Instead of committing to a redesign of their courses, instructors may view technology as an added piece to a portion of their instruction. Instructional designers face a daunting task when working to build an innovative, technology-enhanced online course with teachers who resist technology change. Fortunately, this dilemma is diminished when partnering with a business or industry to create a new short program or course to meet a workforce need.

Biasin (2010) suggests the best instructor role for the adult and continuing education learners is the accompanying relationship. In this role, the instructor performs more as a facilitator in a collaborative learning environment. The instructional design of the learning environment must focus on providing an active experience that transitions training into action for today's lifelong and life-wide learners (Fouts & Mallory, 2010). In many traditional content areas, instructional emphasis is heavy with theory, but adult learners may already have theory and thus require more application with their new knowledge. Instructional designers must work with faculty and administrators to emphasize and design research-based practice training into these noncredit certificate programs.

As the University of Miami expands globally with a focus on partnering with institutions and organizations by implementing short-length noncredit certificates, UM is searching for in-country instructors to supplement and spread the teaching load. Though UM has developed a cultural workshop for its faculty to teach online learners from other cultures, the addition of native instructors could generate stronger connections with learners and a shared strategy with UM faculty. Often in face-to-face instructions, the instructors and learners feel intimidated when they are from different cultures. The design of the cultural workshop is to provide strategies that will bring confidence for all parties in this learning environment. Native instructors also make it easier to offer courses in their languages, which initiates a market to non-English speaking students.

The instructional designers of these programs must also have training and conduct-culture-specific research as part of the effort to provide a stress-free learning environment. Since these programs tend to be new, this benefit means the instructional designers and faculty are constructing new content and blending a few components of a much longer program and redesigning for learning activities. The instructional designer provides the instructional technology knowledge that will support online teaching and learning activities. The instructional designer also works with the subject matter expert to review the desired content and outcomes of the organization (Buus & Georgsen, 2018). Together they pair the elements of digital media, content delivery, and student engagement. By further employing a design for the learning process, they create a learning environment where students are actively engaged in activities and collaborative events that can be applied to their immediate or future employment endeavors (Beetham & Sharpe, 2007; Goodyear & Dimitriadis, 2013).

## 1.4 Transcending Cultures

For decades, educators have investigated the best approach to teaching diversity in the classroom. Initially, they focused on multicultural education, which eventually led to intercultural education (Aldridge, Kilgo, & Christensen, 2014). While these approaches helped to celebrate and interact with various cultures, they often emphasized a family or group rather than the individual learner. Today, educators seek to transcend culture to engage individuals rather than cultures to help students thrive regardless of their cultural upbringing, which is no easy task.

As higher education administration and experts regularly review the optimum methods of reaching international students, they often explore the concern for cultural gaps and knowledge barriers to the experience, which exist for both the students and the instructors (Ryan, 2011). As they address these concerns, they often overlook a significant resource, which is the international students themselves. These students authentically understand their needs and concerns, and their input can be invaluable to develop an effective course design. Although the faculty may be aware of the learning issues involved, including the international students in the discussion who allows everyone to properly address the underlying attitudes and values that may cloud the learning environment, including those of both the students and the instructors. Together they can guide instructional designers to a successful course for all participants.

In addition to considering students' needs, it is important to consider pedagogy that increases student engagement. One of the most effective strategies is to incorporate active learning to encourage international students to creatively interact with content and demonstrate mastery (Takkula, Kangaslahti, & Banks, 2008). As these students actively engage with the lessons, other students as well as the professors can learn from their ideas. Likewise, the instructor can use this student-centered approach to incorporate learning beyond the classroom to include global and social issues.

## *1.5 The Brazil Project*

A recent graduate of the University of Miami's online Master of Sport Administration program informed the College of Education that there are several sports organizations in Brazil that are interested in exploring collaborative opportunities. Among these initial leads are the top volleyball and favorite football clubs in Brazil. The University of Miami's Kinesiology department sent a group to Brazil to meet with these organizations to determine the potential collaborations, and whether faculty research collaborations and student internship opportunities exist. All of the organizations have expressed interest in each of the areas. A plan has been developed to leverage both the UM and club brands by providing several learning options that include both credit and noncredit programs that would be available for their club members only.

Initially, a program that explores five topics was designed in a series of noncredit micro-learning online units in either English or Portuguese language. Topics include

- Sport Marketing and Promotions,
- Public Relations and Event Management,
- Sport Nutrition,
- Sport Conditioning, and
- Sport Injuries.

The program would run for 5 weeks and be taught by adjunct instructors who are employed by each specific club or a University of Miami faculty. The instructors would provide the learners with some content material on a small scale of each specific topic, and then the learners demonstrate their understanding through the completion of a simple project. After successfully completing all 5 weeks, the learners would receive a certificate, displaying both their sports club logo and the University of Miami logo. The micro lessons would simulate the experience of taking a full certificate course online before the learner invests their money and time. The initial program also serves to strengthen ties between the sports club and their membership. The capstone event would be a visit to the club headquarters or training facility with the opportunity to meet the experts in each of the areas as they share their experiences as well as award the certificates. All of the micro topics are offered as 6-week, noncredit certificate courses in an online format.

The partnership between UM and the Brazil sports clubs will provide an adult continuing education program to potentially millions of learners across Brazil as many club members participate in all levels of organizations that support the main professional club. A further extension of the plan is to leverage UM online programs at the graduate level with the popular Master of Sport Administration program. Internship possibilities, club executive guest speakers, and venue tours are part of a potential full-study abroad program option between the education partners.

## 2 Conclusion

In conclusion, the University of Miami anticipates the opportunity to demonstrate its commitment to global education through the continuing education project in Brazil. The university seeks to leverage its brand as an indication of quality to further develop a lasting partnership with businesses and local learners. By incorporating small samples of courses, the university hopes to help learners make informed decisions about potential programs before investing their time and money.

The entire process of partnering, designing courses, and training faculty to better understand cultural differences in an online global world will constantly be evaluated. The University of Miami will consider exploring new partnerships as they learn from processes and results from the initial ones. Future plans will consider many languages depending on the success of early designs and whether the University can meet the needs new partners request. Cost will also play a part in the expansion to other countries and will likely result in different amounts depending on the economy of each country.

## References

Aldridge, J., Kilgo, J., & Christensen, L. (2014). Turning culture upside down: The role of transcultural education. *Social Studies Research and Practice, 9*(2), 107–119.

Beetham, H., & Sharpe, R. (Eds.). (2007). *Rethinking pedagogy for a digital age: Designing and delivering e-learning*. New York: Routledge, London.

Biasin, C. (2010). *Accompaniment: Theories, practices, contexts*. Franco Angeli: Milan, Italy.

Buus, L., & Georgsen, M. (2018). A learning design methodology for developing short learning programmes in further and continuing education. *Journal of Interactive Media in Education, 2018*(1). Retrieved from https://search.ebscohost.com/login.aspx?direct=true&db=eric&AN=EJ1180365&site=ehost-live.

Carnegie Foundation. (2009). *2008 community engagement classification*. Retrieved from https://classifications.carnegiefoundation.org/descriptions/community_engagement.php.

Fouts, S., & Mallory, J. (2010). The credit/non-credit divide: Breaking down the wall. *Journal of Continuing Higher Education, 58*(3), 180–183. Retrieved from https://search.ebscohost.com/login.aspx?direct=true&db=eric&AN=EJ911071&site=ehost-live.

Goodyear, P., & Dimitriadis, Y. (2013). In medias res: Reframing design for learning. *Research in Learning Technology, 21*. CoAction Publication. https://doi.org/10.3402/rlt.v21i0.19909.

Miranda, A. (2018, October 9). *What is blockchain and how can it transform higher education?* Retrieved from https://www.academicimpressions.com/what-is-blockchain-higher-education/.

Moroney, P., & Boeck, D. (2012). Thriving in partnership: Models for continuing education. *Continuing Higher Education Review, 76*, 112–121. Retrieved from https://search.ebscohost.com/login.aspx?direct=true&db=eric&AN=EJ1000657&site=ehost-live.

Olcott, D., Jr. (2009). Global connections to global partnerships: Navigating the changing landscape of internationalism and cross-border higher education. *Journal of Continuing Higher Education, 57*(1), 1–9. Retrieved from https://search.ebscohost.com/login.aspx?direct=true&db=eric&AN=EJ866276&site=ehost-live.

Ryan, J. (2011). Teaching and learning for international students: Towards a transcultural approach. *Teachers and Teaching, 17*(6), 631–648. https://doi.org/10.1080/13540602.2011.625138.

Takkula, H., Kangaslahti, J., & Banks, J. (2008). Teaching transcultural competence: From language learning to experiential education'. *Policy & Practice: A Development Education Review, 7,* 88–95.

Tapscott, D., & Tapscott, A. (2016). *Blockchain revolution: How the technology behind bitcoin and other cryptocurrencies is changing the world.* New York, NY: Penguin Random House LLC.

University of Miami. (2017). *Roadmap to our new century.* Retrieved from https://roadmap.miami.edu/.

Zheng, P., & Harris, M. (2007, August). The university in the knowledge economy: The triple helix model and its implication. *Industry and Higher Education, 21*(4), 253–263.

**Rik Bair** is the Associate Dean of the Distance Learning Institute in the Division of Continuing and International Education at the University of Miami. Dr. Bair has served as a doctoral program chair, instructor, faculty trainer and led instructional design teams in online, hybrid, and classroom instructions in K-12, postsecondary and corporate environments.

**Beth Teagarden Bair** is an Associate Director at the University of Miami. Dr. Bair has developed and delivered traditional and online instructions at all education levels. With a passion for authentic learning, she incorporates innovative online teaching strategies to reach students and give them a voice in lifelong learning.

# Optimizing the Functional Values of Practicum in Nurturing Pre-service Early Childhood Teacher Self-efficacies

**Eunice Pui-Yu Yim**

**Abstract** Students taking an early childhood education teacher training program in an university in Hong Kong ($N = 509$) responded to survey items, before or after practicum, concerning 2 self-efficacy variables (academic self-efficacy and career-specific self-efficacy) and 4 sources of teacher self-efficacy measured by the Self-Efficacy Inventory (SOSI). Confirmatory Factor Analysis (CFA) supported three sources: mastery and vicarious experiences, social persuasions, and physio-emotional arousal. Structural Equation Modeling (SEM) found that practicum had a positive association with mastery and vicarious experiences but near-zero associations with the other sources and the two self-efficacies. The findings called for a dire need to (1) cultivate cognitive competence in the junior year of teacher training programs, (2) review the conventional view on the values and functional role of practicum, and (3) maximize the potentials of the essential sources of self-efficacy during practicum that contribute to pre-service teacher's academic and career-specific self-efficacy.

**Keywords** Early childhood education · Teacher self-efficacy · Academic self-efficacy · Cognitive competence · Practicum · Teacher training

## 1 Introduction

Research has shown that one's competence beliefs influence one's behaviors and accomplishments. For pre-service teachers (i.e., students who are undertaking teacher training programs), two competence beliefs seem equally relevant—academic self-efficacy (i.e., competence in academic work for an academic qualification such as a diploma), and career-specific efficacy (i.e., competence in teaching-related course contents for a professional status). Reviews of research on pre-service teachers

E. P.-Y. Yim (✉)
School of Education and Languages, The Open University of Hong Kong, Ho Man Tin, Kowloon, Hong Kong, China
e-mail: euniceyim@gmail.com



W. W. K. Ma et al. (eds.), *Learning Environment and Design*, Educational Communications and Technology Yearbook,
https://doi.org/10.1007/978-981-15-8167-0_3

suggest that these two types of self-efficacy are closely associated with their academic outcomes and career commitment, but how Early Childhood Education (ECE) teacher training programs may effectively enhance these self-efficacies is a major issue to consider. In many vocational training programs, the role of practicum is highly emphasized as actual experiences in the field with an aim to develop career-related self-efficacy, and therefore professionalism, to maximize his or her expert accomplishments after training. The assumed role of practicum in promoting ECE teachers' self-efficacy, however, has not been rigorously tested. Addressing this gap in the literature, this paper examines the relations of the practicum with academic-focused (academic) and career-focused (teacher) self-efficacies, and their relations with various sources leading to these self-efficacies. The findings are expected to have significant implications for optimizing any gains obtained through practicum in teacher training programs.

## 1.1 Self-efficacy

Self-efficacy, a construct derived from Bandura's social cognitive theory, is defined as the perception or belief of one's capability to execute an action in order to produce a desired outcome under certain circumstances (Bandura, 1977). Self-efficacy is concerned about one's judgment of one's capability in terms of what one can do rather than will do (Bandura, 2006). However, perceived self-efficacy of what one can do is a major determinant of what one will do (intention) with consideration of the involved contextual demands (Bandura, 2006). Such perception influences one's resilience to adversity, the life choices they make and the accomplishments they realize (Perry, DeWine, Duffy, & Vance, 2007). The strength of perceived self-efficacy is not unitary across settings. Rather, it is subject to perceived contextual demands (Zimmerman, Bandura, & Martinez-Pons, 1992). In other words, high perceived self-efficacy in one domain does not necessarily imply the same degree of perceived self-efficacy in a different domain (Bandura, 2006). More importantly, the perceived self-efficacy of each individual within the same domain could vary at a different time due to the complexity of the immediate contextual demands involved (Boyer et al., 2000; Moritz, Feltz, Fahrbach, & Mack, 2000).

**Academic and teacher self-efficacy**. In a vocational training program, two different self-efficacies related to the contextual demands at hand are relevant to pre-service teacher training program participants:

(1) academic self-efficacy refers to one's convictions to perform successfully at designated levels of attainment (Schunk, 1991), which may be increased through effective use of learning strategies and effort (Pintrich & De Groot, 1990). It affects their determination to do well academically in general; and
(2) career-specific self-efficacy refers to the individual's perceived capability to execute a variety of learning strategies such as self-evaluation, goal setting consequences, and environmental restructuring (Zimmerman et al., 1992;

Zimmerman & Martinez-Pons, 1988). It affects their determination to succeed in their career.

These two distinguishable but interrelated constructs suggest that participants need to demonstrate their academic attainment and be confident in achieving a designated academic level (i.e., academic) and their mastery of teacher training contents so as to function well in teaching contexts (i.e., career-specific). The positive relationship between academic and vocational competence beliefs are supported by a number of studies (DeWitz & Walsh, 2002; Zimmerman et al., 1992). In these studies, the term "learning self-efficacy" was used as an inclusive term to highlight the "learning" nature of the construct, regardless of the purposes of learning (Zimmerman et al., 1992; Zimmerman & Martinez-Pons, 1988). In this paper, to clearly differentiate the "learning" construct, which has a vocational purpose, from the "academic" construct, which is essentially learning with an academic focus, we use the terms "academic" and "career-specific" self-efficacy, respectively.

With a high self-efficacy, an individual is confident to deploy effective strategies to cope with immediate contextual demands. Hence, high self-efficacy in academic work may maximize one's likelihood to attain designated levels of attainment (Bong & Clark, 1999; Zimmerman et al., 1992; Zimmerman & Bandura, 1994). Likewise, an individual who has a high self-efficacy in learning career-specific knowledge and skills is likely to deploy effective strategies to achieve competencies directly relevant to the career context such as the classroom. Even under unfavorable circumstances which limit one's performance, those who attribute performance to controllable factors such as their effort or choice of strategies tend to maintain their self-efficacies (Schunk & Gunn, 1986), which keep them functioning in desirable ways. Building pre-service teachers' self-efficacies attributing their performance to their stable and controllable attributes is therefore an important goal in pre-service teacher education programs. Promoting academic self-efficacy will enhance one's learning effectiveness while promoting career-specific self-efficacy will ensure these teachers' career commitment to engage in high-quality teaching and continual improvements in knowledge and skills that promote professionalism.

**Measurement of teacher self-efficacy**. There are a number of scales measuring individuals' teacher self-efficacy (career-specific). The 30-item Teacher Efficacy Scale (TES) developed by Gibson and Dembo (1984) measures teachers' expectancy on both their own academic learning abilities and teaching outcomes (career-specific). Other similar scales include the 12-item Teachers' Sense of Self-Efficacy Scale (TSES) which was validated in five cultural settings (Klassen et al., 2009), and the Teachers' Efficacy Beliefs System-Self (TEBS-Self) which measures teachers' self-beliefs in successfully performing specific teaching-related tasks (Dellinger, Bobbett, Olivier, & Ellett, 2008). Although there are adequate inventories measuring the level of teacher self-efficacy, none of the above-mentioned scales focuses on the sources of teacher self-efficacy, which is central to the present study on nurturing competent teachers in teacher training program. Tschannen-Moran, Hoy, and Hoy

(1998) and Kieffer and Henson (2000) revealed that the development of teacher self-efficacy is positively related to four major sources of self-efficacy, including mastery experience, vicarious experience, social persuasions, and physio-emotional arousal (Schunk & Pajares, 2002). Kieffer and Henson (2000) developed the Self-Efficacy Inventory (SOSI) for measuring sources of teacher self-efficacy. However, the inventory lacks validity as the results of factor analysis were not satisfactory. It therefore calls for a need to modify and validate the SOSI so as to measure the different sources of teacher self-efficacy.

## 1.2 Sources of Teacher Self-efficacy

In order to promote teacher self-efficacy, teacher educators need to know the sources of teacher self-efficacy and how they relate to the two types of self-efficacy (academic and career-specific self-efficacies). In brief, there are four major sources (Kieffer & Henson, 2000).

***Mastery experience*** refers to one's own subjective interpretations of previous attainment (Kieffer & Henson, 2000). Lane (2002) further found that subjective perceptions of one's mastery experiences are better predictors of self-efficacy than are objective results. Successful mastery in a domain can have lasting effects on one's self-efficacy in that particular domain. Individuals who have positive mastery experiences are more likely to overcome obstacles or succeed in challenging tasks, especially in those that are difficult for others (Britner & Pajares, 2006). For pre-service teachers' development in self-efficacy, experiences of successful mastery of knowledge and skills in teaching are expected to promote teacher self-efficacy. How mastery experience is related to academic self-efficacy is, however, unclear. This is because first, a successful mastery of knowledge and skills in teaching is not necessarily academic in nature, and second, self-efficacies in academic work and in the teaching career are likely to be affected by quite different variables at different levels, due to the domain-specific nature of self-efficacy (Gloudemans, Schalk, Reynaert, & Braeken, 2013).

***Vicarious experience*** refers to observing others (Schunk & Ertmer, 1999). In many occasions, pre-service teachers may not have a lot of successful experiences in the field; instead, they may gauge their capabilities in relation to the performance of similar others. They are most likely to alter their perceived teacher self-efficacy following a model's success or failure to the degree that they feel similar to the model in the area in question (Schunk, 1987). Social cognitive theory maintains that the greater the strength of in-group projection, the greater the chance of developing in-group homogeneity and cohesion with perceived in-group members, namely similar others. With this, an individual's in-group identity arises when the individual affiliates himself or herself to a social group and develops an identity of "us" with the group (Robbins & Krueger, 2005). When a need for social connectedness is activated, one's motivation to act like, or compare against, the affiliated others increases (Kelley & Jacoby, 1996).

***Social persuasions*** refer to people receiving comments from others as a source of self-efficacy (Kieffer & Henson, 2000). Encouragement and supportive messages about a teacher's performance from significant others they trust can boost their confidence and competence beliefs in related areas. Particularly, when accompanied by conditions and instructions that help bring about success, social persuasions can be a powerful source of teacher self-efficacy (Hattie & Timperley, 2007).

***Physio-emotional arousal*** refers to people's interpretation of their physiological and emotional arousal as an indicator of personal competence by evaluating their own performances under differing conditions (Kieffer & Henson, 2000). Realistic interpretations of personal competence would elicit corresponding levels of physio-emotional reaction (Schwarzer, 2014). In general, situations that promote students' positive physio-emotional well-being and reducing negative emotional states would lead to positive teacher self-efficacy and favorable self-evaluation.

Bandura (1997) argues that the mastery experience is the most influential among four sources. Generally, a successful past experience strengthens an individual's self-efficacy while experience of failure diminishes it. However, in situations where individuals, such as pre-service teachers, have limited prior experience, the effects of vicarious experience and social persuasions may also be assumed to be prominent. These sources seem to be particularly prevalent especially in situations where individuals can build up self-efficacy through observing successful experiences of similar others or receiving encouragement from others, or both. To provide such experiences, teacher training institutions emphasize the role of practicum which provides novice teachers the opportunities to observe, and be mentored by, more experience teachers and to try out their own teaching in real classroom settings.

## *1.3 Practicum*

The practicum is a strategy in which students gain experiences in a workplace relevant to their program of study and career aims (Houshmand & Papadakis, 1995). It is assumed to provide important benefits for both students and employers in a number of ways. For the students, practicum provides an environment that is close to real-life, but relatively safe and protected for novices to prepare themselves for entering the career (Keller-Schneider, 2010). The experience is also assumed to minimize what some researchers call a "reality shock" (Dicke et al., 2015), primarily resulting from a mismatch between what the pre-service teachers believe they can do and what is actually required of them in real settings. It promotes practical skill development and self-efficacy enhancement at work and in learning to be an effective teacher. Under the supervision and mentoring of more experienced teachers, the novice has a chance to put the theories he or she has learned from taught courses to real practice in the classroom. Learning at the workplace is different from learning at an academic setting such as the university where the link between theory and practice requires to be constructed through experiences and inner cognition. It can be argued that the advantage of the practicum is twofold. First, one's inner cognition skills which include reflection, analysis, integration, and synthesis of learned knowledge can be facilitated and enhanced through practice (mastery), observation of others (vicarious), input from others (social persuasions), and feelings on the practicum experiences (physio-emotional arousal) (Clift & Brady, 2005), Second, the novice also gains social-mediated cognition through guidance provided by mentors and observation of experienced others (Eames & Bell, 2005).

For pre-service teachers engaging in the practicum, they may gain valuable input from the workplace in terms of all four sources of self-efficacy, particularly mastery experience, according to existing literature (Perry et al., 2007; Schunk & Ertmer, 1999; Zimmerman, 2000). However, there are two issues that have not been explored: (1) does practicum directly influence academic and career self-efficacy, and (2) does practicum influence the sources of self-efficacy which in turn improve academic and career self-efficacy, and if so, which sources are most salient? These questions are of utmost significance to teacher training education. If the practicum had no positive effect on teacher self-efficacy, the use of practicum could be challenged. If the practicum in an institution was only effective in facilitating a certain source of self-efficacy, the nature of the practicum would need to be revisited to maximize its benefits. An understanding of the role of practicum and the relations between academic self-efficacy, career-specific self-efficacy and the sources of teacher self-efficacy, teacher educators will be able to tailor-make instructional strategies that are supportive of the specific and maximize the potential of various sources of teacher self-efficacy to become successful learners and effective teachers who are committed to their career and further professional development.

## 1.4 The Present Investigation

The present study focused on a sample of pre-service teachers undergoing a 2-years ECE full-time teacher education program to become professional kindergarten teachers. The program consists of over 800 teaching hours and 320 practicum hours with on-site supervisions. The study examined the relations between the practicum, four sources of teacher self-efficacy, and two related but distinctive self-efficacies: academic and career-specific. Considering the lack of well-validated scale to measure sources of teacher self-efficacy, this study starts with a validation of the SOSI factors based on which the following goals are to be achieved: (1) to adopt and validate the scales of sources of teacher self-efficacy; (2) to examine the relations between different sources of teacher self-efficacy and the two types of self-efficacy (academic and career-specific); and (3) to investigate the effects of practicum on students' experiences of various sources of teacher self-efficacy and how they relate to the two types of self-efficacy.

## 2 Method

### 2.1 Participants

Pre-service early childhood teachers enrolled in a 2-year full-time teacher education program in an university in Hong Kong participated in the survey research ($N =$ 509). All of the participants completed the survey and missing data were minimal (<1%). Of the participants, 96.7% were female, which is similar to the Hong Kong early childhood teacher population (Chen & Rao, 2011). Majority of the participants were between 18 and 24 years old (96.1%), and 3.7% of them were between 25 and 34 years old. The sample consisted of students in higher diploma in early childhood education (Year 1: 58.4%; Year 2: 38.5%) and Year 1 of undergraduate degree in Early Childhood Studies (3.1%). In the present sample, some of the respondents had already participated in the mandatory 320-h teaching practicum arranged by the program. Near the end of the practicum, students are required to eventually take up the role as the regular class teacher for a class of 15 children enrolled in formal kindergarten settings. Among the participants, 368 (71.7%) students had not yet started their teaching practicum, 59 (11.5%) of them were undergoing their practicum, and the rest 86 students (16.8%) had already completed the practicum component of the program.

### 2.2 Materials

The survey was designed to understand pre-service early childhood teachers' academic self-efficacy, career-specific self-efficacy, their relations with various sources that may affect the two types of self-efficacy, and the role of practicum within these relations. The scales and sample items used in the study are presented in the Appendix. Academic self-efficacy items were adopted from the College Self-Efficacy Inventory (CSEI) (Kieffer & Henson, 2000), whereas teacher self-efficacy items were adopted from the Motivated Strategies for Learning Questionnaire (MSLQ) instrument (Pintrich, Smith, Garcia, & McKeachie, 1991). The sources of teacher self-efficacy scale was adopted from Bandura's (1997) Sources of Self-Efficacy Inventory (SOSI).

**Academic self-efficacy**. Five items from the CSEI were used (Kieffer & Henson, 2000). This scale reflects the respondents' perceived competence in academic work. An example is "How confident are you that they could successfully complete the following tasks?".

**Teacher self-efficacy**. Eight items from the MSLQ was used (1990). This scale also reflects the respondents' perceived competence in learning, but not purely in an academic sense. Instead, the items refer specifically to learning the ECE contents relevant to their career in teaching. An example is "If I believe I will receive an excellent grade in this program."

**Mastery experience**. Mastery experience is a scale in the SOSI instrument (Kieffer & Henson, 2000). The items from the SOSI were pilot tested to reduce the length of the instrument while maintaining its reliability (see Appendix). Three items were used for this Mastery scale. An example is "I have developed many of my teaching skills by actually teaching."

**Vicarious experience**. Vicarious experience is a scale in the SOSI instrument (Kieffer & Henson, 2000). The items were pilot tested to reduce the length of the instrument while maintaining reliability. Four items were used for this Vicarious Experience scale. An example is: "I have learned about how to be a teacher by watching other skillful teachers."

**Social persuasions**. The items from the SOSI (Kieffer & Henson, 2000) were pilot tested and three items were used for this Persuasions scale. An example is "I believe I can teach as well as the teachers portrayed in popular movies."

**Physiological and emotional arousal**. Physio-emotional arousal is also a scale in the SOSI (Kieffer & Henson, 2000). After pilot testing, three items were used for this Emotion scale. An example is "When I say the wrong things to a class, I become anxious."

**Practicum**. Practicum was a single-item measure differentiating participants who had not and those who had started or completed the teaching practicum. The dichotomous measure was coded 0 (no practicum) and 1 (practicum started or completed).

## 2.3 *Procedure*

Informed consent was obtained before the survey was administered. Printed surveys and instructions were distributed to students in class. Instructions included information about the purposes of the study, anonymity, and their right to withdraw from the study any time. The survey was a shorter version of an initial piloted survey with another cohort earlier. The completed survey responses were entered and transformed into SPSS for analysis (Field, 2013).

## 2.4 *Data Analysis*

The responses to the survey items were coded such that higher scores reflected more positive responses. Cronbach's alpha reliability was preliminarily estimated for each factor, followed by a series of three principal-axis factoring procedures. The first analysis included items for academic and teacher self-efficacy. The second analysis included the items for the four sources of teacher self-efficacy.

After these analyses, which were exploratory in nature, Confirmatory Factor Analysis (CFA) was conducted (Jöreskog & Sörbom, 2006). A series of models were tested, which started with the first model testing the ability of 27 items to form two self-efficacy factors (academic and teacher), and four sources of self-efficacy (mastery, vicarious, social persuasions, and physio-emotional arousal). The rest of the analyses further tested the factors until the most reasonable factor structure was derived, which will be detailed in the Results section. The reliability of each scale would then be reassessed.

Based on the factors established in the CFA models, a structural equation model examining the paths from practicum to four sources of self-efficacy and from these sources to the two types of self-efficacy was constructed. The relative strengths of the predictive paths from each of the four sources of self-efficacy would give us directions as to which source was most salient in promoting which self-efficacy. The strength of the paths from practicum to all other variables would enlighten our understanding of the role of practicum in yielding any salient source that may promote the two types of efficacy and whether practicum would directly enhance either self-efficacy variable.

The procedures for conducting CFA have been described elsewhere (Jöreskog & Sörbom, 2006), so they are not detailed here. The goodness of fit of models was evaluated by observing the Tucker–Lewis index (TLI, also known as the non-normed fit index) as the primary goodness-of-fit index. However, the chi-square statistics, Root Mean Square Error of Approximation (RMSEA), and Comparative Fit Index (CFI) are also reported. In general, for an acceptable model fit, the values of TLI and CFI should be equal to or greater than 0.90, whereas 0.95 may be taken as an excellent fit. For RMSEA, according to Browne and Cudeck (1993), a value of 0.05 indicates a close fit and values close to 0.08 indicate a fair fit. In short, based on commonly accepted criteria (Browne & Cudeck, 1993; Jöreskog & Sörbom, 2006), an acceptable model would show an acceptable model fit (i.e., TLI and RNI $= 0.90$ or above and RMSEA $< 0.08$), an acceptable factor loading for each item pertaining to its respective factor ($>0.30$), and reasonable correlations among the latent factors.

# 3 Results

## *3.1 Factor Analysis*

**Principal-axis factoring**. Three sets of principal components analyses were conducted. The first analysis included academic self-efficacy and teacher self-efficacy yield two well-defined factors explaining 61.7% of variance. The factors loading, respectively, to each factor were all above 0.50. The second set including items of the presumably four sources (mastery, vicarious, social persuasions, and

Table 1 Goodness of fit of models

| | Model | c2 | *df* | TLI | CFI | RMSEA |
|---|---|---|---|---|---|---|
| 1. | 6 factors (4 sources + 2 self-efficacies) | 893.12 | 309 | Improper | | |
| 2. | 5 factors (3 sources + 2 self-efficacies) | 915.36 | 309 | 0.920 | 0.929 | 0.061 |
| 3. | 3 factors (1 source + 2 self-efficacies) | 1197.67 | 321 | 0.886 | 0.896 | 0.073 |
| 4. | 2 factors (1 source + 1 self-efficacy) | 1927.91 | 323 | 0.793 | 0.810 | 0.099 |
| 5. | 6 factors (practicum + 3 sources + 2 self-efficacies) | 939.67 | 336 | 0.920 | 0.929 | 0.060 |
| 6. | Path model | 939.67 | 336 | 0.920 | 0.929 | 0.060 |

*Note* $N = 509$. CFI = Comparative fit index. TLI = Tucker-Lewis index. RMSEA = Root mean square error of approximation

physio-emotion); however, yielded three factors explaining 55.7% of variance. The factor loadings were all acceptable, with the lowest being 0.46. The items for mastery and vicarious formed one of the factors.

**Confirmatory Factor Analysis**. A series of CFA models were tested (Table 1). Model 1 testing the ability of 27 items to form six factors (two self-efficacy factors: academic and teacher self-efficacy; and four sources: mastery, vicarious, social persuasions, and physio-emotion) yielded an improper solution. The latent factor correlation between the mastery and vicarious factors was 1, indicating that the two factors should be treated as a single factor, which was consistent with the preliminary principal-axis factoring results.

Model 2 (Table 1) tested a five-factor model merging Mastery and Vicarious items into a single factor. Model 2 resulted in a reasonable fit (TLI = 0.92, CFI = 0.93, RMSEA = 0.06). Models 3 and 4 were constructed to further scrutinize the structure of Model 2. Neither Model 3 assuming a single source of self-efficacy (TLI = 0.89, CFI = 0.90, RMSEA = 0.07) nor Model 4 assuming both a single self-efficacy and a single source (TLI = 0.79, CFI = 0.81, RMSEA = 0.10) had a comparable fit. Hence, Model 2 was taken as the best fitting model for further analysis. Based on Model 2, practicum was added as a single-item factor. This model (Model 5) had a reasonable fit (TLI = 0.92, CFI = 0.93, RMSEA = 0.06). The solution of Model 5 is presented in Table 2 in which the factor descriptive statistics and reliability estimates for the validated factors are also presented. Note that the term Mastery hereafter refers to a factor representing both mastery and vicarious experiences.

## 3.2 Correlations of Factors

As can be seen in Table 2, the two self-efficacy latent factors (academic and teacher self-efficacy) were positively correlated ($r = 0.69$). Likewise, the three sources of self-efficacy were all positively correlated ($r$s from 0.43 to 0.81). Between the two

**Table 2** Solution of 6-factor model (Model 5)

| **Mean** | 6.29 | 6.17 | 6.17 | 6.07 | 5.64 | | |
|---|---|---|---|---|---|---|---|
| ***SD*** | 1.38 | 1.14 | 1.13 | 1.06 | 1.23 | | |
| ***Alpha*** | 0.90 | 0.92 | 0.93 | 0.77 | 0.72 | | |
| **Factor loadings** | | | | | | | |
| Academic 1 | 0.67* | | | | | | 0.55* |
| Academic 2 | 0.83* | | | | | | 0.31* |
| Academic 3 | 0.86* | | | | | | 0.27* |
| Academic 4 | 0.87* | | | | | | 0.25* |
| Academic 5 | 0.83* | | | | | | 0.31* |
| Career 1 | | 0.68* | | | | | 0.54* |
| Career 2 | | 0.67* | | | | | 0.55* |
| Career 3 | | 0.74* | | | | | 0.45* |
| Career 4 | | 0.76* | | | | | 0.42* |
| Career 5 | | 0.80* | | | | | 0.37* |
| Career 6 | | 0.74* | | | | | 0.45* |
| Career 7 | | 0.84* | | | | | 0.30* |
| Career 8 | | 0.85* | | | | | 0.27* |
| Mastery 1 | | | 0.77* | | | | 0.41* |
| Mastery 2 | | | 0.82* | | | | 0.32* |
| Mastery 3 | | | 0.69* | | | | 0.53* |
| Vicarious 1 | | | 0.78* | | | | 0.39* |
| Vicarious 2 | | | 0.76* | | | | 0.42* |
| Vicarious 3 | | | 0.82* | | | | 0.32* |
| Vicarious 4 | | | 0.73* | | | | 0.47* |
| Persuasion 1 | | | | 0.70* | | | 0.51* |
| Persuasion 2 | | | | 0.57* | | | 0.67*<br>0.41* |
| Persuasion 3 | | | | 0.77* | | | 0.41* |
| Persuasion 4 | | | | 0.68* | | | 0.54* |
| Emotion 1 | | | | | 0.73* | | 0.46*<br>0.52* |
| Emotion 2 | | | | | 0.69* | | 0.52* |
| Emotion 3 | | | | | 0.61* | | 0.63* |
| Practicum | | | | | | 1.00 | 0.00 |
| **Factor correlations** | | | | | | | |
| Career | 0.69* | | | | | | |
| Mastery | 0.66* | 0.68* | | | | | |

(continued)

**Table 2** (continued)

| | | | | | | | |
|---|---|---|---|---|---|---|---|
| Persuasion | 0.60* | 0.66* | 0.81* | | | | |
| Emotion | 0.37* | 0.43* | 0.59* | 0.63* | | | |
| Practicum | −0.07 | −0.01 | 0.18* | 0.04 | 0.08 | | |

*Note* $N = 509$. $^{*}p < 0.05$. Unique = Uniqueness. Mastery items 1–3 were from the original Mastery scale, whereas items 4–7 were from the original Vicarious scale

**Table 3** Descriptive statistics for key variables by practicum participation

| Variable | | Practicum Participation | |
|---|---|---|---|
| | | No ($n = 366$) | Yes ($n = 143$) |
| Academic self-efficacy | *M* | 6.31 | 6.15 |
| | *SD* | (1.23) | (1.47) |
| Career-specific self-efficacy | *M* | 6.18 | 6.12 |
| | *SD* | (1.11) | (1.21) |
| Mastery and vicarious experiences | *M* | 5.79 | 5.99 |
| | *SD* | (1.19) | (1.17) |
| Social persuasion | *M* | 6.31 | 6.58 |
| | *SD* | (1.09) | (1.29) |
| Physio-emotional arousal | *M* | 5.51 | 5.64 |
| | *SD* | (1.22) | (1.39) |

*Note* * $p < 0.05$. Responses ranged from 1 = low to 10 = high

self-efficacies and the three sources, the correlations were positive (*r*s from 0.37 to 0.68), but were less strong for the physio-emotional arousal source (*r*s = 0.37 and 0.43, respectively; Table 2).

Table 3 shows the means and standard deviations of the two self-efficacies and three sources for pre-service teachers who had participated in the practicum and those who had not. An inspection of the mean scores found that participants of the practicum (*M*s = 6.15 and 6.12, respectively) were no higher than those non-participants in the two self-efficacies (*M*s = 6.31 and 6.18, respectively). In contrast, the mean scores were higher for the practicum participants in all three sources of self-efficacy (*M*s = 5.99, 6.58, and 5.64, respectively, for Mastery, Persuasion, and Emotion) compared to non-participants (*M*s = 5.79, 6.31, and 5.52, respectively). However, these differences were mostly not significant statistically. As shown in Table 2, Practicum was not correlated at all with four of the five variables of interest (*r*s were close to zero) except for Mastery ($r = 0.18$), which was statistically significant. In essence, the practicum did not seem to have benefited the participants in self-efficacies, nor did it empower social persuasions ($r = 0.04$), or physio-emotional arousal ($r = 0.08$). The only apparent benefit was observed in Mastery ($r = 0.18$). These results imply that the practicum added value in pre-service teachers' mastery and vicarious experiences but had little to add to other sources of teacher self-efficacy, or to academic and teacher self-efficacy directly.

## 3.3 Path Model

Based on Model 5, Model was a full path model (Fig. 1) that estimated the relations among practicum participation, three sources of teacher self-efficacy, and the two self-efficacy factors. As shown in Fig. 1, practicum experience was positively associated with Mastery ($\beta = 0.18$), but the paths from practicum to Persuasion ($\beta = 0.04$) and Emotion ($\beta = 0.08$) were not statistically significant, though positive. Practicum was negatively associated with academic self-efficacy ($\beta = -0.18$) and teacher self-efficacy ($\beta = -0.11$). These results suggest that for this sample of pre-service teachers, practicum may not have direct positive input to enhance self-efficacies, both academic and teacher self-efficacy. Instead, practicum tends to increase mastery and vicarious experiences, which in turn contribute to both self-efficacies—academic ($\beta = 0.58$) and teacher ($\beta = 0.50$).

Interestingly, the paths from social persuasions to both self-efficacies were significantly positive—academic ($\beta = 0.20$), and teacher ($\beta = 0.27$). However, although

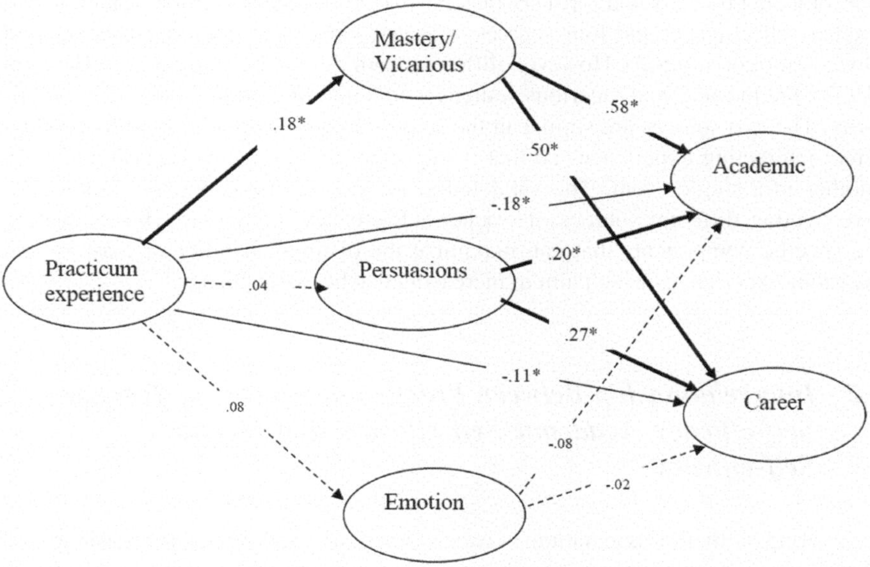


**Fig. 1** Path model. Mastery/Vicarious = mastery and vicarious experiences; Persuasions = Social persuasions; Emotion = physio-emotional arousal; Academic = academic self-efficacy; Career = career-specific self-efficacy. * $p < 0.05$

this source of self-efficacy seems strong in contributing to both types of self-efficacy, the practicum did not add any value to this source, which suggests that the practicum for this particular sample in this particular training program may not have fully used the opportunity of practicum to promote this essential source that has the potential to enhance the two types of self-efficacy.

## 4 Discussion

This study had three aims: (1) to adopt and validate the scale of sources of teacher self-efficacy; (2) to examine the relations between different sources of teacher self-efficacy and the two self-efficacies (academic and career-specific); and (3) to investigate the effects of practicum on students' experiences of various sources of teacher self-efficacy and how they relate to the two self-efficacies.

### *4.1 Validation of Sources of Teacher Self-efficacy Scale*

The present investigation tested the relations between sources of teacher self-efficacy and the two self-efficacies (academic and teacher). The factor structure of 27-item was validated by CFA and form six factors (two self-efficacy factors: academic and teacher self-efficacy; and four sources: mastery, vicarious, social persuasions, and physio-emotion arousal). However, different from the model suggested in Bandura (1977), the mastery and vicarious sources were combined in the scale due to similarity. The two sources are similar in the sense that they were highly influenced by direct interacting experiences so that it might be difficult to distinguish one from another in a single event. The validated scale may be used in future studies for investigating different sources of teacher self-efficacy, particularly for evaluating the specific components that can maximize the distinguishability of mastery and vicarious experiences which aim at increasing teacher self-efficacy.

### *4.2 Interrelationship Between Practicum, Sources of Teacher Self-efficacy, Academic Self-efficacy, and Teacher Self-efficacy*

Comparing with the association between practicum and social persuasions and physio-emotion, the stronger positive association between practicum and mastery and vicarious as a source of teacher self-efficacy shown in the path model (Fig. 1) and the negative association between practicum and the two types of self-efficacies (academic and career) suggest that practicum may not have direct positive input to enhance both types of self-efficacy. Instead, practicum tends to increase mastery and vicarious experiences, which in turn contribute to both types of self-efficacy. The mediating effect of the mastery and vicarious source of self-efficacy on the two types of self-efficacy shed a light on the functional role of practicum as a platform to maximize the most influential source of self-efficacy (as a combination of mastery and vicarious experiences) that has direct impact on both academic and career-specific self-efficacy as oppose to the conventional view of considering practicum itself as exerting influential impacts on both types of self-efficacy.

The path model also indicated that the association between social persuasions to both types of self-efficacy was significantly positive. Social learning theory encompasses that the influences of significant and similar others should not be overlooked (Zimmerman et al., 1992). Domain-specific comments from significant and similar others are powerful sources of academic and career-specific self-efficacy. These sources are particularly strong when there is acknowledgement of the peer group they feel they belong to and the professionalism of those they respect (Hattie & Timperley, 2007).

The positive association between practicum and three sources of self-efficacy and the positive correlations among these three sources imply that the three sources might be developed together with each complementing the others. That is, the event that enhances any one of the sources could also directly or indirectly enhance others positively. Further research will be needed to examine the interplay of these factors in building each source's capacity which in turn boost academic and career-specific self-efficacy.

From the path model (Fig. 1), physio-emotional arousal as a source of self-efficacy was weakly associated with practicum, and with the academic and career-specific self-efficacy factors. The weak relationship suggests that the affective aspect of teacher self-efficacy (physio-emotional arousal) is more closely related to non-academic and non-career-specific self-efficacy such as social self-efficacy which is a distinct construct that focuses on self-expectations for personal skills in performing the specific behavior that underlie personal relationship (Connolly, 1989, p. 259).

## 4.3 *Effect of Practicum Experience*

Practicum students were no higher than those non-practicum in the academic and career-specific self-efficacy. Such finding contradicted with the conventional beliefs about the benefits of practicum on learners' academic and career self-efficacy enhancement. In contrast, apparent benefit was observed in one of the sources of teacher self-efficacy—Mastery (which is mastery and vicarious combined in this analysis). Such a reality shock can be explained by the holistic and general conceptualization of practicum rather than constitutional conceptualization of its critical components that maximizes the power of practicum. These findings are partly coherent with the arguments in Bandura (1997), which stated that mastery experience is the strongest source of teacher self-efficacy, but individuals who have limited prior experience tend to also gain their teacher self-efficacy beliefs from others' encouragement or own observations during practicum in which both teacher and academic self-efficacy can be enhanced. The findings indicated that a practicum without maximizing the potential of essential sources of self-efficacy would have no, or limited if any, impact on developing academic and career-specific self-efficacies in pre-service teachers.

## 5 Implications

**Review and optimize the functional values of practicum arrangement**. The positive association between practicum and mastery and vicarious experiences but zero association with social persuasions and physio-emotional arousal sources, together with the non-positive associations with academic and career-specific self-efficacies imply that the values and role of practicum in this pre-service teacher education program need to be reviewed and redefined. The conventional view assumes that practicum is a powerful channel in developing teachers' professionalism. This study indicated that such an assumption oversimplifies the complexity of the interplay between practicum, sources of self-efficacy, and teacher self-efficacy in terms of academic and career-specific focuses.

The direct and positive association between practicum and mastery and vicarious experiences as a major source of self-efficacy suggests that practicum is effective in providing students with direct experiences in teaching activities or observing teaching models which help students build evidence-based experiences. Only these valuable experiences can lead to the strengthening and development of academic and career-specific self-efficacy. In other words, practicum itself without emphasizing the components of direct teaching and model observation experiences may not benefit teacher efficacy, whether in academic or career-specific terms. It is the direct teaching experience and model observation that provide pre-service teachers opportunities to engage themselves in reflective processes about their own as well as others' teaching practices (Schön, 1991).

**Nurturance of cognitive competence in promoting self-efficacies**. The mediating effect of mastery and vicarious experiences on academic and career-specific self-efficacies via one's inner cognitive process such as reflection and knowledge construction suggests the need in nurturing cognitive competence in pre-service teachers. Cognitive competence include skills for knowledge construction, task completion, problem-solving, decision-making, critical thinking (reasoning and making inferences), and creative thinking (stretching one's spectacles, evaluating multiple ideas and alternatives, and generating novel and practical ideas) (Sun & Hui, 2012). Pre-service teachers who are cognitively competent are more likely to reflect on their teaching experiences at two levels: reflection in action and reflection on action (Schön, 1991). At level one, a pre-service teacher reflects on his or her teaching experiences as well as reflects on the observations acquired from the observation models, systematically analyzes the experiences from multiple sources or perspectives, explores better alternatives and constructs practical new ideas (i.e., reflection in action). At level two, a pre-service teacher makes use of the new knowledge and skills and applies in the practical setting for the betterment of his or her own teaching performance in real classroom settings (i.e., reflection on action). The cycle of reflection in- and on- action leads to continuous professional improvement that would inevitably enhance academic and career-specific self-efficacy.

**Insights to teacher education program design**. Within the teacher training institution, for Year 1 students who have not started practicum yet, the curriculum should place more emphasis on elements that promote cognitive competence via teaching and learning approaches that emphasize case analysis and problem-solving. The problem-based learning approach has been widely used in vocational training programs. It presents novel and challenging learning contexts wherein students are exposed to cognitive conflicts and stimulated to engage in higher order thinking such as reflection, case analysis, and problem-solving (Sun & Hui, 2012). On the one hand, students are required to interpret the presented case with reference to their prior knowledge in the related subject areas and existing levels of thinking skills, and engage in case analysis, critical thinking, and practical solution generation. On the other hand, vicarious and social persuasions sources of self-efficacy can be enriched by providing opportunities for quality contact with good teacher models across different settings. The more similarity perceived with the model, the higher the effect of vicarious experience would have on the individual. These quality contacts include volunteer services in professional teaching contexts, alumni sharing, professional exchange, class observations of good practices, site visits, and teacher or peer evaluations.

In senior years, practicum experience becomes a crucial part of the teacher training curriculum. The emphasis should be to deepen pre-service teachers' abilities in transferring both cognitive and teaching skills to practical settings which lead to continuous enhancement in professionalism. The curriculum should aim at developing students' competence by accumulating successful mastery experiences, reflecting on observation models encouraging knowledge and skills synthesis and transfer across settings, both within and outside the academic institution.

## 6 Limitations

Despite the contributions of the present study illustrating the relationship of practicum experiences to sources of teacher efficacy and to both academic self-efficacy and career-specific self-efficacy outcomes of pre-service teachers, there are a number of limitations that need to be addressed in future studies. First, because the population of the profession is typically female dominant, we are unable to conclude that the same findings can be applied to male teachers in other cultures where early childhood educators show a more gender balance. Second, data were drawn from a one-time survey and the self-reported measures used in this study were adopted from existing inventories developed from non-Asian populations. Cultural aspects (e.g., Chinese cultural values and beliefs on self-evaluation, their virtue of being humble, etc., which may have substantial influences on the current sample) have not been considered in this study. Therefore, the current findings do not represent a comprehensive assessment of generalizable patterns across culture and time. Third, causal relationships between practicum, sources of teacher self-efficacy, and self-efficacy outcomes should be interpreted with caution. Longitudinal data in further studies will benefit interpretations of causal ordering by establishing a clear sequence of variables measured at multiple time points. In essence, a variety of materials, designs, and program characteristics needs to be examined for generalizability of the findings. Although the current findings may not be able to generalize to other teacher training programs, the findings are primarily useful and important to early childhood educators and perhaps also to the wider teacher community that share similar concerns (Banks & Mangan, 1999). Finally, qualitative data should be included to provide richer information to cross-validate the quantitative findings here, and to highlight how exactly practicum experience can be framed so as to benefit teacher efficacies in the most effective way.

**Acknowledgements** I would like to thank Professor Alexander Yeung for his expert advice and encouragement.

## Appendix

### *Scales and Sample Items*

Academic self-efficacy (Alpha = 0.90; 5 items).
How confident are you that they could successfully complete the following tasks?

1. Manage time effectively.
2. Take good class notes.

Self-efficacy for Learning (Alpha = 0.92; 8 items).

1. If I believe I will receive an excellent grade in this program.
2. I'm certain I can understand the most difficult materials presented in the readings for this program.

Mastery Experience (Alpha = 0.78; 3 items).

1. I have developed many of my teaching skills by actually teaching.
2. I have learned a great deal from teaching in classrooms.

Vicarious Experience (Alpha = 0.85; 4 items).

1. I have learned about how to be a teacher by watching other skillful teachers.
2. Watching other teachers make mistakes has taught me how to be a more effective teacher.

Social Persuasion (Alpha = 0.77; 4 items).

1. I believe I can teach as well as the teachers portrayed in popular movies.
2. I often compare my own abilities to other teachers.

Physio-emotional Arousal (Alpha = 0.72; 3 items).

1. When I say the wrong things to a class, I become anxious.
2. The idea of being in a classroom as a teacher makes me nervous.

Practicum
Coded: 0 = no practicum; 1 = practicum experience.

## References


Bandura, A. (1977). Self-efficacy: Toward a unifying theory of behavioral change. *Psychological Review, 84,* 191–215. https://doi.org/10.1037//0033-295X.84.2.191

Bandura, A. (1997). *Self-efficacy: The exercise of control.* New York: W.H. Freeman.

Bandura, A. (2006). Guide for constructing self-efficacy scales. In F. Pajares & T. U. Urdan (Eds.), *Self-efficacy beliefs of adolescents* (pp. 307–339). New York: Information Age Publishing.

Banks, C. K., & Mangan, J. M. (1999). *The company of neighbours: Revitalizing community through action-research.* Toronto, ON: University of Toronto Press.

Bong, M., & Clark, R. E. (1999). Comparison between self-concept and self-efficacy in academic motivation research. *Educational Psychologist, 34*(3), 139–153. https://doi.org/10.1207/s15326985ep3403_1

Boyer, D. A., Zollo, J. S., Thompson, C. M., Vancouver, J. B., Shewring, K., & Sims, E. (2000, June). *A quantitative review of the effects of manipulated self-efficacy on performance.* Poster Session Presented at the Annual Meeting of the American Psychological Society, Miami, Florida.

Britner, S. L., & Pajares, F. (2006). Sources of science self-efficacy beliefs of middle school students. *Journal of Research in Science Teaching, 43*(5), 485–499. https://doi.org/10.1002/tea.20131

Browne, M. W., & Cudeck, R. (1993). Alternative ways of assessing model fit. In K. A. Bollen & J. S. Long (Eds.), *Testing structural equation models* (pp. 136–162). Newbury Park, CA: Sage.

Chen, E. S. L., & Rao, N. (2011). Gender socialization in Chinese kindergartens: Teachers' contributions. *Sex Roles, 64,* 103–116. http://dx.doi.org/10.1007/s11199-010-9873-4.

Clift, R. T., & Brady, P. (2005). Research on methods courses and field experiences. In M. Cochran-Smith & K. M. Zeichner (Eds.), *Studying teacher education: The report of the AERA panel on research and teacher education* (pp. 309–424). Mahwah, NJ: Lawrence Erlbaum Associates Publishers.

Connolly, J. (1989). Social self-efficacy in adolescence: Relations with self-concept, social adjustment, and mental health. *Canadian Journal of Behavioral Science, 21,* 258–269.

Dellinger, A. B., Bobbett, J. J., Olivier, D. F., & Ellett, C. D. (2008). Measuring teachers' self-efficacy beliefs: Development and use of the TEBS-Self. *Teaching and Teacher Education, 24,* 751–766. https://doi.org/10.1016/j.tate.2007.02.010

DeWitz, S. J., & Walsh, W. B. (2002). Self-efficacy and college student satisfaction. *Journal of Career Assessment, 10,* 315–326. https://doi.org/10.1177/10672702010003003

Dicke, T., Parker, P. D., Holzberger, D., Kunina-Habenicht, O., Kunter, M., & Leutner, D. (2015). Beginning teachers' efficacy and emotional exhaustion: Latent changes, reciprocity, and the influence of professional knowledge. *Contemporary Educational Psychology, 41,* 62–72.

Eames, C., & Bell, B. (2005). Using sociocultural views of learning to investigate the enculturation of students into the scientific community through work placements. *Canadian Journal of Science, Mathematics and Technology Education, 5,* 153–169. https://doi.org/10.1080/14926150509556649

Field, A. (2013). *Discovering statistics using SPSS: And sex and drugs and rock 'n' roll* (4th ed.). Los Angeles, CA: Sage.

Gibson, S., & Dembo, M. H. (1984). Teacher efficacy: A construct validation. *Journal of Educational Psychology, 76,* 569–582. https://doi.org/10.1037/0022-0663.76.4.569

Gloudemans, H., Schalk, R., Reynaert, W., & Braeken, J. (2013). The development and validation of a five-factor model of sources of self-efficacy in clinical nursing education. *Journal of Nursing Education and Practice, 3,* 80–87. https://doi.org/10.5430/jnep.v3n3p80

Hattie, J., & Timperley, H. (2007). The power of feedback. *Review of Educational Research, 77,* 81–112. https://doi.org/10.3102/003465430298487

Houshmand, A. A., & Papadakis, C. (1995). *One century of cooperative education in the United States 1906–2006*. Philadelphia, PA: Drexel University.

Jöreskog, K. G., & Sörbom, D. (2006). *LISREL 8.80.* Chicago, IL: Scientific Software International, Inc.

Keller-Schneider, M. (2010). *Entwicklungsaufgaben im Berufseinstieg von Lehrpersonen. Beanspruchung durch berufliche Herausforderungen im Zusammenhang mit Kontext- und Persönlichkeitsmerkmalen.* Münster: Waxmann.

Kelley, C. M., & Jacoby, L. L. (1996). Adult egocentrism: Subjective experience versus analytic bases for judgment. *Journal of Memory and Language, 35,* 157–175. https://doi.org/10.1006/jmla.1996.0009

Kieffer, K. M., & Henson, R. K. (2000, April). *Development and validation of the sources of self-efficacy inventory (SOSI): Exploring a new measure of teacher efficacy.* Paper Presented at the Annual Meeting of the National Council on Measurement in Education, New Orleans, Los Angeles.

Klassen, R. M., Bong, M., Usher, E. L., Chong, W. H., Huan, V. S., Wong, I. Y., & Georgiou, T. (2009). Exploring the validity of a teachers' self-efficacy scale in five countries. *Contemporary Educational Psychology, 34*(1), 67–76.

Lane, A. M. (2002). Relationships between performance towards accomplishment and self-efficacy in amateur boxing. *Perceptional and Motor Skills, 94*(3), 1056–1056.

Moritz, S. E., Feltz, D. L., Fahrbach, K. R., & Mack, D. E. (2000). The relation of self-efficacy measures to sport performance: A meta-analytic review. *Research Quarterly for Exercise and Sport, 71,* 280–294. https://doi.org/10.1080/02701367.2000.10608908

Perry, J. C., DeWine, D. B., Duffy, R. D., & Vance, K. S. (2007). The academic self-efficacy of urban youth. A mixed-methods study of a school-to-work program. *Journal of Career Development, 34,* 103–126. https://doi.org/10.1177/0894845307307470

Pintrich, P. R., & DeGroot, E. V. (1990). Motivational and self-regulated learning components of classroom academic performance. *Journal of Educational Psychology, 82,* 33–40.

Pintrich, P. R., Smith, D. A. F., Garcia, T., & McKeachie, W. J. (1991). *A manual for the use of the Motivated Strategies for Learning Questionnaire (MSLQ).* Ann Arbor, MI: National Center for Research to Improve Post-secondary Teaching.

Robbins, J. M., & Krueger, J. I. (2005). Social projection to ingroups and outgroups: A review and meta-analysis. *Personality and Social Psychology Review, 9,* 32–47. https://doi.org/10.1207/s15327957pspr0901_3

Schunk, D. H., & Gunn, T. P. (1986). Self-efficacy and skill development: Influence of task strategies and attributions. *The Journal of Educational Research, 79*(4), 238–244.

Schunk, D. H. (1987). Peer models and children's behavioral change. *Review of Educational Research, 57,* 149–174. https://doi.org/10.3102/00346543057002149

Schunk, D. H. (1991). Self-efficacy and academic motivation. *Educational Psychologist, 26,* 207–231.
Schunk, D. H., & Ertmer, P. A. (1999). Self-regulatory processes during computer skill acquisition: Goal and self-evaluative influences. *Journal of Educational Psychology, 91,* 251–260. https://doi.org/10.1037/0022-0663.91.2.251
Schunk, D. H., & Pajares, F. (2002). The development of academic self-efficacy. In A. Wigfield & J. S. Eccles (Eds.), *Development of achievement motivation* (pp. 15–31). San Diego, CA: Academic Press.
Schön, D. A. (1991). *The reflective turn: Case studies in and on educational practice.* New York: Teachers College Press.
Schwarzer, R. (2014). *Self-efficacy: Thought control of action.* New York: Routledge.
Sun, R. C. F., & Hui, E. K. P. (2012). Cognitive competence as a positive youth development construct: A conceptual review. *The Scientific World Journal*, 210953. https://doi.org/10.1100/2012/21095.
Tschannen-Moran, M., Hoy, A. W., & Hoy, W. K. (1998). Teacher efficacy: Its meaning and measure. *Review of Educational Research, 68,* 202–248. https://doi.org/10.3102/00346543068002202
Zimmerman, B. J., & Martinez-Pons, M. (1988). Construct validation of a strategy model of student self-regulated learning. *Journal of Educational Psychology, 80*(3), 284–290.
Zimmerman, B. J., & Bandura, A. (1994). Impact of self-regulatory influences on writing course attainment. *American Educational Research Journal, 31,* 845–862.
Zimmerman, B. J. (2000). Self-efficacy: An essential motive to learn. *Contemporary Educational Psychology, 25,* 82–91. https://doi.org/10.1006/ceps.1999.1016
Zimmerman, B. J., Bandura, A., & Martinez-Pons, M. (1992). Self-motivating for academic attainment: The role of self-efficacy beliefs and personal goal setting. *American Educational Research Journal, 29*(3), 663–676.

**Dr. Eunice Yim** is a chartered psychologist and an associate professor at the School of Education and Languages at The Open University of Hong Kong. Her research interests include parent education, special education, early childhood education, behavioral science, and whole person development. Dr. Yim has published book chapters and refereed journal articles.

# Metaphor Analysis of International Doctoral Students' Learning Experiences—A Case Study of Chinese Doctoral Students in Australia

**Xing Xu, Hing Wa Sit, and Shen Chen**

**Abstract** Against the tidal wave of internationalization of higher education is the change of learning environment once student sojourners begin their studies abroad. International doctoral students' perceptions of their lived experiences of learning overseas have attracted wide attention. Yet a few studies have addressed students' assessment of doctoral education via a metaphor analysis. Shedding light on the experiences of Chinese doctoral students in Australia as a case, interviews were conducted in Australian universities. Through metaphor elicitation, their doctoral education experiences within the context of the institutional and academic learning environment were described and discussed. The results showed that the conceptualization of the experience resided in the intersection of two pairs of dichotomies: (1) outcome orientation (goals accomplishment) versus process orientation (experiential exploration) and (2) internal factors (educational efforts and cross-cultural adaptation) versus external factors (learning environment and educational change). This study bears wide implications for international Chinese PhD students registered around the globe.

**KeyWords** Overseas learners · Learning environment/context · Learning analytics · Internationalization of doctoral education · Metaphor analysis · The role of educators

X. Xu (✉)
School of International Relations, Sichuan International Studies University, 400031 Chongqing, China
e-mail: xing.xu@uon.edu.au

H. W. Sit · S. Chen
School of Education, the University of Newcastle, Callaghan Campus, 2308 Newcastle, NSW, Australia
e-mail: helena.sit@newcastle.edu.au

S. Chen
e-mail: Shen.chen@newcastle.edu.au



W. W. K. Ma et al. (eds.), *Learning Environment and Design*, Educational Communications and Technology Yearbook,
https://doi.org/10.1007/978-981-15-8167-0_4

## 1 Introduction

The internationalization of doctoral education has been a global phenomenon as governments strive to build their respective knowledge economies. Subsequently, scholarly attention has been paid to the international doctoral student cohort. In this research, a group of overseas Chinese doctoral students were examined as a case study to expand the pertinent scholarship. Although there is a growing interest among scholars to explore the lived experiences of international research students in Australia (e.g., Son & Park, 2014; Soong, Thi Tran, & Hoa Hiep, 2015; Yu & Wright, 2016), which stands as one of the top four importing countries of doctoral students, a few have focused specifically on the single biggest learner group—namely, the Chinese cohort—the exact number of which is, however, not readily accessible due to the unavailability of official data (Shen, Wang, & Jin, 2016). Despite recent efforts to address this gap (e.g., Ai, 2017; Xu & Hu, 2019), there remains a scarcity of Chinese doctoral students' emic assessment of their lived experiences of learning in the global educational context (Xu, Sit, & Chen, 2020). This study aspired to give substance to that gap. Grounded in metaphor analysis, the research endeavored to explore how students conceptualize the doctoral trajectory that might be culturally and/or contextually bound. Commencing with an overview of pertinent research on the Chinese learner as a whole and Chinese doctoral students specifically, theoretical and empirical underpinnings of metaphor analysis are examined, followed by a description of the research design. Building upon an analysis of the findings, the discussion highlights from a cultural linguistics view how, most notably, the perceptions of overseas doctoral students on their learning experiences manifested salient cultural schemas that were entrenched in Chinese culture.

## 2 An Overview of Chinese Learners

Members of different cultural communities may have different perceptions, expectations, values, and beliefs about learning and how to learn. To unpack Chinese doctoral students' perceptions of their overseas study experiences, it is necessary to cast some light on this particular cohort.

The term Chinese learner was first used by Watkins and Biggs (1996) to refer to Chinese students from Confucian heritage cultures. Since their pioneering scholarships, many researchers have paid attention to the learning and teaching of Chinese students across the globe, with a focus on demystifying the paradox of the Chinese learners who are censured for their outward demeanor of passivity and yet excel in their studies with remarkable achievements (e.g., Marton, Dall'Alba, & Tse, 1996; Mok et al., 2001). For example, challenging this cohort as surface learners, Dahlin and Watkins (2000) argued that both rote memorization and understanding are employed by the Chinese learner as complementary learning strategies which lead to higher cognitive outcomes. Kember (2016) contended that in most Asian countries where education is motivated by family and societal orientation, students are pressured to obtain good results in examination-oriented elite educational systems.

Regardless of the paradox debate, others have criticized the notion of the Chinese learner as inappropriately reductionist. On the one hand, it is suspicious that the so-called "Asian values" that define this cohort's Chineseness would be exclusive only to Confucian heritage cultures and not other cultural societies (Ryan & Louie, 2005). On the other hand, the changing sociohistorical context complicates interpretation of these original values and contributes multiple ideological sources to shape the Chinese learners' learning styles (Ryan, 2010). Such views are encapsulated in Holliday and Amadasi's (2020) perception of culture as decentering unbounded hybrid personal narratives that are formed in everyday experiences rather than grand narratives imposed upon a group that work as structures to confine and prescribe them. We as authors although disapproving of an essentialist approach to the Chinese learners, do however corroborate that they might share regularities and patterns of features in their perceptions of learning since "human mental functioning … is inherently social, or sociocultural, in that it incorporates socially evolved and socially organized cultural tools" (Wertsch & Tulviste, 1992, p. 551). As noted by Li (2016), there are cultural models that function as conceptual maps, scripts, or ethnotheories pertaining to human learning in specific cultures. An array of such models in explaining the Chinese learner are widely discussed. For example, the value of education for upward social mobility (Rao, Ng, & Sun, 2016), an emphasis on the importance accorded to effort (Rao & Chan, 2010), and beliefs of moral obligation in relation to academic achievement (Tao & Hong, 2014) are recognized common features that have their cultural embeddedness in Confucianism.

Nevertheless, despite a few exceptions (e.g., Ingleby & Chung, 2009; Hu, van Veen, & Corda, 2016), the current literature on the Chinese learner is centered around cohorts at the graduate level or below, and only a small number of studies have probed into Chinese doctoral students, not least on those studying overseas whose learning experiences might (or might not) reflect Chinese cultural models due to their transnational educational mobility. With an influx of Chinese doctoral students to further their education abroad, pertinent research on their learning experiences has flourished. These studies cover a wide range of topics, such as their persistent motivation (Zhou, 2014), academic socialization (Wu, 2017), identity formation (Ye, 2018), student–supervisor relationship (Shen, 2018), and career decisions (Lee, McMahon, & Watson, 2018). These efforts have enriched pertinent scholarship grounded in a plethora of theoretical underpinnings such as cross-cultural, sociological, and psychological perspectives. However, as an important composite of the Chinese learner, the scholarly light cast on the Chinese doctoral student is disproportionately scant given the huge number in this cohort. Besides the most important agents of their learning trajectory, there are limited voices from learners themselves to assess their lived experiences of studying overseas. This is the lacuna that this study aims to address via metaphor analysis, a sociolinguistic tool that unearths people's ways of structuring meaning, both culturally and socially. Elicited metaphor analysis in educational discourse encourages a participant to "recognise, discuss and evaluate one's educational beliefs and working practices in metaphoric terms" (Wan & Low,

2015, p. 1). It has been claimed as an effective strategy for increasing self-recognition and enhancing learning during one's learning journey. Bian's (2010) metaphor analysis of social sciences PhD students based in a Chinese university revealed four major factors influencing the participants' academic research experiences. Insightful as the study was, it did not take into account the background of the participants as Chinese learners and offered no deep investigation into hidden cultural forces behind these factors. To date, no such research has been conducted in the Australian higher education context. The present study, therefore, attempts to expand the scholarship pertinent to this particular cohort, by delving into the sociocultural schemas embedded in the metaphor conceptualization of Chinese PhD students. It is hoped that these findings will prompt further research into other cultural cohorts, thus contributing to practice and inquiry in the internationalization of doctoral education.

# 3 The Study

## *3.1 Method: Metaphor Analysis*

In general, metaphor is "a figure of speech in which one thing is compared to another by saying that one is the other" (Kövecses, 2002, p. vii). Metaphor as a mediating psychological tool offers researchers access to participants' conceptualization of a broader repertoire of cognitive models that coherently represent multimodal conceptual entities (Evan, 2006) in which their attitudes, beliefs, and goals are embedded. Uncovering veiled assumptions of awareness, it affords a means to understand how people make sense of their lives, including how they think, speak, and act. As a research method, it is suitable for the current study, which delves into an emic understanding of lived learning experience.

Since Lakoff and Johnson's (1980) initiated conceptual metaphor theory (CMT), it has dominated scholarship pertinent to the systematic analysis of metaphor, despite the fact that criticisms persist with regard to "the explanatory value of 'conceptual metaphor' construct" (McGlone, 2007, p. 109). According to CMT, analyzing metaphors that people employ to perceive an experience or object will give insight into the mental constructs configuring their cognition of the world. Relating an abstract concept, action, or phenomenon to concrete sensorimotor experiences via metaphorical mapping (Amin, Jeppsson, & Haglund, 2015), people actively produce thoughts, including those hidden at the subconscious level. Investigating the consensus of semantic entailments embedded in a group of metaphors may reveal an overarching knowledge gestalt or an image schema predominating the construction of shared meanings in terms of an experience (Kövecses, 2010). That being so, our conceptualization of the nuance and complexity of a given phenomenon underpinned by metaphorical concepts can be articulated herein.

There has been some pertinent research illuminating doctoral education referring to metaphor analysis, including supervisor–student relationship (e.g., Mackinnon,

2004; Grant, 2008), doctoral students' competences (e.g., Baptista & Huet, 2012), ontological understanding of supervision (e.g., Lee & Green 2009), and attributes of supervisors (e.g., Grant, Hackney, & Edgar, 2014). Nevertheless, there is still a rarity of empirical work that examined doctoral students' experiential reflection on their PhD trajectories based on the narrative of metaphors. To the authors' best knowledge, no research has yet specifically focused on the Chinese cohort. The current study therefore attempts to bridge this gap.

## 3.2 Research Design and Process

This study forms part of a larger research project undertaken by the researchers. It was carried out in 11 institutions located in different states, such as Western Australia, New South Wales, Victoria, and Queensland, throughout Australia. A qualitative-oriented research method through semi-structured interviews was used to gain Chinese overseas PhD students' perceptions of their doctoral journeys in the new Australian educational discourse. This method is appropriate for the current study as its flexibility to promote fruitful reflection by the participants (Mill, 2001) can ensure a collection of data necessary to understand the phenomenon under investigation (Holstein & Gubrium, 1995). After obtaining ethics approval, the recruitment of participants was launched via social media by using a snowball strategy. Students who grew up in China and had come to Australia for doctoral study were invited for a 15-minute, one-on-one, face-to-face interview. By means of WeChat and QQ groups of Chinese Student Associations across universities in Australia, the recruitment secured 25 participants. The inclusion of more than one individual as well as time and space as sources of data was executed following what Denzin (1978) termed data triangulation to understand a social phenomenon under a variety of conditions. Despite a relatively small sample size, the participants came from a wide range of disciplines and from across different years into their candidature at their respective universities (see Table 1). The university names have been coded from U1 to U11 and the students from S1 to S25.

The overriding research question guiding this study was: "How do Chinese doctoral students use metaphors to make sense of their learning experiences in the Australian academic environment?" All participants were invited to express their feelings and viewpoints. Depending on the participants' availability and preference, either face-to-face (11 out of 25) or telephonic (14 out of 25) interviews were conducted. In these interviews, they were encouraged to provide a metaphor best capturing their experiences of their PhD trajectories. In order to prompt them to flesh out their elicitations, questions asked were open-ended, such as "What metaphor would you employ to depict your PhD experience?", "Why do you think your PhD experience is analogous to it?", and "Can you be more specific regarding the similarities between your PhD experience and the metaphor?" Most participants responded by starting with "I think my PhD is…" and further elaborated their rationale with a short explanation.

**Table 1** Demographic Information of Participants

| Participant | Gender | Academic discipline | Stage of PhD at time of interview (year) | University |
|---|---|---|---|---|
| S1 | Male | Social Science | 2 | U5 |
| S2 | Male | Medicine | 2 | U1 |
| S3 | Male | Social Science | 1 | U7 |
| S4 | Female | Medicine | 2 | U1 |
| S5 | Male | Engineering | 3 | U1 |
| S6 | Female | Social Science | 1 | U3 |
| S7 | Male | Engineering | 3 | U8 |
| S8 | Male | Engineering | 2 | U4 |
| S9 | Male | Science | 3 | U9 |
| S10 | Male | Social Science | 2 | U8 |
| S11 | Male | Business | 4 | U1 |
| S12 | Male | Humanities | 1 | U8 |
| S13 | Male | Engineering | 3 | U1 |
| S14 | Female | Medicine | 2 | U6 |
| S15 | Male | Engineering | 3 | U10 |
| S16 | Female | Humanities | 2 | U8 |
| S17 | Female | Medicine | 1 | U5 |
| S18 | Female | Social Science | 4 | U1 |
| S19 | Male | Engineering | 1 | U1 |
| S20 | Male | Engineering | 2 | U11 |
| S21 | Female | Engineering | 4 | U1 |
| S22 | Female | Business | 2 | U1 |
| S23 | Female | Humanities | 2 | U5 |
| S24 | Female | Science | 4 | U2 |
| S25 | Male | Medicine | 2 | U1 |

The participants' responses were recorded, transcribed, and translated verbatim. All data were then imported into NVivo 12 for data analysis. To boost credibility, before analysis, texts were returned to some participants to verify if the account was accurate, impartial, and complete. Also, as suggested by Mathison (1988), it is good practice to involve more than one investigator in the research process; thus investigator triangulation was adopted. This involved rounds of discussions and revisions among the investigators (namely, the authors) to reach a consensus prior to the finalization of codes elaborated below, which ensured that dialogical intersubjectivity (Benjamin, Williams, & Maher, 2017) mitigated bias thus generating validity. It is noteworthy that the data analysis was inductive, viz., the findings were categorized into a hierarchy of conceptualization emerging from within, although it was unavoidably informed by theoretical underpinnings and epistemological positions taken by the authors. Data analysis was conducted within a six-phased thematic analysis, argued by Braun and Clarke (2006) as a foundational method for qualitative analysis. After rounds of reading, we familiarized ourselves with the data, which were segmented into basic meaning units that were subjected to initial codes. Themes were then generated and reviewed with a cross-code examination. After discussion, the research team reached consensus on themes which were then defined and named. The last phase was to organize these findings into a final report, as below.

# 4 Results

All 25 metaphors were mapped onto each category not so much for the mere literal sense suggested by each metaphor per se as for an analytic judgement of a dominant metaphorical concept illuminated by the participants' response. Further, although the ensuing discussion was developed within two dichotomies, metaphors were often not binary opposites of one or the other; they were allocated to a category more for the degree than for the nature of taking possession of that particular pattern. In the results section, metaphors and segments of elicitations were interspersed to ensure that analysis was grounded in actual expressions intended by the users. The next section delineates the results of data analysis.

## *4.1 Outcome orientation versus Process orientation*

In this study, a dichotomy of outcome and process orientations was firstly highlighted by the participants regarding their experiential explorations of their doctoral journeys and the process of accomplishing the common goal of obtaining their PhD degrees.

### 4.1.1 Outcome orientation

Eleven participants used metaphors to entrench a nucleus of seeing the doctorate process as a goal-accomplishing activity. Participants were clear about their goal of completing their studies; thus, adopting an outcome-oriented approach may be most effective in motivating them and in sustaining their learning interests. As McCulloch (2013) suggests, the objects pursued may take various forms as participants longed for their objects in different combinations and to varying degrees from their doctoral education. For example, six participants perceived their doctoral education as a final-goal working purpose, though tough but still worthwhile to accomplish. They demonstrated a propensity to view the doctorate as working toward specific goals.

> I think PhD is a lengthy war, consuming a lot of my time, energy and health … I will be extremely happy the day I win the war when I get my degree. (S3)
>
> It's like the rebirth of a phoenix rising from the ashes. There is a whole range of tests to go through from which will be a rebirth. Further, there are a lot of struggles and torments; but once you overcome them, your life will have a new beginning. (S5)
>
> My PhD is … fighting for success despite a series of wars defeated.… The process is arduous and full of setbacks … but I am looking forward to my success in the future.… I am so desirous of getting my first payslip upon graduation. (S13)

Be it "win[ing] a war", "a rebirth", or "fighting for success", the goals they set were of great significance and value, denoting high expectations they attached to (the process of obtaining) the title "Doctor" and projecting a strong sense of honor as well as achievement once their goals are accomplished. On the other hand, expressions like "consuming a lot of my time, energy, and health", "a lot of struggles and torments", and "arduous and full of setbacks" supported the other common feature of one's acute self-awareness regarding how onerous and back-breaking it is to finally acquire such a precious and desirous result.

By contrast, another five interviewees held less challenging perceptions. Here are some representative viewpoints:

> My PhD study is a flight journey …. There is a clear departure and destination. During the whole journey, I must guarantee a sufficient provision of fuel, lest the plane drops down. It is just like I need to continuously enrich myself so that I can finish the journey with a safe landing. (S4)
>
> It is a process of climbing a mountain. Although I tend to think of it as a painstaking process, it is still rewarding as I obtain a lot. Even though I feel very exhausted, I continue to climb bit by bit, ever reaching the peak. (S14)

Despite the fact that these participants found their PhD journeys "exhaust[ing]", and there was a need of "sufficient provision of fuel", they planned to "finish the journey with a safe landing" and to "reach the peak" if they could "continuously enrich self" and "climb bit by bit". For these participants, the attainability was more predictable as the process was a routinized journey with a known departure and destination mapped out in advance. Self-driven perseverance, persistence, and patience to overcome weaknesses were considered vital in the process of learning.

### 4.1.2 Process orientation

In comparison, process-oriented perspectives were expressed by another nine participants who considered the doctorate journey as an experiential exploration. They cared more for developing a plan of action that could guide them to carry out their research projects and sail through their doctorate routes. In contrast to the participants who were outcome-oriented, those with a process orientation valued the qualities of "doctorateness" or more broadly of "personness" and tended not to focus on pursuing tangible and/or intangible objects (i.e., credentials, publications, jobs, reputation, and self-actualization).

With regard to their day-to-day PhD-related tasks, five participants noted their understanding of the attributes, such as solitude and ambiguity, that impact on their PhD education trajectories. The general feedback of being lonely learners was noted:

> I am a raft adrift in the sea, feeling helpless. Because you know very few understand your research and very few can help you. The sea is changeable and I don't know where I am floating to. I have no idea if I will remain in academia. (S16)
>
> Like (running) a marathon in a mountain forest .... In sum, it is lonely because you are all alone, although there are family and friends; yet, on more occasions doing research is independent. (S22)

Further, the PhD study process was also associated with ambiguity for these participants whose responses ranged from enthusiasm to intolerance. These participants manifested that doing a PhD entailed an exploration into the boundless unknown. Below are the reflected points regarding their ambiguity levels to the doctorate study:

> This metaphor should be an adventure because there are new things happening, day by day. Every day is full of the unknown. I think the most adventurous part is that there is no destination because every day is a new day. (S1)
>
> Just a raft adrift in the sea. I have a sense of fear that it is impossible for me to learn all that I need to know so I feel helpless and vulnerable .... Like Zheng Jun's lyrics: "In the foreign land that I am familiar with, I send myself into exile, year after year". I feel more or less like that. Even now, I do not have any destination. (S9)

As shown explicitly, doctoral education was a milieu where uncertainty, unpredictability, and blurred boundaries played out at all levels (Tesch et al. 2015). Confronting its ambiguity, students harbored opposing perceptions, either as desirable or threatening. As PhD students are usually expected to be critical thinkers and new knowledge creators, uncertainties may prevail the process of producing originality. It is consistent with what Kumaravadivelu (2008a, b) notes, "change produces anxiety, particularly if it involves a move from the comfortable climate of familiarity to an unpredictable arena of uncertainty" (p. 225). However, he further argues that these changes are less disorientating when being developed within a context in which one can play a decision-making role, implement their decisions, and succeed in their overseas doctorate learning. As highlighted below, one student commented:

> Studying abroad is not as smooth as it would be at home due to language challenges and cultural differences. If using a metaphor to explain this, I think this is like a process of cultivating immortals. First of all, you will experience all kinds of difficulties, grit your teeth to work even harder and it will be relatively smooth .... It really is a process of cultivation. (S25)

## *4.2 Internal Factors versus External Factors*

The other level of conceptualization revealed that both internal and external forces were inclined to either positively or negatively affect these doctorate students' learning journey. Discussing their experiences in metaphorical terms enabled the participants to further describe their self-motivated learning traits and to identify specific issues as well as possible ways to overcome challenges that they may face in the current and future educational changes and/or trends.

### 4.2.1 Internal Factors

Twelve participants felt their doctoral education required intrinsic learning motivation and tremendous self-directed learning efforts. These students expressed that their inner strengths and adaptabilities affected their long-term academic outcomes:

> Like a growing seedling.... The seedling has very good genes but what it will turn into is subject to external factors. But I am hopeful regarding it because it is ever growing. (S24)

> I think it is just like once an arrow is discharged from the bow, there is no turning back. Once an arrow is discharged, the direction it moves towards and the landing point are circumscribed within a boundary. The running course is subject to the adaptation of external forces such as wind. Similar to the PhD, there are many limitations from the outside. All I can do is, with the joint efforts of my supervisor, to set a reasonable target in the beginning when I choose my research topic and try the utmost to shoot the arrow towards the target accurately. (S11)

As supported by Bronfenbrenner and Morris (2006), the findings showed that "human development takes place through processes of progressively more complex reciprocal interaction between an active, evolving biopsychological human organism and the persons, objects, and symbols in its immediate external environment" (p. 797). The juxtaposition of internal forces, viz., "good genes" and personal efforts, and external forces, such as "wind", navigates the doctoral trajectory. At the end of the spectrum, students who have experienced greater amounts of engaged concentration are more likely to achieve the preset goals.

### 4.2.2 External Factors

Eight students also suggested that the doctoral journey was circumscribed by external factors such as a new institution and learning environment, teachers' and supervisors' guidance, and changing technology and multimodality involved in their doctorate studies.

I am like Sisyphus who rolls an immense boulder up a hill. Whenever I try to fix a problem, feeling like a solution is to be found, a bigger problem comes to the surface and defeats me all the way to the start.... Just like Sisyphus, I feel I am pushing the boulder up and not strong enough to move it. But I keep doing it. (S7)

The quote above reveals the struggles experienced by this participant, whose doctoral journey has been fraught with unexpected challenges that feel insurmountable. Nevertheless, the participant enacted strong agency to "keep doing it" although knowing "not strong enough to move it". This finding provides supporting evidence to show that becoming a doctor is primarily dependent upon the student, the leading actor, who "when faced with apparently insurmountable problems, finds both the courage and means to solve, detour around, or design them out" (McCulloch, 2013, p. 62).

Moreover, along with the digital world, the need to cope with the context of educational change and multimodality was also identified, as seen by the metaphors "information explosion era", "AI-mediated experiments", and "the 72 polymorphic transformations" given by S25, whose doctoral study was in the field of medicine. PhD education is co-shaped by both internal and external influences that have the potential to impact their learning outcomes. The participant viewed curriculum design involving digital tools as an integrated part of their continuous learning practices and of the future landscape of the chosen field.

> [L]iving in the era of information explosion, our current learning operations have been computerized and technologized, and we also need to adapt to different changes, master and update new technological knowledge and apply it to medical education …. For example, the use of artificial intelligence is about to be applied in medical clinical operations and practice soon. Usually, the CT images, MRI results, and X-rays are read and judged by doctors and specialists themselves. However, the use of artificial intelligence can help analyze more objective results, improve result accuracy, and provide clinicians with good data reference …. if using a metaphor to express my current feeling, I think that the impact of adapting new technology to the current learning environment and working environment, we need to become Sun Wukong [a main character, Monkey King, in the Chinese classical novel *Journey to the West*, who experiences 72 polymorphic transformations] with 72 changes. (S25)

These elicitations demonstrate that the participants had been considering such questions as what is the role of the teacher/supervisor or of the students themselves? How is the learning instruction? What is the relationship between the learning trajectory and the kinds of twenty-first century global citizens in this artificial intelligence and digital era? The participants revealed they needed to quickly identify any possible changes required in their everyday learning environments. Not only did they possess determination, learning spirit, and strength of willpower, they also demonstrated a willingness to endure arduous effort and be adaptive to institutional policies, supervisor–student relationships, and research communities.

## 5 Discussion

Having presented and reviewed the metaphorical description and analysis of interview data, four emerging points will be discussed in further detail below.

First, a close scrutiny of these findings gives weight to what Sharifian (2013, 2014, 2015) contended from a cultural linguistics view regarding the interplay between language attainment, cultural environment, and conceptualization. The participants' metaphorical conceptualizations resonate with Sharifian (2017) who described the source of metaphors as being typically derived from shared bodily experiences that are grounded in cultural schemas and abstracted from the collective cognitions associated within a cultural group. For example, in metaphors such as "war" (S3, S13, S16), "the rebirth of a phoenix rising from the ashes" (S5), and "Sisyphus who rolls an immense boulder up a hill" (S7) the doctoral journey is analogous to a certain type of bodily experience. Although the participants did not personally undergo them, by imagining themselves as the agent, they projected these experiences onto their own bodies. As Yu (2015) noted, the only way humans know, understand, think, and reason is from and within our body whose sensorimotor systems set up the contours of what we perceive and understand. Via these imaginary bodily experiences involving a variety of organs, the participants illuminated their feelings, emotions, mood, etc., of their doctoral education.

Second, metaphor analysis revealed that the participants' general perception was that doing their PhDs was not only effortful but also valuable. From a cultural linguistics view, this notion is a typical cultural schema often associated with the Chinese education context. As localized conceptualizations of how the world works, cultural schemas are shared cognitive structures distributed within a particular culture. Heavily influenced by its historical and philosophical underpinnings, it is a cultural schema embedded in the Chinese culture that education is of supreme significance and must be achieved through hard work. Since the enlightening of education, sayings such as "Everything else is low-grade, only study is above all" and "In books one finds golden mansions and maidens as beautiful as jade" have infiltrated the Chinese psyche as a deep-seated culturally constructed conceptualization. Be it the thousand-year-old Chinese Imperial Examinations initiated in the Sui Dynasty or its modern variation, the college entrance examination, they are tools that are widely adopted to secure socially upward mobility in the eyes of the Chinese. As Kipnis (2011) noted, the origin that Chinese people give high priority to education lies in the notion that education equates to a system of prestige and those with educational accomplishments are thus superior to the non-educated. Given that the doctorate is the highest academic degree that one can possibly obtain in the current education system, it comes as no surprise that the supremacy of importance was indicated by participants through such notions as "rebirth" and "triumph of a war".

There is also cultural embeddedness in regarding education to be effortful. Contrary to Western philosophy which holds that it is the individual's innate ability or capacity to learn, Chinese education is entrenched in Confucian philosophy that values effort, hard work, and endurance as determinants of learning outcome (Lee, 1996). While Western cultures support a manifestation of self-worth in personal competence (Covington & Omelich, 1979) that seems a stable and uncontrollable disposition, Chinese philosophy emphasizes human perfectibility through enhancing ability that is "malleable and under one's control" (Salili, 1996, p. 93). As discussed earlier, this cultural schema was markedly echoed by participants who were committed to accomplishing their mission with strenuous effort and sacrifices. Also, some other metaphorical elicitations equating the doctoral education to an exploration into qualities of "personness", exemplarily manifested in metaphors such as "trip" (S2) and "growing tree" (S10), have their cultural embeddedness in Confucianism that views the learning process of becoming a full human as the primary concern (Tu, 1985). In these and similar metaphors, the doctoral trajectory is considered to be an unfolding and formative process of self-cultivation that humans as an open system continuously engage in and devote to. This line of thought embodies a classic Confucian concept which considers learning as self-cultivation that endorses the broadening of the self to be fuller, more expansive, and receptive to the world (Tu, 1985).

Third, the findings revealed the impact of both internal and external forces on the learning journey, as what S11 and S24 described as "wind", and "with the joint efforts with my supervisor … towards the target accurately". Chen and Sit (2017), in their study on the supervision of research students in Australian universities, pointed out that while relationships between international students and their Australian supervisors are developing and changing, they are still influenced by students' and teachers' home cultures. Like many other Western countries, the role of a teacher in the Australian learning environment is regarded as a facilitator who believes it is his/her basic responsibility to help students develop and foster their abilities to be critical thinkers, independent learners, and voice-makers of their own areas of research interests. In addition to being independent learners and critical thinkers, Brick (2011) further identified Australian students to be apprentices and problem solvers. She explained: "the essential thing about an apprentice is that they (students) learn by doing … and are constantly expected to apply what they are learning in real situations" (p. 54). Rather than providing the students direct solutions to a given problem, it is not surprising to see that most Australian supervisors are more inclined to "raise questions without answering them" (Brick, 2011, p. 53) because teachers would expect students to evaluate various approaches to a question and encourage them to decide the most useful approach to fulfill the research purpose on their own. This helps us understand why some of the study participants likened their doctoral education experiences to "roll[ing] an immense boulder up a hill". As such, it is instructive for teachers/supervisors to help their international students be aware of the cultural differences between their home countries and Australia before supervising students' PhD studies.

In the meantime, the participants were motivated by divergent goals entailing complexities and subtleties circumscribed by what they look for in the upcoming trajectory. Their decision-making was subjected to practical considerations and shaped by various social agents and situational factors in different subsystems within the bioecological system in which they were situated. While perceptions vary, they manifested a high degree of autonomy in the academic context and/or in everyday life. Bringing different knowledge, competency, values, and expectations, coupled with varying performability in affective, behavioral, and cognitive aspects, the subjects were supposed to enact their navigation of the doctoral journey in (dis)similar fashions. From this perspective, the adjustment of the relationships between international students and their Australian supervisors would require more mutual efforts.

Fourth, digital humanities as an emerging global educational trend has had an impact on the students' PhD studies. PhD trajectory is more self-directed, research-based, and directly relevant to students and their inquiry-oriented learning practices. Aligned with the rapidly changing educational environment, for example, in medicine and healthcare education, as expressed by S25, the trend of using technology-enhanced learning and diverse pedagogical practices has continued to evolve throughout their doctoral studies and clinical training. Guze (2015) stated that "the trend in the use of technology has primarily developed in response to the challenges facing medical education" (p. 260). Over the years, the use of digital tools and learning management systems has been increasingly incorporated at all educational levels, allowing teachers to transform their instructional activities while giving learners the opportunity to engage with or access virtual communities. Whether we like or dislike it, people need to realize the full potential of technology's impact on teaching and learning across the curriculum. There is an assumption that learners, especially youngsters, are digital natives; however, their possible anxiety, disorientation, or varying understanding of the value of computer-assisted or digital-medicated learning design and the training environment need to be considered (Yang, 2001). This is also echoed by Lotherington (2004) who suggests that younger learners today "were born into a world complete with digital gizmos. For them ... their history is one of computerisation" (p. 317). As a result, this paper also represents the authors' attempts to throw light upon any noteworthy explorations of new possible ways of teaching and supervising international doctorate students in the twenty-first century.

## 6 Conclusion

Metaphors mirror a variety of thoughts, images, and feelings. According to this study, the learner's education experience in an alien environment is never linear but involves multifarious dynamics—saturated in complexity and uncertainty as well as trial and error, subjected to interior tenacity and exterior impacts, and brimming with emotional ups and downs. The above metaphor analysis brought to light Chinese doctoral students' conceptualization of their learning experiences in Australia within two pairs of dichotomies, namely, outcome orientation versus process orientation, and internal forces versus external forces. As evinced by the findings above, in the broadest sense, those "psychologically held understandings, premises, or propositions" (Richardson, 1996, p. 103) constitute cognitive aspects of what the participants believe the doctoral experience is and communicate an indication of their behavioral approach toward it. From a cultural linguistics aspect, the metaphors indexing the perception of PhD study manifest some salient cultural schemas entrenched in the Chinese culture. They cast light on how the Chinese doctoral students' assessment of their learning experiences is subjected to cultural impacts and sociohistorical constraints. The study reveals that despite how the participants differed with regard to gender, discipline, and years into candidature, these variables did not seem to significantly affect their perception of learning trajectory, as metaphors were shared across cases although they undoubtedly featured divergent academic experiences that were subjected to these variables. This, presumably, is one limitation of this small-size study that yet allowed for a meaningful search for subtle patterns shaped by different variables, which is anticipated to be tackled by future studies with large datasets involving more participants.

Backdropped by the growing trend of internationalization of education, this study suggests practical implications for stakeholders in education in the global context.

In an era when educational internationalization presents itself as a major shift in the changing landscape of education, it is imperative to probe more into the emic conceptualization of learners' assessment of learning experiences in diverse learning environments. For example, it warrants more cultural sensitivity from supervisors who should enact pedagogical principles to cater for the needs of international students who may encounter additional challenges studying in an alien environment. Also, operationally and mentally, supervisors should adopt a two-way street perspective where international students are valued as assets contributing to an interconnected world due to their unique cultural capital that could potentially be harnessed. Furthermore, as significant constituents of external forces, policy-makers at institutional and national levels should heed the strengthening of formal and informal organizational socialization (Rhoads, Zheng, & Sun, 2016) within and among campuses in order to cultivate a diversified learning environment.

This study does not imply any generalization of the cohort of Chinese doctoral students given that their intrinsic heterogeneity resulting from shaping forces other than a shared culture may differentiate one member from the others. Nevertheless, it is the authors' hope that this piece has filled the lacuna proposed in the introduction and that additional research can be conducted to explore more sociocultural schemas entrenched in the metaphorical conceptualization of Chinese students and other student cohorts around the world.

## References


Ai, B. (2017). Constructing an academic identity in Australia: An autoethnographic narrative. *Higher Education Research & Development, 36*(6), 1095–1107.

Amin, T. G., Jeppsson, F., & Haglund, J. (2015). Conceptual metaphor and embodied cognition in science learning. *International Journal of Science Education, 37*(5–6), 745–758.

Baptista, A. V., & Huet, I. (2012). Making sense of metaphors about doctoral students' competences: Analysis of supervisors' voices. *Procedia - Social and Behavioral Sciences, 47,* 930–937.

Benjamin, S., Williams, J., & Maher, M. A. (2017). Focusing the lens to share the story: Using photographs and interviews to explore doctoral students' sense of well-being. *International Journal of Doctoral Studies*, 12, 197–217. Retrieved from http://www.informingscience.org/Publications/3862

Bian, G. Y. (2010). Academic research experiences of PhD students in social sciences: A metaphor analysis. *Modern University Education, 3,* 28–33.

Braun, V., & Clarke, V. (2006). Using thematic analysis in psychology. *Qualitative Research in Psychology, 3*(2), 77–101.

Brick, J. (2011). *Academic culture: A student's guide to studying at university* (2nd ed.). Melbourne, Australia: Macmillan Education.

Bronfenbrenner, U., & Morris, P. A. (2006). The bioecological model of human development. In W. Damen & R. M. Lerner (Eds.), *Handbook of child psychology: Theoretical models of human development* (pp. 793–828). New York, NY: Wiley.

Chen, S., & Sit, H. W. (2017). Developing a "3-I model" for supervising international students' PhD research at Australian universities. *International Journal of Learning in Higher Education, 24,* 13–21.

Covington, M. V., & Omelich, C. L. (1979). It's best to be able and virtuous too: Student and teacher evaluative responses to successful effort. *Journal of Educational Psychology, 71,* 686–700. https://doi.org/10.1037/0022-0663.71.5.688.

Dahlin, B., & Watkins, D. A. (2000). The role of repetition in the processes of memorizing and understanding: A comparison of the views of German and Chinese secondary school students in Hong Kong. *British Journal of Educational Psychology, 70,* 65–84.

Denzin, N. K. (1978). *The research act: A theoretical introduction to sociological methods.* New York, NY: McGraw-Hill.

Evan, V. (2006). Lexical concepts, cognitive models and meaning construction. *Cognitive Linguistics, 17,* 491–534.

Grant, B. M. (2008). Agonistic struggle: Master slave dialogues in humanities supervision. *Arts and Humanities in Higher Education, 7*(1), 9–27.

Grant, K., Hackney, R., & Edgar, D. (2014). Postgraduate research supervision: An "agreed" conceptual view of good practice through derived metaphors. *International Journal of Doctoral Studies, 9,* 43–60.

Guze, P. A. (2015). Using technology to meet the challenges of medical education. *Transactions of the American Clinical and Climatological Association, 125,* 260–270.

Holliday, A., & Amadasi, S. (2020). *Making sense of the intercultural: Finding decentred threads* (1st ed.). London, UK: Routledge.

Holstein, J. A., & Gubrium, J. F. (1995). *Qualitative research methods: The active interview.* Thousand Oaks, CA: Sage.

Hu, Y., van Veen, K., & Corda, A. (2016). Pushing too little, praising too much? Intercultural misunderstandings between a Chinese doctoral student and a Dutch supervisor. *Studying Teacher Education, 12*(1), 70–87.

Kember, D. (2016). Understanding and teaching the Chinese learner: Resolving the paradox of the Chinese learner. In R. B. King & A. B. Bernardo (Eds.), *The psychology of Asian learners* (pp. 173–187). Singapore: Springer.

Kipnis, A. B. (2011). Subjectification and education for quality in China. *Economy and Society, 40*(2), 289–306.

Kövecses, Z. (2002). *Metaphor: A practical introduction.* Oxford, UK: Oxford University Press.

Kumaravadivelu, B (2008a). *Understanding Language Teaching: from Method to Post-Method.* Taylor and Francis e-Library.

Kumaravadivelu, B. (2008b). *Understanding language teaching: From method to postmethod*. New York, NY: Routledge

Ingleby, R., & Chung, M. (2009). Cultural issues in commencing the supervision of Chinese research students. *Australian Universities' Review, 51*(2), 42–48.

Lakoff, G., & Johnson, M. (1980). *Metaphors we live by.* Chicago, IL: University of Chicago Press.

Lee, A., & Green, B. (2009). Supervision as metaphor. *Studies in Higher Education, 34*(6), 615–630.

Lee, M. C. Y., McMahon, M., & Watson, M. (2018). Career decisions of international Chinese doctoral students: The influence of the self in the environment. *Australian Journal of Career Development, 27*(1), 29–39. https://doi.org/10.1177/1038416217743023.

Lee, W. O. (1996). The cultural context for Chinese learners: Conceptions of learning in the Confucian tradition. In D. A. Watkins & J. B. Biggs (Eds.), *The Chinese learner: Cultural, psychological and contextual influences* (pp. 25–41). Hong Kong/Melbourne: Comparative Education Research Centre, The University of Hong Kong/Australian Council for Educational Research.

Li, J. (2016). The indispensable role of culture in shaping children's learning beliefs. In R. B. King & A. B. Bernardo (Eds.), *The psychology of Asian learners* (pp. 37–51). Singapore: Springer.

Lotherington, H. (2004). Multiliteracies and inner-city kids: Finding our way into the loop. *TEL & CAL, 4*(1), 18–26.

Mackinnon, J. (2004). Academic Supervision: Seeking metaphors and models for quality. *Journal of Further and Higher Education, 28*(4), 395–405.

Marton, F., Dall'Alba, G. A., & Tse, L. K. (1996). Memorizing and understanding: The keys to the paradox? In D. A. Watkins & J. B. Biggs (Eds.), *The Chinese learner: Cultural, psychological and contextual influences* (pp. 69–83). Hong Kong/Melbourne: Comparative Education Research Centre, The University of Hong Kong/Australian Council for Educational Research.

Mathison, S. (1988). Why triangulate? *Educational Researcher, 17*(2), 13–17.

McCulloch, A. (2013). The quest for the PhD: A better metaphor for doctoral education. *International Journal for Researcher Development, 4*(1), 55–66.

McGlone, M. (2007). What is the explanatory value of a conceptual metaphor? *Language & Communication, 27,* 109–126.

Mill, J. (2001). Being bilingual: Perspectives of third-generation Asian children on language, culture and identity. *International Journal of Bilingual Education and Bilingualism, 4,* 383–402.

Mok, I., Chik, P. M., Ko, P. Y., Kwan, T., Lo, M. L., Marton, F., ... Szeto, L. H. (2001). Solving the paradox of the Chinese teacher? In D. A. Watkins & J. B. Biggs (Eds.), *Teaching the Chinese learner: Psychological and pedagogical perspectives* (pp. 161–180). Hong Kong/Melbourne: Comparative Education Research Centre, The University of Hong Kong/Australian Council for Educational Research.

Rao, N., Ng, S. S. N., & Sun, J. (2016). Early learning experiences of young Chinese learners in Hong Kong: The role of traditional values and changing educational policy. In R. B. King & A. B. Bernardo (Eds.), *The psychology of Asian learners* (pp. 635–649). Singapore: Springer.

Rao, N., & Chan, C. K. K. (2010). Moving beyond paradoxes: Understanding Chinese learners and their teachers. In C. K. Chan & N. Rao (Eds.), *Revisiting the Chinese learner: Changing contexts, changing education* (pp. 3–32). Hong Kong/Dordrecht, NL: Comparative Education Research Centre, University of Hong Kong & Springer.

Richardson, V. (1996). The role of attitudes and beliefs in learning to teach. In J. Sikula (Ed.), *Handbook of research on teacher education* (pp. 102–119). New York, NY: Simon & Schuster.

Rhoads, R. A., Zheng, M., & Sun, X. Y. (2016). The methodological socialization of social science doctoral students in China and the USA. *Higher Education, 73*(2), 335–351.

Ryan, J. M. (2010). "Chinese learners": Misconceptions and realities. In J. Ryan & G. Slethaug (Eds.), *International education and the Chinese learner* (pp. 37–56). Hong Kong: Hong Kong University Press.

Ryan, J., & Louie, K. (2005). Dichotomy or complexity: Problematising concepts of scholarship and learning. In M. Mason (Ed.), *Proceedings of the 34th annual conference of the Philosophy of Education Society of Australia, "Critical thinking and learning: Values, concepts and issues"* (pp. 401–411). Hong Kong: Philosophy of Education Society of Australia.

Salili, F. (1996). Learning and motivation: An Asian perspective. *Psychology and Developing Societies, 8*(1), 55–81.

Sharifian, F. (2013). Cultural linguistics. In C. A. Chapelle (Ed.), *The encyclopedia of applied linguistics* (pp. 1590–1596). Chichester, UK: Wiley-Blackwell.

Sharifian, F. (2014). Advances in cultural linguistics. In M. Yamaguchi, D. Tay, & B. Blount (Eds.), *Approaches to language, culture, and cognition: The intersection of cognitive linguistics and linguistic anthropology* (pp. 99–123). London, UK: Palgrave Macmillan.

Sharifian, F. (2015). Cultural linguistics. In F. Sharifian (Ed.), *The Routledge handbook of language and culture* (1st ed., pp. 473–492). Abingdon Oxon, UK: Routledge.

Sharifian, F. (2017). *Cultural linguistics: Cultural conceptualisations and language.* Amsterdam, The Netherlands: John Benjamins.

Shen, W. (2018). Transnational research training: Chinese visiting doctoral students overseas and their host supervisors. *Higher Education Quarterly, 72*(3), 224–236.

Shen, W., Wang, C., & Jin, W. (2016). International mobility of PhD students since the 1990s and its effect on China: A cross-national analysis. *Journal of Higher Education Policy and Management, 38*(3), 333–353.

Son, J. B., & Park, S. S. (2014). Academic experiences of international PhD students in Australian higher education: From an EAP program to a PhD program. *International Journal of Pedagogies and Learning, 9*(1), 26–37.

Soong, H., Thi Tran, L., & Hoa Hiep, P. (2015). Being and becoming an intercultural doctoral student: Reflective autobiographical narratives. *Reflective Practice, 16*(4), 435–448. https://doi.org/10.1080/14623943.2015.1023276.

Tao, V. Y. K., & Hong, Y. Y. (2014). When academic achievement is an obligation: Perspectives from social-oriented achievement motivation. *Journal of Cross-Cultural Psychology, 45*(1), 110–136.

Tesch, J., Baptista, A., Frick, L., Holley, K. A., Remmik, M., & Åkerlind, G. (2015). The doctorate as an original contribution to knowledge: Considering relationships between originality, creativity, and innovation. *Frontline Learning Research, 3*(3), 55–67.

Tu, W.-M. (1985). *Confucian thought: Selfhood as creative transformation.* New York, NY: State University of New York Press.

Wan, W., & Low, G. (2015). *Elicited metaphor analysis in educational discourse.* Amsterdam, The Netherlands: John Benjamins.

Watkins, D. A., & Biggs, J. B. (Eds.). (1996). *The Chinese learner: Cultural, psychological and contextual influences.* Hong Kong/Melbourne: Comparative Education Research Centre, The University of Hong Kong/Australian Council for Educational Research.

Wertsch, J. V., & Tulviste, P. (1992). LS Vygotsky and contemporary developmental psychology. *Developmental Psychology,* 28, 548–557. Retrieved from https://pdfs.semanticscholar.org/9362/96a075caefccd6af941420d3322887cb61aa.pdf.

Wu, R. (2017). Academic socialization of Chinese doctoral students in Germany: Identification, interaction and motivation. *European Journal of Higher Education, 7*(3), 276–290.

Xu, L., & Hu, J. (2019). Language feedback responses, voices and identity (re)construction: Experiences of Chinese international doctoral students. *Innovations in Education and Teaching International*, 1–12.

Xu, X., Sit, H. H. W., & Chen, S. (2020). *The eastern train on the western track: An Australian case of Chinese doctoral students' adaptation.* Singapore: Springer.

Yang, S. C. (2001). Integrating computer-mediated tools into the language curriculum. *Journal of Computer Assisted Learning, 17*(1), 85–93.

Ye, L. L. (2018). *Intercultural experience and identity: Narratives of Chinese doctoral students in the UK.* London, UK: Palgrave Macmillan.

Yu, B. H., & Wright, E. (2016). Socio-cultural adaptation, academic adaptation and satisfaction of international higher degree research students in Australia. *Tertiary Education and Management, 22*(1), 49–64.

Yu, N. (2015). Embodiment, culture and language. In F. Sharifian (Ed.), *The Routledge handbook of language and culture* (1st ed., pp. 227–239). Abingdon Oxon, UK: Routledge.

Zhou, J. (2014). Persistence motivations of Chinese doctoral students in science, technology, engineering, and math. *Journal of Diversity in Higher Education, 7*(3), 177–193.

**Xing Xu** is a lecturer at Sichuan International Studies University, China, and a PhD candidate at the University of Newcastle, Australia. Her areas of interest are cross-cultural adaptation, international education, and sociolinguistics. She has publications in these research areas in both Chinese and international refereed journals.

**Dr. Hing Wa (Helena) Sit** is a Senior Lecturer and PhD supervisor in the School of Education at the University of Newcastle. She holds a PhD in Education and an MEd in TESOL from the University of Newcastle. Her research expertise includes Second Language Education, International Education, Higher Education, and Teacher Education.

**Dr. Shen Chen** is a multi-lingual teacher educator in School of Education at University of Newcastle, Australia. His publication record includes 30 funded research projects with successful completions and 86 publications including 8 books with high quality in wide areas as well as professional consultancy provided in Australia and abroad.

# Structural Models of Statistics Learning Within an IT Environment

**Ken W. Li and Merrilyn Goos**

**Abstract** Students' motivation to learn, which is a common topic for discussion, is the theme of this paper in a specific context, i.e., statistics learning within an IT environment that was organized to promote social interaction among students as well as between students and a teacher in statistical computing laboratory. Data were gathered from a questionnaire-based survey on students' perceptions of and attitudes toward the learning environment. The data were analyzed by using a sophisticated statistical technique, Structural Equation Modeling to study under what conditions, settings or situations students would be motivated to learn. It was found that supportive interaction with the teacher has a direct effect on social interaction between students, which ultimately enhances students' motivation to learn. Communication is the core element in the above learning environment that motivated students to learn, through maintaining verbal exchanges, promoting social interaction, fostering positive peer relationship or enhancing interpersonal relationships.

**Keywords** Peer relationships · Social interaction · Statistical computing laboratory · Students' motivation · Teacher–student relationships · Vocational education

## 1 Introduction

Many research studies report that students find it difficult to learn statistics, for a number of reasons. The language of statistics is difficult to learn (Dunn, Carey, Richardson, & McDonald, 2016). Statistical concepts are hard to grasp, with

K. W. Li (✉)
Hong Kong Institute of Vocational Education, Department of Information and Communication Technology, Tsing Yi Island, Hong Kong, China
e-mail: dr_ken_li@yahoo.com.hk

M. Goos
School of Education, University of Limerick, Limerick, Ireland
e-mail: merrilyn.goos@ul.ie



W. W. K. Ma et al. (eds.), *Learning Environment and Design*, Educational Communications and Technology Yearbook,
https://doi.org/10.1007/978-981-15-8167-0_5

some researchers (e.g., Pfannkuch & Budgett, 2016; Wild, Pfannkuch, Regan, & Parsonage, 2017) investigating the enhancement of statistics teaching and learning through various pedagogical approaches and the use of technology. Some teachers lack the necessary background to teach statistics owing to limited statistical training, and so some other researchers have developed and evaluated short courses for teachers that offer underpinning statistical knowledge (Biehler, 2016). Some students also hold negative attitudes toward statistics learning owing to unpleasant learning experiences or lacking control of learning (Bude et al., 2007). To address students' negative attitudes, various contexts of motivation to statistics learning have been spelt out by Bude et al. but without touching on socio-contextual influences. In fact, the influences on students and teacher are significant in mathematics classroom (Watt & Goos, 2017). To achieve an understanding of these influences, educational psychologists (e.g., Bempechat & Shernoff, 2012; Pianta, Hamre, & Allen, 2012; Skinner & Pitzer, 2012; Wentzel, 2009, 2012) give an explicit account of interpersonal relationships, social interaction, family and peer group dynamics, peer influence, as well as achievement of social goals, social relationship goals, and social responsibility goals in learning through interactions with peers and the teacher. All these social process factors interact in complex ways and can be explained by socio-cultural theories of learning (Watt & Goos, 2017).

## 2 Literature Review

Socio-cultural theories of learning are extensively applied to the contexts of mathematics education research (e.g., Galbraith, Goos, Renshaw, & Geiger, 2001; Goos, 2004, 2009; Steele, 2001) as building social environments in classroom teaching and learning. This social environment is characterized by caring and support offered by peers or teachers that would motivate students to learn (Brand, Felner, Shim, Seitsinger, & Dumas, 2003). Goos reported an account of teaching and learning practices using technology in secondary mathematics classrooms where the teacher took up both facilitating and managing roles of student learning, such as clarifying students' misunderstandings or misconceptions, stimulating and structuring their thinking, moderating discussions among students, and so forth. Through discussions, the students made a concerted effort to construct knowledge and derive mathematical meanings, achieved more or better understanding of problem setting, as well as proposed and refined approaches to problem solving. All these outcomes would arise under the assumption that they were motivated to participate in learning activities. Besides, educational psychology (e.g., Juvonen, Espinoza, & Knifsend, 2012; Pianta et al., 2012; Skinner & Pitzer, 2012; Wentzel, 1991, 2012) sheds the light on probable causes under which students' motivation to learning would be enhanced.

Everybody is expected to achieve social responsibility goal as fulfilling social roles, norms, and rules for being adaptive and cohesive within a peer group, a family, and a society. This goal is also a prerequisite for classroom learning during which students behave and interact with peers and a teacher so as to build a positive interpersonal relationship. The relationship has a positive orientation toward students' motivation to learn (Wentzel 1991).

Student perceptions of teacher's responsiveness, enthusiasm, and classroom management skills have an impact on their own attitudes or behavior, in turn affecting teacher–student relationships as well as interactions. Hence, the teacher should play a scaffolding role for cognitive development, create well-structured and caring classroom environments with students' autonomy of learning approach and strategies, and manage physical and social resources (Pianta et al., 2012). Skinner and Pitzer (2012) stressed the importance of social resources for fostering pleasant learning atmosphere, maintaining group cohesiveness, and mediating group conflicts.

Peer relationship concerning social interaction among students has mutual influence on one another. The influence is characterized by the nature and quality of inputs and responses, where inputs can be verbal or nonverbal expressions, time and effort devoted, considerate attitudes, and burden-sharing. A positive or negative response given by peers like an endorsement or a challenge of academic or social goals, an appreciation or a criticism, a positive or negative feedback, and so forth is dependent upon how they perceive and interpret the inputs. If positive, the interpersonal relationship is enhanced; both parties would thus be motivated to work collaboratively (Juvonen et al., 2012).

All these research articles focus on the study of students' motivation either from students' viewpoint or a teacher's roles raising all the psychological issues are within socio-cultural contexts (Wentzel, 2012). The theoretical framework for the present study was thus derived from socio-cultural theories of learning and relevant theories of educational psychology to address the research question, "Under what probable causes would students be motivated to learn in the social organization of a statistics classroom within an IT environment?" In accordance with socio-cultural theories of learning, students in the present study were divided into small groups to do exercises demanding the analysis, design, or implementation of the solutions in a statistical computing laboratory, thus facilitating collaborative learning. Data were gathered from the students about their perceptions of and attitudes toward learning with IT. Statistical analysis of data was performed, with educational psychology theories hinting at constructing arguments to justify statistical findings.

## 3 Empirical Study

The class of all 58 full-time students enrolling in Year 2 of the three-year Higher Diploma in Applied Statistics Course (HDASC) voluntarily participated in the study. This cohort of students was selected because they knew each other more after completing Year 1, and they were offered more opportunities for peer learning and collaboration in a statistics module taught by the teacher (the first author of the paper).

A questionnaire was designed to solicit feedback from the students about their perceptions of and attitudes toward the use of IT in the learning environment where they could have more peer interactions and student–teacher interactions. The questionnaire consisted of three major parts (see Appendix). The first part, Qs 1–15 focused on investigating students' perceptions of the educational use of IT when working alone, whereas the second part (Qs 16–42) and the third part (Qs 43–53) explored how well students' learning was constructed through social interaction among students and between students and their teacher in an IT environment, respectively. These three parts, which were initially assumed independent based on the question contexts were verified by Li and Goos (2017) using factor analysis, but each part asked a series of related questions in accordance with socio-cultural perspective. Apparently, the last two parts of the questionnaire also adhere to the theories of motivation.

In Part 1, Qs 1–4 were set to check students' motivation to learning in terms of their interest, confidence, and belief. Qs 5–12 were intended to evaluate how digital tool, IT would facilitate student learning. Qs 13–15 provided students with an opportunity to report how they progressed when learning with IT.

In the second part, Qs 16–18 were designed to collect students' general perceptions of peer learning. Qs 19–21 were to check peer interaction. Qs 22–27 looked at cognitive issue relating to the types and the ways of knowledge that was shared between partners. Qs 28–31 studied the working relationship among the student group members. Qs 32–35 examined the importance of communication between the students and their learning partners. Qs 36–38 asked students to elaborate on how workload was distributed. Qs 39–42 investigated the social relationship among the members of the group.

In the last part, Qs 43–46 aimed to gather students' general views about the teacher's intervention. Qs 47–48 enabled students to express the role and significance of the teacher in the learning process. Qs 49–51 sought to discover whether or not IT was regarded as a vehicle of education delivery that could replace human teaching. The prime purpose of Qs 52–53 was to learn students' experience of interacting with their teacher.

The students were interviewed individually; their responses given to each of the questions were recorded and then were inputted into computer for data validation and analysis. Twenty-nine of these questions provided multiple response categories so as to enable the interviewees to indicate the extent of their agreement with a proposition; the remaining twenty-four questions were open-ended, as enabling them freely to supplement or explain their responses given to the multiple-choice questions.

# 4 Data Analysis and Results

## *4.1 Preliminary Analysis*

Preliminary analysis of the survey data was conducted to investigate students' perceptions of the educational use of IT under these three circumstances: (i) when learning on their own, (ii) when learning with peers, and (iii) when learning with the teacher that corresponds to Parts 1–3 in the questionnaire. The multiple-choice responses given by the students were analyzed using frequency tables, whereas different wordings but similar nature or context of students' responses given to the open-ended questions were summarized into more meaningful findings that were subsequently used to substantiate the establishment of the structural models of learning in Subsection 4.3.

Among these fifty-eight students' perceptions of learning with IT on their own (i.e., Qs 1–15 in the first part of the questionnaire), forty-seven (81.0%) found learning with IT interesting or very interesting and eleven (19.0%) gave a neutral response but none of them found learning with IT boring or very boring ($x_1$). Twenty-four students (41.4%) were either confident or very confident, thirty-one (53.4%) gave a neutral response, and the remaining three students (5.2%) felt unconfident ($x_2$). A little over two-thirds of students (40, 69.0%) opined that IT helped them make sense of what they were learning, and eighteen students (31.0%) and none gave neutral and negative responses, respectively ($x_3$). Most students (51, 87.9%) saw learning activities in the computer laboratory as relevant, while seven students (12.1%) and none gave neutral and negative responses, respectively ($x_4$). Most students (52, 89.7%) believed that IT widened or much widened their scope of learning, whereas five and one had neutral and negative responses, respectively ($x_5$). Most students reported IT widening their scope of learning as appreciating applicability of statistics in real world and developing thinking and reasoning beyond conceptual understanding of statistics ($x_6$); a few explained that IT narrowed down their scope of learning because of slow information traffic limiting exploration of or access to the practice of statistics ($x_7$). Thirty-seven students (63.8%) believed they had control of their learning process, and twenty students (34.5%) and only one student (1.7%) gave neutral and negative responses, respectively ($x_8$).

Regarding which type of learning materials, IT-based or non IT-based, helped students focus best on the learning task ($x_9$), forty-eight students (82.8%) selected IT-based learning materials. Only one student (1.7%) chose non IT-based learning materials, and nine students (15.5%) gave a neutral response. The majority of students (46, 79.3%) had better or much better learning progress when learning with IT, and twelve (20.7%) gave neutral responses, but no student gave negative responses, i.e., neither worse nor much worse ($x_{10}$). The reason given by them for having better learning progress is that they exploited the use of information technology as an important role for enhancing student understanding through hands-on animation and visualization tools, chat rooms for seeking learning assistance outside classrooms, and data experimentation using statistical software ($x_{11}$). Apart from the positive comments in positive or neutral responses, there were also some negative comments about hardware problems and unmanageable software tools and computer syntax ($x_{12}$).

Many students reported coping practice with learning problems outside the classroom through academic support from peers or teacher via the Internet or finding reference materials from libraries or the Internet ($x_{13}$). Students' responses ($x_{14}$ and $x_{15}$) given to both open-ended questions, Qs 14 and 15 were similar as shown below. They could understand more about the world through statistical exploration of data, make more attempts to experiment with data using handy graphing and software tools, and utilize logical and statistical thinking as well as analytic skills.

All responses given by students to the second part of the questionnaire (Qs 16–42) show affirmative attitudes toward collaborative learning within an IT environment. When working with a learning partner, most students (50, 86.2%) found learning with IT helpful or very helpful, whereas five (8.6%) students had a neutral response and three (5.2%) found it unhelpful ($x_{16}$). A large majority of students (52, 89.7%) had better or much better learning progress when working with their learning partners, while five (8.6%) gave a neutral response and only one student (1.7%) had worse learning progress ($x_{17}$). The reasons the students held for having better learning progress were concerned with intellectual exchanges like getting help from a more able learning partner, exchanging views for co-construction of knowledge, and clarifying or correcting misconceptions. On the other hand, the reasons for having worse learning progress were about a lack of mutually held goals and difficulties in resolving discrepant opinions ($x_{18}$).

The majority of students (47, 81.0%) found their interaction with their learning partners collaborative, while ten (17.2%) students gave a neutral response and only one student (1.7%) found the interaction disruptive ($x_{19}$). Around three-quarters of students (44, 75.9%) found their interaction produced agreement, none found their interaction produced disagreement, whereas fourteen (24.1%) students gave a neutral response ($x_{20}$). Many students would resolve disagreement by soliciting teacher's or a third party's opinion ($x_{21}$).

Most students (52, 89.7%) shared much or very much knowledge with their learning partners, while five students (8.6%) gave a neutral response and only one (1.7%) shared little knowledge ($x_{22}$). Similarly, many students (48, 82.7%) said that their learning partners shared much or very much knowledge with them, respectively, whereas eight (13.8%) students gave a neutral response and two (3.5%) claimed their partner shared little knowledge ($x_{23}$). Either the students or their learning partners shared conceptual, procedural, technical, tactical, or analytic knowledge as well as knowledge around statistical communication skills ($x_{24}$ and $x_{26}$). However, only one student did not want to share approaches to problem solving with his learning partner ($x_{25}$) and one student said her learning partner did not share concepts and knowledge of statistical applications with her ($x_{27}$).

The largest proportion of students (35, 60.3%) thought that an IT environment helped students foster a better or a much better interpersonal relationship ($x_{28}$). Twenty-one students (36.2%) gave a neutral response and two students (3.5%) had a negative opinion. A large majority of students (49, 84.5%) thought that learning partners helped them make sense of what they were learning and eight (13.8%) students gave a neutral response, whereas only one student (1.7%) thought that learning partners did not help him make sense of what he was learning ($x_{29}$). Almost all students (54, 93.1%) said they co-learnt rather than competed with their learning partners ($x_{30}$), and many (51, 87.9%) thought their learning partners co-learnt with them ($x_{31}$). Co-learning here refers to the development of learning strategies or achievement of common learning goals with concerted effort and displays a positive working relationship between the students and their partners.

Most students (53, 91.4%) believed that communications with their learning partners were beneficial or very beneficial to their learning process and four (6.9%) students gave a neutral response, whereas the remaining student (1.7%) had a negative opinion ($x_{32}$). The students gave an account of communications beneficial to their learning process relating to social and cognitive dimensions. To collaborate on learning tasks efficiently, they built interpersonal relationship, shared personal views, clarified misconceptions and/or misunderstanding, and resolved conflicting views. Upon fulfilling the above social conditions mandatory for learning, they monitored peers' learning progress, determined goals, formulated strategies, and stimulated one another to think ($x_{33}$). Few students reported unbeneficial communications when less competent learning partners could not assist their learning ($x_{34}$).

More than a half (33, 56.9%) of the fifty-eight said they had good communications and found no need for improvement. Those students wanted to improve the communications with their learning partners by taking the initiative in chatting to them, listening attentively to peers, being more empathetic about their situations and feelings, and providing moral support where necessary. All of these approaches aim to improve communication in the social context ($x_{35}$).

Students mentioned the extent of their involvement with their learning partners ($x_{36}$), in terms of the focus of their joint activity that characterized the hierarchy of statistical thinking from high to low levels: strategic, tactical, and operational. Almost all students reported positively about the contribution they made to collaborative learning ($x_{37}$). For example, they aroused interest and built teamwork spirit, focused discussion on tasks, and gave constructive ideas. The order of contribution (i.e., from social to cognitive) they gave is rational, as students would find a social environment in which they feel secure prior to learning. Students also reported that learning tasks that were equally shared within collaborating groups ($x_{38}$) were cognitive rather than socially oriented, such as problem formulation, approaches to problem solving, graphing and computing tasks, and statistical reporting.

A little over half of the students (30, 51.7%) reported having much interaction, nearly as many gave a neutral response (26, 44.8%), and only two students (3.5%) reported having little interaction ($x_{39}$). The great majority (47, 81.0%) found this partnership harmonious; some students (11, 19.0%) gave a neutral response, but none found this partnership conflictive ($x_{40}$). Most students (41, 70.7%) said their learning partners made learning more fun, while seventeen (29.3%) gave neutral responses and none gave negative responses ($x_{41}$). Students described the feeling of learning with partners in a number of common ways ($x_{42}$), mostly related to development of rapport, good interpersonal relationships, and a pleasant learning climate.

The research findings from Part 3 ($x_{43}$–$x_{53}$) outlined the importance of the teacher's intervention in students' development of statistical understanding. The majority of students (46, 79.3%) said they needed the intervention, and twelve (20.7%) gave neutral responses ($x_{43}$). A little over two-thirds (40, 68.9%) found the timing of teacher's intervention appropriate, and fifteen students (25.9%) and three students (5.2%) gave neutral and negative responses, respectively ($x_{44}$). Almost all students (54, 93.1%) thought the intervention was beneficial and none gave a negative response, while four students (6.9%) were neutral ($x_{45}$). Circumstances under which students found the teacher's intervention beneficial to their learning process ($x_{46}$) were summarized using Tharp and Gallimore's (1988) means of offering learning assistance. The intervention is a social process within which the teacher and students were engaged in various assisting means, feeding back, modeling, instruction, questioning, and cognitive structuring. Apart from these categories, some students mentioned that the teacher maintained an active dialogue with them to encourage their participation and involvement, share their views as well as ideas, and respond to assistance they sought. Conversely, those few students who found the teacher's intervention unbeneficial to their learning process felt uncomfortable when their mistakes were pointed out.

Students gave a full description of how the teacher orchestrated the learning activities in computing laboratory within social and cognitive contexts. The teacher fostered an amusing climate for learning and utilized teaching and learning resources, for example, handouts as well as computer hardware and software. He organized meaningful learning activities, established a community of practice and inquiry, provided prompt feedback to students, and motivated students to excel at tasks ($x_{47}$). These responses also appeared in students' responses ($x_{48}$) about how well students had learnt from the teacher; they also put in additional comments as shown below. The teacher was conscientious in organizing learning in the classroom students felt confident, competent, and autonomous. He was responsive in facilitating group discussions, maintaining group process, or resolving group conflicts.

To examine preferences for interaction with the teacher, learning partners, and IT, students gave the following indications ($x_{49}$–$x_{51}$). A little over one-third of students (23, 39.7%) preferred learning with a teacher to IT, only one student (1.7%) preferred learning with IT to a teacher, and thirty-four students (58.6%) gave a neutral response. When being asked to choose between learning partners and IT, about half (30, 51.7%) preferred the former, while twenty-four students (41.4%) had no preference and very few preferred the latter (4, 6.9%). Students were evenly divided in choosing to learn with a partner (14, 24.1%) or the teacher (12, 20.7%), and about half (32, 55.2%) gave a neutral response. To sum up $x_{49}$–$x_{51}$, students would like to interact with humans rather than IT.

A large majority of students (50, 86.2%) had better or much better learning progress when working with their teacher in an IT environment; eight (13.8%) gave a neutral response, but no students gave a negative response ($x_{52}$). Students who considered they had better learning progress with the teacher gave a number of reasons ($x_{53}$), which were very similar to both responses, $x_{47}$ and $x_{48}$, albeit with the following supplements. The teacher was attuned and responsive to their questions and also created and maintained a positive and warm classroom atmosphere conducive to learning. Apart from the positive reasons shown in neutral responses in which negative reasons spelt out that the teacher did not give direct instruction and communications with the teacher slowed down learning progress.

All the responses, $x_1$–$x_{53}$, which were summarized from Li (2012) as well as Li and Goos (2015), had a rich social context. The responses undergone correlation analysis showed whether there were relationships among them, as clueing us up to propose certain causal relationships among the responses that were due to be verified by the structural equation modeling in Subsection 4.3.

## 4.2 Correlation Analysis

Nevertheless, the results of preliminary analysis showed that the survey data skewed to positive responses to all the multiple-choice and open-ended questions with indication of most students appraising the value of educational use of IT highly. It would be of interest to know what elements of the IT environment motivated students to learn. Motivation here is concerned about how to arouse and maintain students' interest in learning. The data were further analyzed by means of Spearman correlation to examine the relationship among survey responses, particularly $x_1$ (students' interest in learning) versus $x_2$–$x_5$, $x_8$–$x_{10}$, $x_{16}$, $x_{17}$, $x_{19}$, $x_{20}$, $x_{22}$, $x_{23}$, $x_{28}$–$x_{32}$, $x_{39}$–$x_{41}$, $x_{43}$–$x_{45}$, and $x_{49}$–$x_{52}$. These are responses measured on an ordinal scale (also known as manifest variables in context of structural equation modeling in the following section) given to all the multiple-choice questions in the questionnaire, Qs 1–5, 8–10, 16, 17, 19, 20, 22, 23, 28–32, 39–41, 43–45, and 49–52.

Correlations shown in Table 1 are weak to moderate. More importantly, at the 5% level of statistical significance, $x_1$ (students found learning with IT interesting) positively correlated with $x_9$ (IT-based learning materials helped them best focus on their learning tasks), $x_{10}$ (having better learning progress when learning with IT on their own), $x_{17}$ (having better learning progress when working with their learning partners), $x_{22}$ (they shared much knowledge with their learning partners), $x_{23}$ (their learning partners shared much knowledge with them), $x_{32}$ (communication beneficial to learning process), $x_{39}$ (having much interaction with partners), and $x_{41}$ (their learning partners made learning process more fun), while $x_1$ correlated with $x_{44}$ (appropriate timing of teacher's intervention) at the level of statistical significance slightly higher than 5%. These relationships imply that students were motivated to learning with IT when it involved social interaction with peers based on communication, sharing of knowledge, enjoyable learning atmosphere, IT-based materials, and appropriate timing of teacher's intervention. These are cognitive as well as affective

**Table 1** Correlation analysis results

| | $x_2$ | $x_3$ | $x_4$ | $x_5$ | $x_8$ | | $x_9$ | $x_{10}$ | $x_{16}$ | $x_{17}$ |
|---|---|---|---|---|---|---|---|---|---|---|
| $x_1$ | 0.092 | −0.039 | −0.164 | 0.133 | −0.059 | | $0.285^b$ | $0.294^b$ | 0.240 | $0.326^b$ |
| | $x_{19}$ | $x_{20}$ | $x_{22}$ | $x_{23}$ | $x_{28}$ | | $x_{29}$ | $x_{30}$ | $x_{31}$ | $x_{32}$ |
| $x_1$ | 0.094 | 0.145 | $0.458^c$ | $0.381^c$ | 0.200 | | 0.125 | −0.170 | 0.044 | $0.363^c$ |
| | $x_{39}$ | $x_{40}$ | $x_{41}$ | $x_{43}$ | $x_{44}$ | $x_{45}$ | $x_{49}$ | $x_{50}$ | $x_{51}$ | $x_{52}$ |
| $x_1$ | $0.345^c$ | 0.089 | $0.411^c$ | 0.161 | $0.252^a$ | 0.092 | −0.112 | −0.039 | 0.200 | 0.177 |

*Note* $^a p = 0.056$, $^b p < 0.05$, $^c p < 0.01$

base for developing students' interest, perceiving as a positive orientation toward learning (Ainley, 2012).

## 4.3 Structural Equation Modeling

Correlation analysis is confined to checking the noncausal relationship between two variables at a time while assuming no interrelationship among variables. If relaxing this assumption, another statistical tool, structural equation modeling would deepen our understanding or extend the implications of results from the correlation analysis because the tool can illustrate causal relationships among these latent variables corresponding to the theme of three parts of the questionnaire in the present study, i.e., $y_1$ (students' motivation to learn) in Part 1, $y_2$ (social interaction among students) in Part 2, and $y_3$ (supportive interaction with teacher) in Part 3 through their effects on these manifest (observable) variables: $x_1$–$x_5$, $x_8$–$x_{10}$, $x_{16}$, $x_{17}$, $x_{19}$, $x_{20}$, $x_{22}$, $x_{23}$, $x_{28}$–$x_{32}$, $x_{39}$–$x_{41}$, $x_{43}$–$x_{45}$, and $x_{49}$–$x_{52}$. The first eight, the middle fourteen, and the last seven manifest variables based upon their contexts compose of the themes of $y_1$, $y_2$, and $y_3$, respectively, namely $x_2$ (student's confidence), $x_3$ (student's belief), $x_4$ (relevant learning activities), $x_5$ (IT widened scope of learning), $x_8$ (student's autonomy of learning), $x_9$ (IT-based learning materials helped students focus best on the task), $x_{10}$ (better learning progress when learning with IT), $x_{16}$ (finding learning with IT helpful when working with their learning partners), $x_{17}$ (having better learning progress when working with their learning partners), $x_{19}$ (collaborative interaction), $x_{20}$ (peer interaction produced agreement), $x_{22}$ (students shared much knowledge with their learning partners), $x_{23}$ (their learning partners shared much knowledge with them), $x_{28}$ (IT fostered a better interpersonal relationship in peer learning), $x_{29}$ (peer helped make sense of what they were learning with IT), $x_{30}$ (students co-learnt with peer), $x_{31}$ (peer co-learnt with student), $x_{32}$ (communication beneficial to learning process), $x_{39}$ (much interaction), $x_{40}$ (learning harmony), $x_{41}$ (peer made learning process fun), $x_{43}$ (needing teacher's intervention), $x_{45}$ (teacher's intervention beneficial to learning process), $x_{49}$ (liking to learn with teacher), $x_{50}$ (liking to learn with peer), $x_{51}$ (liking to learn with peer and teacher), and $x_{52}$ (better learning progress when working with teacher). These manifest variables were derived from the contexts of educational psychology, such as interpersonal relationships, social interaction, peer group dynamics, peer influence, and so on but not including family dynamics because the research participants are adults, ranging in age from 19 to 22 under little influence of family.

The first conceptual model of statistics learning within an IT environment is set up in accordance with socio-cultural theories of learning and correlation results by displaying probable casual relationships among the latent variables, $y_1$, $y_2$, and $y_3$ in the path diagram, Fig. 1 in which arrows indicate both $y_2$ and $y_3$ playing a part in $y_1$ as well as $y_3$ having effects on $y_2$. The rationale for the hypothesized relationship between these three latent variables evolves from educational psychologists, Wentzel (1998) as well as Skinner and Pitzer (2012). Students' motivation to learn

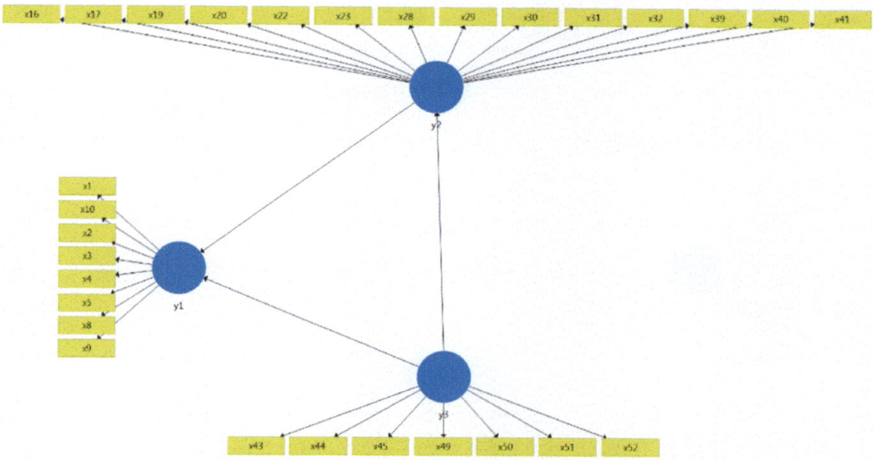

**Fig. 1** Path diagram of a full conceptual model

($y_1$) is hypothesized to be under the influence of social interaction among students ($y_2$) as well as supportive interaction with teacher ($y_3$) (Wentzel, 1998), especially when their teacher is enthusiastic about teaching (Metcalfe & Game, 2006). Their teacher played important roles in student learning; the student–teacher relationship is significant because the students would be motivated to learn when a positive learning environment was fostered and learning assistance with student autonomy was offered by their teacher (Skinner & Pitzer, 2012) so that $y_3$ would directly influence $y_2$ or perhaps influence indirectly $y_1$ via $y_2$. As far as the causal relationship is concerned, the corresponding path is established when achieving statistical significance, i.e., $p < 0.05$ with a relatively large path coefficient (Hair, Black, Babin, & Anderson, 2014). By establishing this path, it would show the evidence in supporting the hypothesized relationship (Schum, 2001). However, causal relationship is only a statistical jargon within the context of structural equation modeling, but here it shows the conditions or situations given rise to students' motivation to learn.

After the structural equation analysis, Fig. 2 presents the causal model with path coefficients, 0.499, 0.183, and 0.470, and together with their corresponding *p*-values, 0.106, 0.397, and 0.012, respectively. Three positive coefficients confirm the head of the arrow, signifying the direction of hypothesized causal relationship, i.e., $y_2$ has positive and moderate effect on $y_1$, $y_3$ has positive and weak effect on $y_1$, and $y_3$ has positive and moderate effect on $y_2$, but the first two causal relationships cannot be substantiated as not fulfilling the prescribed statistical significance, $p < 0.05$. On the contrary, the last causal relationship is statistically significant ($p = 0.012$) with a path coefficient, 0.470; supportive interaction with the teacher ($y_3$) thus directly influences social interaction between students ($y_2$). This argument has factual basis on students' responses, i.e., $x_{46}$, $x_{48}$, and $x_{52}$ are summarized as scaffolding, facilitating, and managing roles the teacher devoted to teaching and learning. According to socio-cultural theories of learning (Goos, 2004), scaffolding assistance was offered

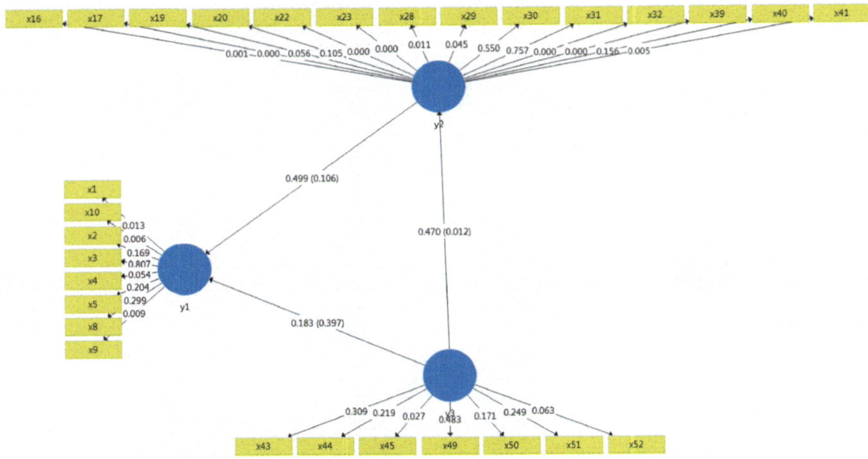


**Fig. 2** A causal model showing path coefficients and *p*-values

to restructure learning tasks to meet students' individual needs and rate of progress to develop shared experience between teacher and students, thus eventually establishing students' zone of proximal development. Student learning was facilitated with high autonomy of learning approaches and strategies. Resources were managed for fostering a pleasant learning atmosphere (Skinner & Pitzer, 2012).

No more causal relationships are statistically significant but the path connecting $y_1$ and $y_2$ illustrated by the Fig. 2 seems to resemble social interaction between students ($y_2$) has a direct impact on students' motivation to learn ($y_1$) to a certain extent owing to a relatively large path coefficient, 0.499. This is evident from students' responses when being asked to choose for learning assistance in these two scenarios: (1) a teacher versus IT and (2) learning partners versus IT. They chose a teacher and learning partners in scenarios (1) and (2), respectively, implying that they would prefer interacting with humans, probably demanding social interaction. In addition, probable correlation between $x_1$ (students found learning with IT interesting) and $x_{44}$ (teacher's intervention beneficial to learning process) hinted that the teacher's intervention was a direct or indirect probable cause under which students would be motivated to learn. All these show a causal chain involving $y_1$, $y_2$, and $y_3$ and would deduce that the teacher's intervention ($y_3$) may eventually influence students' motivation to learn ($y_1$) via $y_2$ because the scaffolding, facilitating, and managing roles have relevance to social contexts as developing a rapport with students and helping them build confidence and competence, thus in turn enhancing students' motivation to learn (Skinner & Pitzer, 2012).

The deduction just made calls our attention to tracing the effects of the first eight manifest variables ($x_1$–$x_5$ and $x_8$–$x_{10}$) on $y_1$, the effects of the next fourteen manifest variables ($x_{16}$, $x_{17}$, $x_{19}$, $x_{20}$, $x_{22}$, $x_{23}$, $x_{28}$–$x_{32}$, and $x_{39}$–$x_{41}$) on $y_2$, as well as the effects of the last seven manifest variables ($x_{43}$–$x_{45}$ and $x_{49}$–$x_{52}$) on $y_3$. The effect here refers to the favorable conditions for being congenial to the three latent variables,

$y_1$ (students' motivation to learn), $y_2$ (social interaction among students), and $y_3$ (supportive interaction with teacher). To achieve an understanding of the effects, we can check each manifest variable associated with its corresponding *p*-value in Fig. 2 to assess its significance. Students' interest ($x_1$), IT-based learning materials ($x_9$), and IT environment ($x_{10}$) are statistically significant at the 0.05 level linking with students' motivation to learn ($y_1$).

Statistical significance reflected on the *p*-values, 0.001, 0.000, 0.000, 0.000, 0.045, 0.000, 0.000, and 0.005 corresponding to these manifest variables, $x_{16}$ (finding learning with IT helpful when working with peers), $x_{17}$ (better learning progress when working with peers), $x_{22}$ (much sharing of knowledge with peers), $x_{23}$ (much knowledge shared by peers), $x_{28}$ (IT fostered a better interpersonal relationship in peer learning), $x_{29}$ (peers helped make sense of what they were learning with IT), $x_{32}$ (communication beneficial to peer learning), $x_{39}$ (much interaction), and $x_{41}$ (peers made learning process fun) connecting to $y_2$ (social interaction among students) gives rise to a belief that these manifest variables have effect on $y_2$. Only the teacher's intervention being beneficial to a learning process ($x_{45}$) has an effect on supportive interaction with teacher ($y_3$).

A question of peculiar interest is raised to show whether there are intermediate links in between latent variables, y's and manifest variables, x's concealing eleven potential factors to form more meaningful contexts. The contexts were derived from $x_1$–$x_4$, $x_5$ and $x_8$–$x_{10}$, $x_{16}$–$x_{17}$, $x_{19}$–$x_{20}$, $x_{22}$–$x_{23}$, $x_{28}$–$x_{31}$, $x_{32}$, $x_{39}$–$x_{41}$, $x_{43}$–$x_{45}$, $x_{49}$–$x_{51}$ as well as $x_{52}$ in order to postulate factors: $f_1$ (students' motivation to learn), $f_2$ (IT as digital tool), $f_3$ (peer collaboration), $f_4$ (peer interaction), $f_5$ (cognitive engagement), $f_6$ (working relationship), $f_7$ (communication), $f_8$ (social relationship), $f_9$ (teacher's intervention), $f_{10}$ (collaboration with human), and $f_{11}$ (interaction with teacher), respectively. To illustrate lucid models, the full model is broken down into three submodels in Figs. 3, 4, and 5 for further examination.

It is envisaged that students' interest ($x_1$), confidence ($x_2$), and belief ($x_3$) as well as the relevancy of learning contents ($x_4$) are crucial to $f_1$ (students' motivation to learn) (Ainley, 2012; Skinner & Pitzer, 2012), but this argument cannot be well substantiated owing to *p*-values ranging from 0.072 to 0.721 are not statistically significant (see Fig. 3). Perhaps, $x_1$, $x_2$, and $x_4$ achieving marginal significance deserve consideration of their contribution to $f_1$. Significant correlations between some variables embedded in $f_1$ (students' motivation to learn) and certain variables linking with $f_2$ (IT as digital tool) lead us to believe that a direct path would demonstrate the effect of $f_2$ on $f_1$ with $p = 0.027$. This naturally draws us a close examination of which of $x_5$, $x_8$, $x_9$, and $x_{10}$ would contribute to $f_2$. It was found that only $x_9$ (IT-based learning materials) and $x_{10}$ (IT environment) are statistically significant at 0.05 linking with $f_2$ (IT as digital tool). To summarize, $f_2$ (IT as digital tool) has an effect on $f_1$ (students' motivation to learn), resulting from the causal chain involving $x_9$ (IT-based learning materials) and $x_{10}$ (IT environment).

Prior to tracing the relations among these factors, $f_3$–$f_8$, it would be better to study how the manifest variables $x_{16}$, $x_{17}$, $x_{19}$, $x_{20}$, $x_{22}$, $x_{23}$, $x_{28}$–$x_{32}$, and $x_{39}$-$x_{41}$ contribute to each of the factors in Fig. 4. Based on the context of each of the variables and

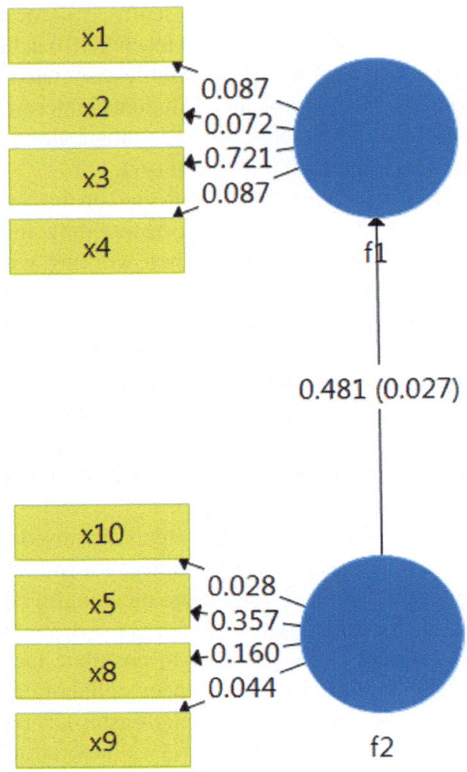


**Fig. 3** Submodel 1 relating to students' motivation and digital tools

statistical significance at 5% level, the variables were grouped so as to generate a common theme of a factor, i.e., factor labeling.

The first two variables, $x_{16}$ and $x_{17}$ are attached to $f_3$ (peer collaboration); $x_{19}$ represents $f_4$ (peer interaction). Both $x_{22}$ and $x_{23}$ form $f_5$ (cognitive engagement). These three variables, $x_{28}$, $x_{29}$, and $x_{31}$ are recruited to $f_6$ (working relationship). A single variable, $x_{32}$ is equivalent to $f_7$ (communication).

The last three variables, $x_{39}$ (peer interaction), $x_{40}$ (having harmony in learning), and $x_{41}$ (peer made learning fun) are statistically significant and are incorporated into $f_8$ (social relationship). These are echoed in students' responses to Q42 asking to describe the feeling of learning with peers which are mostly related to development of rapport, good interpersonal relationships, and a pleasant learning climate that in turn would alleviate stress and pose a favorable condition for student motivation (Juvonen et al., 2012).

To examine the relations among the five factors, $f_3$–$f_8$ based on educational psychology, it calls our direct attention to account of how they stand in relation to each other in these five cases: (1) $f_4$–$f_8$ on $f_3$, (2) $f_5$–$f_7$ on $f_4$, (3) $f_7$ on $f_5$, (4) $f_7$ on $f_6$, and (5) $f_5$ on $f_8$ (see Fig. 4).

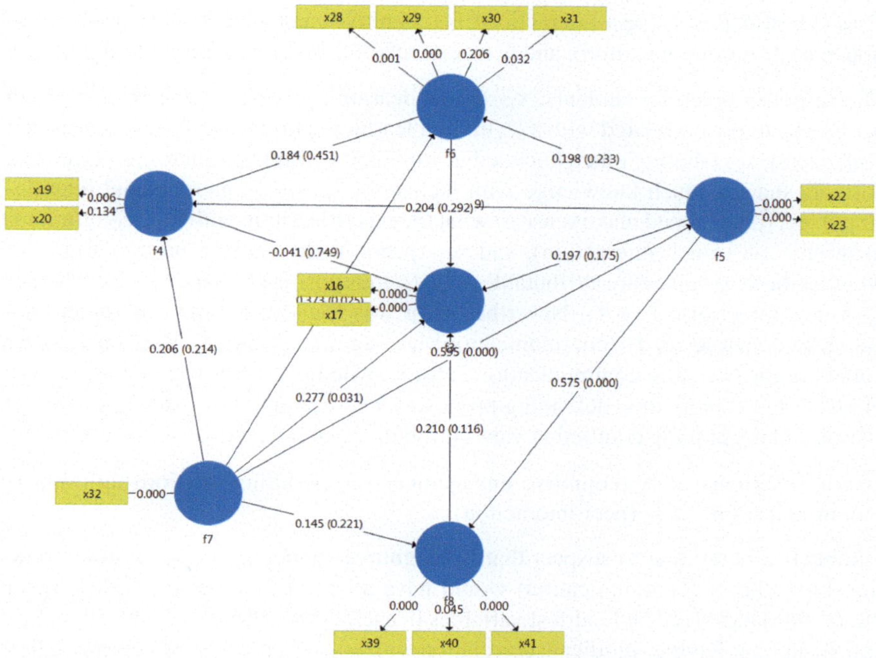


**Fig. 4** Submodel 2 relating to social interaction among students

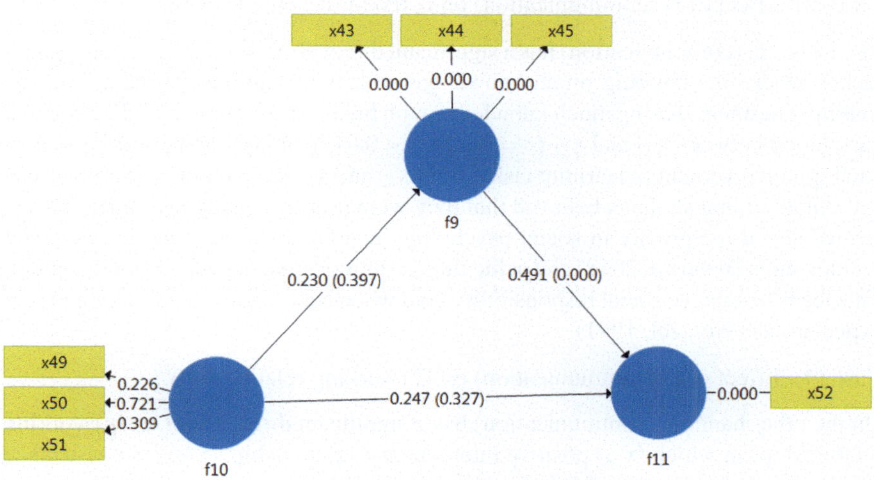


**Fig. 5** Submodel 3 relating to supportive interaction with the teacher

Case (1): Effects of $f_4$ (peer interaction), $f_5$ (cognitive engagement), $f_6$ (working relationship), $f_7$ (communication), and $f_8$ (social relationship) on $f_3$ (peer collaboration).

The response given by students, $x_{17}$ (better learning progress when working with peers) positively correlated with $x_{16}$ (finding learning with IT helpful when working with peers), $x_{22}$ (sharing much knowledge with their learning partners), $x_{23}$ (learning partners sharing much knowledge with them), $x_{28}$ (positive interpersonal relationships), $x_{29}$ (peer helped make sense of what they were learning with IT), $x_{32}$ (communication), $x_{39}$ (much interaction), and $x_{41}$ (peers made learning process fun), but whether these variables are attributable to $f_3$–$f_8$, thus posing clues to examine whether $f_4$–$f_8$ has direct effect on $f_3$. Nevertheless, it was found that only one direct path could be constructed, $f_7$ (communication) has significant direct effect on $f_3$ (peer collaboration) because communication is for knowledge co-construction in the way of clarifying standpoints, defending proposals, exchanging views, debating critical issues, and negotiating conflicting views through social interaction (Emerson, 1996).

Case (2): Effects of $f_5$ (cognitive engagement), $f_6$ (working relationship), and $f_7$ (communication) on $f_4$ (peer interaction)

Although it is rational to suspect that $f_5$ (cognitive engagement), $f_6$ (working relationship), and $f_7$ (communication) would have a causal relationship with $f_4$ (peer interaction) based on the manifest variables in each of the factors, $x_{23}$ in $f_5$, $x_{29}$ in $f_6$, and $x_{32}$ in $f_7$, as having significant correlation with $x_{19}$ in $f_4$, none of the corresponding paths is statistically significant.

Case (3): Effect of $f_7$ (communication) on $f_5$ (cognitive engagement)

The factor, $f_7$ (communication) has a significant direct effect on $f_5$ (cognitive engagement), where $x_{22}$ (sharing much knowledge with their learning partners) and $x_{23}$ (learning partners sharing much knowledge with them) are implicit in $f_5$. The positive correlation between $x_{22}$ and $x_{23}$ ($r = 0.608$, $p < 0.01$) displays their contributions to and high involvement in learning tasks. Both $x_{22}$ and $x_{23}$ were reciprocal obligations and duties so that students believed that they were treated equally and fairly. This is termed social reciprocity in social psychology conducive to building interpersonal relationships (Wentzel, 2012) and achieving a common goal of task accomplishment. This leads to achieve social responsibility goal when adhering to social rules and role expectations (Wentzel, 1991).

Case (4): Effect of $f_7$ (communication) on $f_6$ (working relationship)

On the other hand, $f_7$ (communication) has a significant direct effect on $f_6$ (working relationship) in which ($x_{28}$) positive interpersonal relationship, ($x_{29}$) peer assistance in the way making sense of learning, and ($x_{31}$) co-learning play a significant part. This is understandable as communication is not merely for cognitive exchanges but also serves for articulating expectations of peer's roles (Wentzel, 2012), maintaining social interaction, and expressing personal feelings. The last two purposes are contextualized in Kumpulainen's classifications of talk (1994) as responsive and affectional speeches.

Case (5): Effect of $f_5$ (cognitive engagement) and $f_7$ (communication) on $f_8$ (social relationship)

Nevertheless, $f_7$ (communication) has no direct effect on $f_8$ (social relationship) but a causal path connecting from $f_5$ (cognitive engagement) to $f_8$ (social relationship) is established, with a path coefficient 0.575 and $p < 0.01$. This path established is also on the basis of students' responses given to Q18 about how to achieve better learning progress when working with peers; they were mostly concerned with getting assistance from a more able partner; exchanging views for knowledge construction, clarifying or correcting misconceptions; as well as formulating goals and strategies for problem solving. It is worth noting that the three manifest variables, $x_{39}$ (much interaction), $x_{40}$ (learning harmony), and $x_{41}$ (peer made learning process fun) were incorporated into $f_8$ (social relationship). These are echoed in responses given to open-ended question, Q42 asking to describe the feeling of learning with peers which is mostly related to development of rapport, good interpersonal relationships, and a pleasant learning climate that in turn would alleviate stress and pose a favorable condition for student motivation (Juvonen et al., 2012).

The above comprehensive analysis shows that there is a causal path connecting from $f_5$ (cognitive engagement) to $f_8$ (social relationship). Cognitive engagement here was associated with both $x_{22}$ and $x_{23}$, which were interrelated. The implicit meaning of cognitive engagement is about contributions to and high involvement in learning tasks so as to build interpersonal relationships (Wentzel, 2012) and achieve social responsibility goal (Wentzel, 1991). This would affect students' attitudes and emotions at the time they are studying; cognitive engagement is thus intrinsic motivation.

More importantly, $f_7$ (communication) has an effect on $f_3$ (peer collaboration), $f_5$ (cognitive engagement), and $f_6$ (working relationship); thus, communication is the core element conducive to $y_2$ (social interaction among students). This line of argument was substantiated from students' responses given to the open-ended question, Q33 asking about circumstances in which they found communications beneficial to learning. The students would rather express their own ideas than merely listening to their peers as communication was an effective means of clarifying misconceptions and/or misunderstanding, stimulating thinking, monitoring peer's learning progress, determining goals, formulating strategies, sharing personal views, enhancing interpersonal relationship, and resolving conflicting views. The first four response items were in a cognitive context, whereas the remaining three items were social in nature. In accordance with Laible and Thompson (2007), good interpersonal interaction is derived from trust, sharing, affection, pleasant tone, as well as the content and quality of communication, thus building positive peer relationship that has positive influence on students' motivation to learn (Wentzel, 2005, 2009). Another open-ended question, Q35 asked how students thought they could improve communication with their peers; they said they would be more empathetic about peer's situations and feelings, and more patient to listen to what they said, and would provide moral support where necessary. All of these approaches aim to improve communication in the social context. A point worth emphasizing here is that there is a close link between communication and student's motivation to learn.

In Fig. 5, $f_9$ (the teacher's intervention) impacted positively on $f_{11}$ (interaction with the teacher) with path coefficient, 0.491 and $p < 0.05$ where $f_9$ is composed of $x_{43}$ (needing teacher's intervention), $x_{44}$ (appropriate timing of the teacher's intervention), and $x_{45}$ (the teacher's intervention beneficial to learning process); and $x_{52}$ (better learning progress when working with the teacher) is inclusive of $f_{11}$. This finding is sustained from students' responses to a follow-up question, Q53 asking why they are having a better learning progress. The responses can be summarized as learning assistance was offered for their development of statistical understanding; learning process was structured by initiating classroom discussions to engage students with peers and learning tasks; a positive and warm classroom atmosphere generating feelings of confidence, competence, and autonomy was fostered; as well as the teacher's conscientiousness, enthusiasm, and devotion to student learning were appreciated. The responses show adherence to the theories of education psychology (Metcalfe & Game, 2006; Skinner & Pitzer, 2012; Wentzel, 2005, 2009), quality relationship with the teacher, pedagogical caring, positive interactions with social and physical environments; the teacher's enthusiasm has positive influence on students' motivation to learn.

## 5 Limitations

By establishing paths to claim causal relationships, it is arguable because of not being certain of the claim, but it would be credible (Schum, 2001). Some caution is also needed in interpreting results of structural equation modeling because causal or latent relationships are most likely to exist by using probabilistic assessment at the 5% level of statistical significance, and the relationship is assumed to be linear, which would not give a full picture portraying the real phenomena. Technically speaking, it seems better to construct a single three-level structural equation model to illustrate the causal relationship between latent variables to show how the factors would link with the manifest variables, and how the manifest variables are correlated. However it would overgeneralize the results, without giving a detailed account of latent factors relating to the manifest variables.

## 6 Conclusion

Probable correlations among the manifest variables involving $x_1$, $x_9$, $x_{10}$, $x_{17}$, $x_{22}$, $x_{23}$, $x_{32}$, $x_{39}$, $x_{41}$, and $x_{44}$ illustrate that students are motivated to learning with IT when it involved social interaction with peers based on communication, sharing of knowledge, enjoyable learning atmosphere, IT-based materials, and appropriate timing of the teacher's intervention. These provide a cognitive as well as affective basis for developing students' interest, perceived as a positive orientation toward learning.

Theoretically speaking, there should be dynamics among these three parts, students' motivation to learn ($y_1$), social interaction between students ($y_2$), and supportive interaction with the teacher ($y_3$), but no such dynamics exists in this empirical study owing to all positive causal path coefficients are positive in the first model (see Fig. 2). Indeed, these positive coefficients confirm the one-way path direction as proposed in the model. The first structural model shows that $y_3$ directly influences $y_2$ and indirectly influences $y_1$, i.e., via $y_2$. The first argument involving $y_2$ in Submodel 2 is based upon five factors posing conditions for student motivation: $f_3$ (peer collaboration), $f_5$ (cognitive engagement), $f_6$ (working relationship), $f_7$ (communication), and $f_8$ (social relationship), where $f_6$ (working relationship) is under the influence of interpersonal relationship, peer assistance, and co-learning. Besides, $f_8$ (social relationship) is under the influence of peer interaction, learning harmony as well as learning process being made fun by peers. The latent variable, $y_3$ (supportive interaction with teacher) probably plays a certain role of posing favorable conditions for student motivation and may be due to supportive interaction with the teacher ($y_3$) in Submodel 3 involving two factors, $f_9$ (teacher's intervention) as well as $f_{11}$ (interaction with teacher). The first factor is dependent upon three manifest variables, $x_{43}$ (the significance of the teacher's intervention), $x_{44}$ (appropriate timing of the teacher's intervention), and $x_{45}$ (the teacher's intervention beneficial to learning process). Student motivation is dependent upon the positive experience of interacting with peers and/or the teacher and is mostly social in nature.

Empirical justification of the applicability and relevancy of socio-cultural theories to statistics learning within an IT environment is valid given that the structural model was constructed incorporating a number of factors to portray the influence on student motivation in conjunction with the theory of educational psychology. Implications for classroom teaching and learning that emerge from the submodels with the factors relating to each respective manifest variable provide insight into ways that students can be supported to engage with classroom learning.

## Appendix

Questionnaire

1. Do you find learning with IT interesting or boring?
2. Do you feel confident or unconfident when learning with IT?
3. Does IT help you make sense of what you are learning?
4. Do you see learning activities in computer laboratory as relevant or irrelevant?
5. Does IT widen or narrow down your scope of learning?
6. In what ways does IT widen your scope of learning?
7. In what ways does IT narrow down your scope of learning?
8. When learning with IT, do you have control or no control of your learning process?

9. What type of learning materials help you focus best on the task, IT-based or non IT-based materials?
10. Do you have a better learning progress learning with or without IT?
11. In what ways do you find you have better learning progress when learning with IT on your own?
12. In what ways do you find you have less learning progress when learning with IT on your own?
13. Outside the classroom, how do you cope with a learning problem related to this subject, "Applied Statistical Methods"?
14. Can you describe what you have learnt from this lesson?
15. Can you describe how well you feel you have learnt this material?
16. Do you find learning with IT helpful or unhelpful when working with your learning partners?
17. When learning with IT, do you have a better learning progress when working alone or working with your learning partners?
18. For what reasons do you find you have a better or a worse learning progress when working with your learning partners in an IT environment?
19. Is your interaction with your learning partners collaborative or disruptive?
20. Does your interaction with your learning partners usually produce agreement or disagreement?
21. If disagreement, how do you resolve the discrepancy/dispute or compromise between one another?
22. How much knowledge do you want to share with your learning partners when learning with IT?
23. How much do your learning partners share knowledge with you when learning with IT?
24. What type of knowledge do you want to share with your learning partners when learning with IT?
25. What type of knowledge do you not want to share with your learning partners when learning with IT?
26. What type of knowledge do your learning partners want to share with you when learning with IT?
27. What type of knowledge do your learning partners not want to share with you when learning with IT?
28. Compared with a non-IT environment, does an IT environment help you foster a better or a worse interpersonal relationship when working with your learning partners?
29. Do your learning partners help you make sense of what you are learning with IT?
30. Are you co-learning or competing with your learning partners when working in an IT environment?
31. Are your learning partners co-learning or competing with you when working in an IT environment?
32. Is the communication between you and your learning partners beneficial or unbeneficial to your learning process?

33. In what circumstances, do you find the communication between you and your learning partners are beneficial to your learning process?
34. In what circumstances, do you find the communication between you and your learning partners are unbeneficial to your learning process?
35. How do you think you can improve the communication between you and your learning partners in order to achieve mutual benefit from learning?
36. Can you describe the extent of your involvement when learning with your learning partners in IT environment?
37. Do you make any contribution to your learning partners when learning with IT?
38. How do you share the workload with your learning partners?
39. Do you have much or little interaction with your learning partners about the learning tasks?
40. Do you have harmony or conflict between you and your learning partners when learning with IT?
41. Do your learning partners make the learning process more fun or more threatening?
42. How well do you feel you have learnt with your learning partners?
43. Do you need or not need teacher's intervention in your learning process?
44. Do you find the timing of the teacher's intervention in your learning process appropriate or inappropriate?
45. Do you find the teacher's intervention beneficial or unbeneficial to your learning process?
46. In what circumstance, do you find the teacher's intervention beneficial/unbeneficial to your learning process?
47. How does your teacher orchestrate the learning activities in the computing laboratory?
48. How well do you feel you have learnt from a teacher?
49. Would you like to learn with a teacher or IT?
50. Would you like to learn with your learning partners or IT?
51. Would you like to learn with learning partners or a teacher?
52. Do you have a better or a worse learning progress when working with your teacher in an IT environment?
53. For what reasons do you find you have a better or a worse learning progress when working with your teacher in an IT environment?

## References

Ainley, M. (2012). Students' interest and engagement in classroom activities. In S. L. Christenson, A. L. Reschly, & C. Wylie (Eds.), *Handbook of research on student engagement* (pp. 283–302). Heidelberg, Germany: Springer.

Bempechat, J., & Shernoff, D. J. (2012). Parental influences on achievement motivation and student engagement. In S. L. Christenson, A. L. Reschly, & C. Wylie (Eds.), *Handbook of research on student engagement* (pp. 315–343). Heidelberg, Germany: Springer.

Biehler, R. (2016). Professional development for teaching probability and inferential statistics with digital tools at upper secondary level. *Proceedings of the 13th international congress on mathematical education*, Hamburg, Germany: International Commission on Mathematical Education. Retrieved from http://iase-web.org/documents/papers/icme13/ICME13_I1_Biehler.pdf.

Brand, S., Felner, R., Shim, M., Seitsinger, A., & Dumas, T. (2003). Middle school improvement and reform: Development and validation of a school-level assessment of climate, cultural pluralism, and school safety. *Journal of Educational Psychology, 95*(3), 570–588.

Bude, L., Van De Wiel, M. W. J., Imbos, T., Candel, M. J. J., Broers, N. J., & Berger, M. P. F. (2007). Students' achievements in a statistics course in relation to motivational aspects and study behaviour. *Statistics Education Research Journal, 6*(1), 5–21.

Dunn, P. K., Carey, M. D., Richardson, A. M., & McDonald, C. (2016). Learning the language of statistics: Challenges and teaching approaches. *Statistics Education Research Journal, 15*(1), 8–27.

Emerson, C. (1996). The outer word and inner speech: Baktin, Vygotsky, and the internalization of language. In H. J. Daniels (Ed.), *An introduction to Vygotsky* (pp. 245–264). New York, USA: Routledge.

Galbraith, P., Goos, M., Renshaw, P. & Geiger, V. (2001). Integrating technology in mathematics learning: What some students say. In J.M. Truran & K.M. Truran (Eds.), *Proceedings of the 24th conference of mathematics education research group of Australasia*, Sydney, Australia: MERGA. Retrieved from: https://merga.net.au/Public/Publications/Annual_Conference_Proceedings/2001_MERGA_CP.aspx.

Goos, M. (2004). Learning mathematics in a classroom community of inquiry. *Journal for Research in Mathematics Education, 35*(4), 258–291.

Goos, M. (2009). A sociocultural framework for understanding technology integration in secondary school mathematics. In M. Tzekaki, M. Kaldrimidou, & H. Sakonidis (Eds.), *Proceedings of the 33rd conference of the international group for the psychology of mathematics education* (vol. 1, pp. 113–120), Thessaloniki, Greece: Psychology of Mathematics Education.

Hair, J. F., Black, W. C., Babin, B. J., & Anderson, R. E. (2014). Structural equation modeling overview. In J. F. Hair, W. C. Black, B. J. Babin, & R. E. Anderson (Eds.), *Multivariate analysis* (pp. 541–598). Essex, UK: Pearson.

Kumpulainen, K. (1994). Children's talk during collaborative writing at the computer. *Reading, 28*(2), 6–9.

Juvonen, J., Espinoza, G., & Knifsend, C. (2012). The role of peer relationships in student academic and extracurricular engagement. In S. L. Christenson, A. L. Reschly, & C. Wylie (Eds.), *Handbook of research on student engagement* (pp. 387–401). Heidelberg, Germany: Springer.

Laible, D., & Thompson, R. A. (2007). Early socialization: A relationship perspective. In J. Grusec & P. Hastings (Eds.), *Handbook of socialization: Theory and research* (pp. 181–207). New York, USA: Guilford.

Li, K.W. (2012). A study on computer-supported collaborative learning in statistics classroom. In P. Lam (Ed.), *Proceedings of the 7th international conference on e-learning* (pp. 259–268). Sonning Common, UK: Academic Publishing International Limited.

Li, K. W., & Goos, M. (2015). A study on students' attitudes towards teacher's intervention in statistical computing laboratory. In K. C. Li, T. L. Wong, S. K. S. Cheung, J. Lam, & K. K. Ng (Eds.), *Transforming educational practices with technology* (pp. 116–126). Heidelberg, Germany: Springer.

Li, K. W., & Goos, M. (2017). Factors influencing social processes of statistics learning within an IT environment. *International Journal of Science, Mathematics, and Technology Learning, 24*(2), 21–33.

Metcalfe, A., & Game, A. (2006). The teacher's enthusiasm. *The Australian Educational Researcher, 33*(1), 91–106.

Pfannkuch, M., & Budgett, S. (2016). Markov processes: Exploring the use of dynamic visualizations to enhance student understanding. *Journal of Statistics Education, 24*(2), 63–73.

Pianta, R. C., Hamre, B. K., & Allen, J. P. (2012). Teacher-student relationships and engagement: Conceptualizing, measuring, and improving the capacity of classroom interactions. In S. L. Christenson, A. L. Reschly, & C. Wylie (Eds.), *Handbook of research on student engagement* (pp. 365–386). Heidelberg, Germany: Springer.

Skinner, E. A., & Pitzer, J. R. (2012). Developmental dynamics of student engagement, coping and everyday resilience. In S. L. Christenson, A. L. Reschly, & C. Wylie (Eds.), *Handbook of research on student engagement* (pp. 21–44). Heidelberg, Germany: Springer.

Schum, D. A. (2001). *Evidential foundations of probabilistic reasoning*. Illinois, USA: Northwestern University Press.

Steele, D. F. (2001). Using Sociocultural theory to teach mathematics: A Vygotskian perspective. *School Science and Mathematics, 101*(8), 404–416.

Watt, H. G., & Goos, M. (2017). Theoretical foundations of engagement in mathematics. *Mathematics Education Research Journal, 29*(2), 133–142.

Wentzel, K. R. (1991). Social competence at school: Relation between social responsibility and academic achievement. *Review of Educational Research, 61*(1), 1–24.

Wentzel, K. R. (1998). Social relationships and motivation in middle school: The role of parents, teachers, and peers. *Journal of Educational Psychology, 90*(2), 202–209.

Wentzel, K. R. (2005). Peer relationships, motivation and academic performance. In A. J. Elliot & C. S. Dweck (Eds.), *Handbook of competence and motivation* (pp. 279–296). New York, USA: Guilford.

Wentzel, K. R. (2009). Students' relationships with teachers as motivational contexts. In K. R. Wentzel & A. Wigfield (Eds.), *Handbook of motivation at school* (pp. 301–322). New Jersey, USA: Routledge.

Wentzel, K. R. (2012). Part III commentary: Socio-cultural contexts, social competence, and engagement at school. In S. L. Christenson, A. L. Reschly, & C. Wylie (Eds.), *Handbook of research on student engagement* (pp. 479–488). Heidelberg, Germany: Springer.

Wild, C., Pfannkuch, M., Regan, M., & Parsonage, R. (2017). Accessible conceptions of statistical inference: Pulling ourselves up by the Bootstraps. *International Statistical Review, 85*(1), 84–107.

**Dr. Ken W. Li** is Associate Editor of Electronic of e-Learning, UK and serves on various editorial boards of international academic journals as well as conference proceedings. He was General Secretary of the Hong Kong Statistical Society. His research interests include computer simulation, big data analytics, and the use of technologyin education. He has published research papers in the areas of computer science, statistics as well as education and co-authored an undergraduate textbook.

**Prof. Merrilyn Goos** is Professor of STEM Education and Director of EPI*STEM—the National Centre for STEM Education at the University of Limerick. She is Vice President of the International Commission on Mathematical Instruction and a former President of the Mathematics Education Research Group of Australasia. In addition to her current research interest in numeracy across the curriculum, she fosters collaborations between mathematicians and mathematics educators to support interdisciplinary research and development in university mathematics teaching and pre-service teacher education

# Online Learning and Blended Learning

# Effective Scaffolds (Supported by ICT) for Improving Student Skills: Dyslexic Students' Experiences with Their Peers

Michele Della Ventura

**Abstract** In recent years, flipped learning has attracted the attention of an increasing number of teachers and schools and has been used as a tool to enhance student learning. It is a useful tool (mainly based on cooperative learning) to help the dyslexic student to improve his/her knowledge and skills, participating on equal terms and without passive behaviour. This document analyses the effects on the concept of "positive interdependence" that characterizes a teamwork in the presence of a dyslexic student: student who does not remain an observer of the work done by other students but who leads the group work. The data were obtained from 50 students (of which 5 dyslexic students) of the third year of the Musical High School and showed a positive impact on motivation, attitude and involvement towards learning. Educational implications and research suggestions are provided based on the results of the study.

**Keywords** Creativity · Critical thinking · Dyslexia · Higher education · ICT · Student skill

## 1 Introduction

In the information age, one of the main research fields that is being developed is the one related to how to improve the quality of the learning process in order to help students to reach the learning goals. This has been justified by the fact that the diffusion of the web has produced an increasingly marked detachment of a large part of the school world from the needs of society and the skills and desires of students and their families. It follows that student interests are increasingly developed outside the school walls (Muganga, 2018; Muganga & Ssenkusu, 2019) and the teacher should implement actions that aim to better orient the student (La Marca & Longo, 2016).

M. D. Ventura (✉)
Music Academy "Studio Musica" – Department of Information Technology, Via Andrea Gritti, 25, 31100 Treviso, Italy
e-mail: michele.dellaventura@tin.it



W. W. K. Ma et al. (eds.), *Learning Environment and Design*, Educational Communications and Technology Yearbook,
https://doi.org/10.1007/978-981-15-8167-0_6

In this perspective, digital technologies can represent a valid link between learning and teaching, only if the teacher is able to deal with alternative didactic perspectives. This entails the construction of methodologies and strategies that are attentive to the multimedia reality, and which by meeting the interests of the students can increase the learning motivation. Research has shown that active learning is promising to this end (Freeman et al., 2014), especially considering an increasingly widespread reality in school classrooms: the presence of dyslexic students, that is, students with specific learning disabilities (SLD). These are disorders that do not allow complete self-sufficiency in learning since the difficulties concern some specific activities that are used for learning: reading, writing and counting. Active learning methodology (such as the flipped classroom) has become the preferred way to rethink the traditional classroom teaching model (Roach, 2014) to increase learning motivation (Sergis, Sampson, & Pelliccione, 2018). It is based on activities that include the use of technologies (such as digital video or audio files) in group activities: favourable characteristics for dyslexic students.

This paper proposes a learning approach, based on group work, in which the dyslexic student is an active part of the learning process, leading the other students in reaching the learning goals. The doubts and questions of the dyslexic student become guidelines for the working group, to compare ideas, to share information and experiences, to structure activities and to create artefact. This is possible if the communication takes place using a simple and effective language that takes into consideration the needs of each student, particularly the needs of the dyslexic student. It has been widely demonstrated that responsible learning improves students' performance (Hake, 1998; Freeman et al., 2007; Chaplin, 2009), increases students' involvement, improves their critical thinking [10], improves students' attitude and their ability to interact and cooperate (O'Dowd & Aguilar-Roca, 2006). This approach can positively affect the learning motivation, attitudes and engagement of all students.

This paper is organized as follows:

Section 2 analyses the concept of "dyslexia" and possible teaching strategies useful in a learning process. Section 3 describes the strengths of the working group, related to all students: dyslexic and non-dyslexic students. Section 4 shows a list of learning process indicators useful for the teacher for monitoring the learning process. Section 5 shows an experimental test. Finally, in Section 6 the paper ends with concluding remarks on the current issues and future research possibilities with respect to the efficient enhancement of educational practices and technologies.

## 2 Understanding Dyslexia

Dyslexia is defined as a specific learning disorder (SLD) and it is characterized by difficulties with reading and writing that often lead the dyslexic students to stay behind their classmates (Akinoglu & Tandogan, 2006). Over time, this poor performance may lead to a negative attitude towards learning and the activities that require reading and writing (Lyon, Shaywitz, & Shaywitz, 2003). Students might lose confidence in their abilities and develop a low sense of self-esteem (Graham, Berninger, & Fan, 2007; Orth, Robins, Trzesniewski, Maes, & Schmitt, 2009).

Within the classroom, the dyslexic students are often perceived as "different" by teachers and/or students. They are labelled as lazy, inattentive and people who "don't commit enough". Teachers often use the same teaching methodology with all students (dyslexic students and non-dyslexic students) without looking for alternative solutions that can meet the needs of all students, because some of them are rigid personality types or may be wedded to a single teaching method; some of them fear that the search for alternative solutions requires more demands on time and energy.

In view of the presence of dyslexic students, some teachers consider more useful and productive to use an individualized teaching (Della Ventura, 2019a, b), supporting the didactic activity with the use of compensatory tools and dispensatory measures (Martan, Skočić Mihić, & Matošević, 2018), on the base of the most recent laws (adopted in many countries) about the inclusive education of students with special educational needs. Even though it appears to be necessary, it does not mean that it alone can be effective in helping dyslexic students to reach the learning goals.

The *compensatory tools* are tools that dyslexic students can use in order to compensate the functional weakness due to their disorder (Fogarolo & Scapin, 2010). On the one hand, computers and software could be good compensatory tools only if students know how to use them competently; on the other hand, given that the competence is the result of a formative process (Della Volpe, 2016), teachers should not leave students alone in the use of computers and software, but guide them for an effective and conscious use: this implies knowledge in their use by teachers.

The *dispensatory measures* are interventions that allow dyslexic students not to carry out some activities without modifying learning objectives (Lombardi, 2012). The reduction of the number of questions or the extra time for doing a test represent a solution but at the same time a problem (Reid, 2009). Teachers, often, don't take care of the language used for the questions of the test: they don't use a language suitable for every student (Kudek Mirošević, 2016). Teachers must not only think about what they write in a question, but how they write it: they must consider the clarity of their writing and adapt it to the students. It is important that the teacher uses a didactic method that supports not only dyslexic students, but in which everyone, even the best students, learns in a satisfactory and functional way: what is good for a dyslexic student is good for everyone, but the opposite is not true (Stella, 2001).

## 3 Why Active Learning?

These considerations, referred to some examples of *compensatory tools* and *dispensatory measures*, underlie the research presented in this paper. The weaknesses of the dyslexic student were considered as the basis of the learning process, considering them not as a threat but as an opportunity of improvement for all students: dyslexic and non-dyslexic students. In order to do this, active learning typical of group work was used. In group work, students share a common goal and share the responsibility for completing the task, avoiding that some students do all the work or make all the decisions (Johnson & Johnson, 2018). The elements that characterize the group work are: positive interdependence, individual accountability, promotive interaction, social skills and group processing (Johnson & Johnson, 1989).

First of all, the members of a working group must understand what dyslexia is in order to learn how to relate within the group because the dyslexic can help other group members thanks to his/her skills. The dyslexic student has interpersonal strengths (Chathurika, Jerome, Sowmya, & Rosie, 2018): he/she replies more effectively if the task involves action, active learning and interaction, together with group work and discussion; the dyslexic student is more curious than average and is highly intuitive; the dyslexic student thinks outside the box (Della Ventura, 2019a, b): he/she is fast problem solver and can think laterally, providing unorthodox methods for solving problems.

At the same time, the dyslexic student can indirectly help other group members by means of his/her weaknesses: what may seem like a threat can become an opportunity. The dyslexic student has difficulties with vocabulary acquisition, with retrieving words, with applying the grammar rules and consequently in formulating a discourse (Adlof & Hogan, 2018): this implies that the other members of the group must use a simple language to explain a concept or an idea (Della Ventura, 2019a, b; Adlof & Hogan, 2018). It is necessary to reduce the complexity of the technical language: the student who doesn't speak the technical language in a fluent way has difficulty to make sense of a discourse or a text. To communicate effectively, it is not enough to have well-organized ideas expressed in complete and coherent sentences (Horkoff & McLean, 2015). It is necessary to think of the style and the clarity of text (or discourse) and adapt these elements to the other students. If the students with satisfactory academic performances think that the explanation to the other group members is not necessary, they are wrong: complete mastery of a topic, in fact, is obtained when he/she knows how to transfer the key contents to others. Only in this case, he/she can argue that he/she has understood and internalized the topic in question. In addition, transferring content to those who don't know it requires greater clarity both mentally and in exposure. Finally, this transfer of knowledge helps both the transmitter and the receiver to store it persistently (Ellis, 2016).

The dyslexic student has difficulties with reading, writing and with taking notes. It is possible to go beyond these difficulties by using text messages with as few words as possible, using simple and punctual language that includes descriptions that can create tangible images with details that the dyslexic student can visualize: in fact, dyslexics think mainly in pictures instead of words and are highly skilled at remembering pictures they have seen. They are able to connect different concepts to identify similarities and/or affinities that others students may not see (Ramus et al., 2003).

It follows that it is useful:

- to verify that the dyslexic student has understood the meaning of what is said or written by the members of the group: with simple questions or asking him/her to create a conceptual map or to find images that can support the explained concept;
- to pay attention to the questions of the dyslexic student to understand how to proceed in group work. It could be necessary to explain a concept again using

different words to better understand the concept; to make connections with different and work-related concepts that could help to understand a concept; reconsider all the work done or part of it, for not having considered it from a different and more objective point of view.

In view of the above-mentioned considerations, it is easy to understand the importance of the role of ICT (information and communication technologies) in the learning process. Nowadays, the web provides new useful tools for teaching and different from traditional learning management systems (like Moodle). The social network (like Facebook, WhatsApp and Twitter) allows users to exchange images, videos, audio, text messages or voice notes; the possibility to send and listen to audio messages rather than text messages helps the students who have trouble with reading; the possibility to listen several times to the same message helps them to learn; the possibility to dictate a message rather than write it down also compensates the writing difficulties (Della Ventura, 2016). The conciseness of the messages helps the students who have a hard time reading and it does not tire them when learning the content; the possibility to write short messages helps them to develop the capacity to formulate a question or an answer (moreover, the automatic spell checker, already present in all devices allows them to avoid possible spelling errors) (Della Ventura, 2018).

It is therefore important to monitor and analyse the learning process through a series of indicators that allow support interventions for all students. In the following paragraph, a set of learning process indicators derived from all of the above considerations, and useful for the teacher for monitoring the learning process, are presented.

## 4 Process Indicators

In a learning process, the central role is played by the operations of monitoring and analysis that assume an indispensable role for the development of quality processes. It is essential to identify process indicators as a support for the teacher, in order to be able to interpret the learning objectives.

This paragraph presents a series of indicators (Table 1) divided in turn on the basis of different aspects and elements that characterize a learning process: indicators referred to the internal process, indicators referred to the learning and growth process and indicators referred to the user's perspective.

## 5 Application and Analysis: Research Method

The main aim of the present research was to analyse the effects on learning brought by the implementation of the group work in the learning process, in the presence of dyslexic students: their doubts and questions were considered working group

**Table 1** Learning process indicators

| Learning outcome and/or process requirement | Indicators |
|---|---|
| *Indicators referred to the internal process* | |
| Social competences | Respect of the ideas of all group members |
| | Give value to the student diversity |
| | Increase student retention |
| | Increase of the empathy among the group members |
| | All the group members are aware of other team members' competences, experience and preferred working styles |
| Critical thinking | Critical discussion by providing arguments |
| | Construction of simple sentences for the message |
| | Describe ideas clearly |
| | Use of the technical terms in simple sentences |
| | Understand questions, analysis and evaluation of this aspect of the learning experience thinking |
| | Provide an explanation for a problem in an appropriate way that permits to guide the other group members |
| | Understand what knowledge has been acquired and what not yet |
| Inclusiveness of all students | Take part in different dialogues |
| | Identify a need or opportunity from the context and problem indicated |
| | Discourses of personal experiences |
| | Increase of students' curiosity |
| *Indicators referred to the learning and growth process* | |
| Communication | Number of text messages |
| | Number of audio messages |
| | Number of images used in the messages (to support explanation and discourses) |
| Attitudes | Increase of the ability to autonomously search for relevant information |
| | Increase of the ability to recognize relevant information and use it effectively in the activity |
| | Increase of the ability to elaborate and share ideas verbally and written |
| | Increase of the ability to organize and make meaning from visual information |

(continued)

**Table 1** (continued)

| Learning outcome and/or process requirement | Indicators |
|---|---|
| | Increase of the ability to relate ideas from multiple topics in different contexts |
| | Help group members to make inquiries in a logical way |
| | Help group members to correct misconceptions |
| | Increase of the capacity to use again resources and knowledge |
| | Ability to use the information in different activities |
| | Ability to identify a need or opportunity from the context and problem indicated |
| Acquisition of Knowledge | Creation of concept maps |
| | Formulate a hypothesis |
| | Identify what is still necessary to know |
| | Critical discussion of the findings by providing arguments |
| | Provide feedback on the results of others group members |
| | Identification of what needs to be done next |
| | Evaluation of potential results in terms of identified concepts |
| | Description of potential results |
| | Improvement of communication skills between group members |
| | Reduction of confusion in connecting different concepts |
| | Number of new technical words used during communication |
| | Number of connections between different concepts |
| Communication skills | Use of different types of results visualization |
| | Presentation of the results in clear written form |
| | Description of potential results, through drawings, texts or verbally |
| | Use words as per context |
| Engagement | Increase of the group approaches to problem solving |
| | Number of approaches with a new perspective |

(continued)

**Table 1** (continued)

| Learning outcome and/or process requirement | Indicators |
|---|---|
| *Indicators referred to the user's perspective* | |
| Feelings and perceptions | Increase of the awareness of the group work |
| | Increase of the perception of reinforcement |
| | Increase of the perception of the opportunity to learn |
| | Increase of the satisfaction with the learning experience |
| | Increase of enthusiasm in students |
| Attitudes and perceptions | Interactions with group members |
| | More positive attitudes towards learning |
| | Increase students' awareness of the value of the group work experience |
| | Increase of the student's ability to persist in a task despite failure and difficulty |
| | Reduction of the number of complaints from the group members |

guidelines (threat vs opportunity). The students had to decide how to respond using simple and effective language (strength vs improving skills), which included descriptions that could create tangible images for the dyslexic student (strength vs improving skills). If necessary, they had to explain a concept again using different words to better understand the concept (threat vs opportunity), making connections with different and work-related concepts (strength vs improving skills), reconsidering the previous work or part of it (threat vs opportunity), for not having considered it from a different and more objective point of view. Students were invited to fill out an online diary of their study activities (twice a week), specifying the time spent studying (the number of days and the average number of hours), writing the concepts analysed during the working group, the doubts expressed by the group members, the words (or technical terms) that required an intervention to understand a concept and the solution found to solve a problem or a doubt. It was possible to include an audio file with the explanations of the concepts and the answers to the doubts: in this way it was possible for the dyslexic student to listen to (several time) the audio rather than read the text messages. Students were allowed to use WhatsApp for the personal study, under the supervision of the teacher.

The research was conducted by analysing the learning process of two groups of students of the third year of two different Musical High Schools: the first group, consisting of 24 students of which two dyslexic students, involved in a *Traditional Course* (TC), and the second group of 26 students of which three dyslexic students, involved in a *Work Group Course* (WGC).

The project was conducted for a time period of 10 weeks in the second semester (February–April 2019) and covered the same discipline (Music Technologies) and the same topic (the case study of a sound recording) using the same book. Both groups had the same teacher.

Both courses included:

- classroom lessons each one divided into two parts: the first one dedicated to the explanation of the teacher and the second one dedicated to the analysis of a practical recording event under the supervision of the teacher (for the WGC, the second part of the lesson and the home study were conducted with working group, while the TC was conducted with a personal study);
- homework that included the study of the concepts explained by the teacher and the analysis of other events.

At the end of the project an examination was passed in the two groups, identical for dyslexic and non-dyslexic students: dyslexic students were allowed to use compensatory measures.

## 6 Results

The results of the WGC were very satisfactory demonstrating the effectiveness of the working group. Figure 1 shows the grade point average of the first and second semesters of each student. The main factors that determined the results of the learning process, also derived from reading the student's diaries, are summarized in Figs. 2 and 3.

Figure 2 shows that WGC students spent more time studying each week than the TC students. This is explained by the fact that in WGC the doubts were addressed step-by-step (also using WhatsApp), while in the other case the doubts were only resolved during the lesson in the classroom: this means that during the lessons the teacher was able to face more examples with the WGC students, helping them to consolidate their knowledge/skills with more opportunities to compare similar or different events.

The increased time required for the homework, which could be perceived as a threat (because it reduced the time for other personal activities), translated into an opportunity for skills improvement.

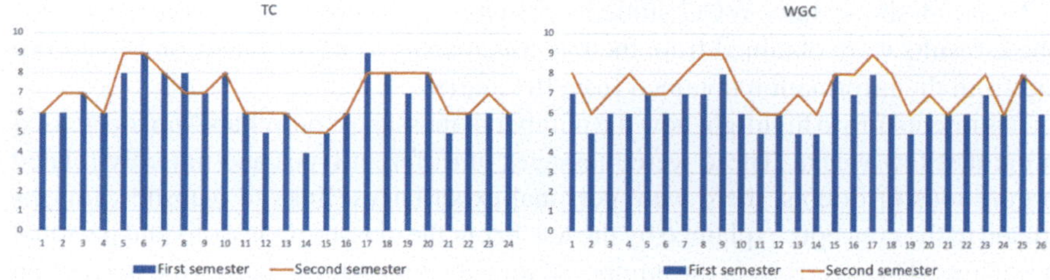


**Fig. 1** Grade point average

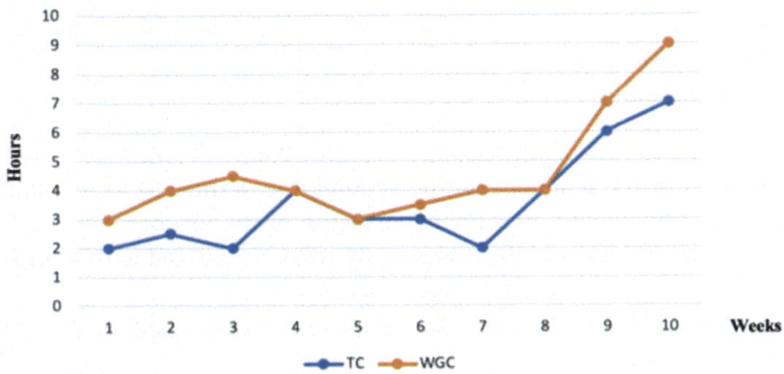


**Fig. 2** Average of time studied each week

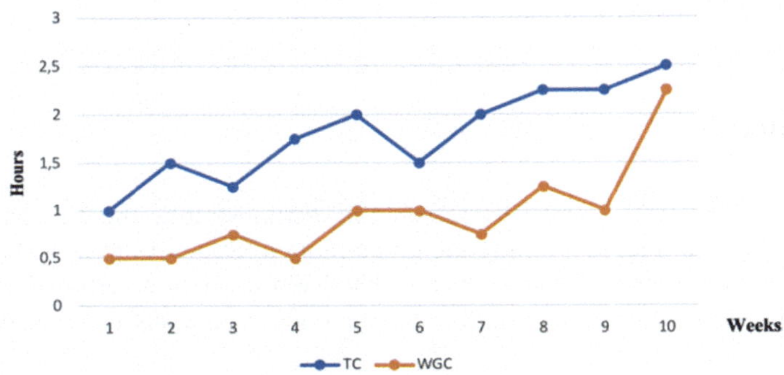


**Fig. 3** Average of time studied each week

Figure 3 shows that the WGC students spent less time studying lecture notes each week than the TC students. This is explained by the fact that the WCG students shared all the work on WhatsApp, where they could: find information related to the concepts explained by the teacher in the classroom (with simpler language), share personal experiences and concrete examples (sometimes related by links to websites), find links between different concepts and get quick answers to personal doubts. Only in the last week, WGC students spent more time studying the lecture notes, to make sure they hadn't omitted any concepts.

Figure 4 shows how WCG students activities took place during the 10 weeks. These results were obtained from the integrated analysis of messages on WhatsApp and from the information obtained from the diaries.

It is interesting to highlight how the number of messages (on WhatsApp) decreased in the first few weeks. These were messages of different types and not only related to questions or doubts, there were personal experiences, links to websites, images related to the concepts explained by the teacher in the classroom lessons and personal curiosities. This decrease in the number of messages was due in part to the perception of the waste of time for personal activities, and in part to the need to write an effective

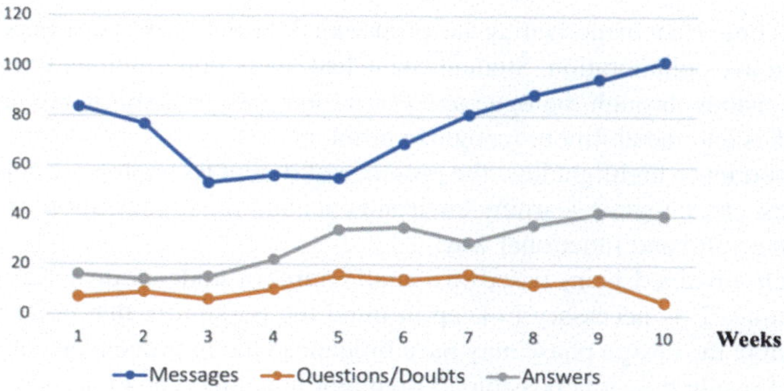


**Fig. 4** Conduct of activities

message in a language suitable for all members of the group. In the last few weeks, the number of messages has increased, thanks to the fact that students became aware of the possibilities offered by this method of learning.

Answers to questions and/or doubts gradually increased over the weeks, as the active participation of all students in research and analysis activities increased as well. Students learned to propose effective feedbacks and follow the relationship between their activities to be aware of their progress.

Finally, it is interesting to point out that for every doubt expressed by a dyslexic student, there were always more than one message of support, not just from students with high academic performances.

An improvement was noted in the papers produced by dyslexic students: in their autonomy to perform a task, in the use of technical language and example to support the explanation, and in the analysis of the event. These improvements also emerged for non-dyslexic students with low and high academic performances. It is very important to make all students aware that even if dyslexic students have difficulties with reading and writing, they can achieve excellent results with the help of their peers and they can help their peers to improve their skills. This is important so that they accept and support the dyslexic learners instead of ridiculing them.

From the point of view of involvement and motivation, the activity certainly proved to be a success, given that about 80% of the WGC students declared that they wanted to repeat this experience also with different disciplines.

## 7 Discussion and Conclusions

The research presented in this document takes into consideration students with heterogeneous learning styles (because also with dyslexic students), and analyses the potentialities of the use of the active learning in the learning process and its impact on students' motivation, attitudes and commitment of all students (dyslexic and non-dyslexic students). The involvement of students during the entire learning process was important in motivating and preparing them to follow the project's phases with a proactive collaboration. Students have had many opportunities for reflection on their decisions through dialogue and discussion, self-evaluation, recognition of their mistakes and the ability to recognize problems.

It is important to highlight how the presence of digital technologies, in particular mobile ones, can integrate learning environments and the organization of teaching activities in a valid and functional way.

The results obtained, being based on a small sample of students, do not allow undue generalizations. It is necessary to keep in mind the possibility that other variables not foreseen in the design phase may have influenced the improvements observed in the group. Despite this, it is possible to highlight how this type of activity has tried to make the concepts that underlie the learning process experiential and directly tangible on a cognitive and metacognitive level.

The modalities with which the project presented in this document was carried out can represent a stimulus and a guide for other teachers who have didactic objectives similar to those pursued.

## References

Adlof, S. M., & Hogan, T. P. (2018). Understanding dyslexia in the context of developmental language disorders. *Language Speech and Hearing Services in Schools, 49*(4), 762–773.

Akinoglu, O., & Tandogan, R. (2006). The effects of problem-based active learning in science education on student's academic achievement, attitude and concept learning. *Eurasia Journal of Mathematics, Science & Technology, 3,* 71–81.

Chaplin, S. (2009). Assessment of the impact of case studies on student learning gains in an introductory biology course. *Journal of College Science Teaching, 39,* 72–79.

Chathurika, K., Jerome, C., Sowmya, P., & Rosie, A. (2018). Not all those who wander are lost: Examining the character strengths of dyslexia. *Global Journal of Intellectual & Developmental Disabilities, 4*(5), 555–648.

Della Ventura, M. (2019a) Exploring the impact of artificial intelligence in music education to enhance the Dyslexic student's skills. In L. Uden, D. Liberona, G. Sanchez, & S. Rodríguez-González (Eds), *Learning technology for education challenges* (Vol. 1011). LTEC 2019. Communications in Computer and Information Science. Springer, Cham.

Della Ventura, M. (2019b) Between research and action: The generative sense of technology. In L. Rønningsbakk, T. T. Wu, F. Sandnes, & Y. M. Huang (Eds), *Innovative technologies and learning* (Vol. 11937, pp. 754–763). ICITL 2019. Lecture Notes in Computer Science. Cham: Springer.

Della Ventura, M. (2016). Creating inspiring learning environments by means of digital technologies: A case study of the effectiveness of WhatsApp in music education. In *Proceedings of the 3rd international conference on e-learning, e-education, and online training* (pp. 36–45). Dublin, Ireland: Springer.

Della Ventura, M. (2018). Using Twitter to enhance the students skills: Motivation a disregarded factor in educational design. In *Proceedings of the 4th international conference on e-learning e-education and online training (eLEOT 2018).* Shanghai: People's Republic of China, Springer.

Della Volpe, V. (2016). Study of compensatory tools and dispensatory devices in Italian inclusive education. *Journal of Research & Method in Education, 6*(5), 7–13.

Ellis, A. W. (2016). *Reading, writing and dyslexia (classic edition): A cognitive analysis.* Psychology Press.

Fogarolo, F., & Scapin, C. (2010). *Competenze compensative: Tecnologie e strategie per l'autonomia scolastica degli alunni con dislessia e altri DSA*. Trento, Erickson 2013.
Freeman, S., O'Connor, E., Parks, J. W., Cunningham, M., Hurley, D., Haak, D., et al. (2007). Prescribed active learning increases performance in introductory biology. *CBE Life Science Education, 6,* 132–139.
Freeman, S., Eddy, S. L., McDonough, M., Smith, M. K., Okoroafor, N., Jordt, H., & Wenderoth, M. P. (2014). Active learning increases student performance in science, engineering, and mathematics. *Proceedings of the National Academy of Sciences of the United States of America*
Graham, Steve, Berninger, Virginia, & Fan, Weihua. (2007). The structural relationship between writing attitude and writing achievement in young children. *Contemporary Educational Psychology, 32,* 516–536. https://doi.org/10.1016/j.cedpsych.2007.01.002.
Hake, R. R. (1998). Interactive-engagement versus traditional methods: A six-thousand student survey of mechanics test data for introductory physics courses. *American Journal of Physics, 66*(1), 64–74.
Horkoff, T., & McLean, S. (2015). *Writing for success* (1st Canadian ed.). Canada: Open Textbook.
Johnson, D. W., & Johnson, R. T. (2018). Cooperative learning: The foundation for active learning. *Active Llearning - Beyond the Future, Sílvio Manuel Brito, IntechOpen.* https://doi.org/10.5772/intechopen.81086.
Johnson, D. W., & Johnson, R. T. (1989). *Cooperation and competition: Theory and research.* Edina, MN: Interaction Book Company.
Kudek Mirošević, J. (2016). The assessment of the competences of students at the department of teacher education studies and of the teachers for inclusive practice. *Croatian Journal of Education, 18*(1), 71–86.
La Marca, A., & Longo, L. (2016). Addressing student motivation, self-regulation, and engagement in flipped classroom to decrease boredom. In *Proceedings of 4th international conference on information and education technology (ICIET)* (pp. 230–235). Los Angeles.
Lombardi, G. F. (2012). *La legislazione internazionale in materia di dislessia: Uno sguardo d'insieme e un'analisi della legislazione britannica e americana.* AP Generale.
Lyon, G. R., Shaywitz, S. E., & Shaywitz, B. A. (2003). *A definition of dyslexia. Springer, 53,* 1–14. [Note: This entry should be placed after the next, which starts with the surname 'Lombardi']
Martan, V., Skočić Mihić, S., & Matošević, A. (2018). *Teachers' attitudes toward teaching students with Dyslexia.* 19.
Muganga, L. (2018). *You can't make "Fish Climb Trees": Overcoming educational malpractice through authentic Learning.* Victoria, BC: FriesenPress.
Muganga, L., & Ssenkusu, P. (2019). Teacher-centered vs. student-centered. *Cultural and Pedagogical Inquiry, 11,* 16–40. https://doi.org/10.18733/cpi29481.
O'Dowd, D. K., & Aguilar-Roca, N. (2009). Garage demos: Using physical models to illustrate dynamic aspects of microscopic biological processes. *CBE Life Science Education, 8,* 118–122.
Orth, U., Robins, R. W., Trzesniewski, K. H., Maes, J., & Schmitt, M. (2009). Low self-esteem is a risk factor for depressive symptoms from young adulthood to old age. *Journal of Abnormal Psychology, 118,* 472–478.
Ramus, F., Rosen, S., Dakin, S. C., Day, B. L., Castellote, J. M., White, S., et al. (2003). Theories of developmental dyslexia: Insights from a multiple case study of dyslexic adults. *Brain, 126*(4), 841–865.
Reid, G. (2009). *Dyslexia: A practioner's handbook.* Chichester: Wiley.
Roach, T. (2014). Student perceptions toward flipped learning: New methods to increase interaction and active learning in economics. *International Review of Economics Education, 17,* 74–84.
Sergis, S., Sampson, D. G., & Pelliccione, L. (2018). Investigating the impact of Flipped Classroom on students' learning experiences: A Self-Determination Theory approach. *Computers in Human Behavior, 78,* 368–378.
Stella, G. (2001). *In class with a learning disorder affected student* (1st ed.). Florence, Italy: Fabbri.

**Michele Della Ventura** Professor of Music Technology, is a learning expert, researcher and instructional designer. His research interests include: correlation between music and mathematics; assessment for learning and strategies and models for the effective integration of technology into the curriculum at all academic levels. He is the author of several articles presented at many International Conferences and published in international science magazines. He proofreads articles and is a member of scientific committees in International Conferences. He was invited as keynote speaker to International Conferences and events world-wide.

# Study on Instructional Designers' Various Perspectives of Different Learning Modes in Blended Learning

**Ying Hui Chan and Wing Sum Cheung**

**Abstract** Many organizations want to implement blended learning. Many have invested resources in their efforts of implementing blended learning. This study is aimed to find out the appropriate situations of using face-to-face activity, asynchronous online activity, and synchronous online activity in a blended learning environment. We explore the reasons why these are the appropriate situations. In this qualitative study, seven instructional designers, who have at least three years blended learning design experience were interviewed. Findings from this study show that there are key situations and potential challenges that instructional designers opine the usage of face-to-face activities, asynchronous online activities, and synchronous online activities in blended learning. Based on the various perspectives from the seven instructional designers, blended learning design guidelines on various situations of using face-to-face activities, asynchronous online activities, and synchronous online activities, and preventing some potential challenges have been postulated, which serve as a guide for instructional designers.

**Keywords** Blended learning · Asynchronous online activities · Synchronous online activities

## 1 Introduction

Educational institutions have invested in lots of resources—effort, time, and money, to make technology infrastructure available to instructors and learners for e-learning. Despite the numerous benefits of blended learning, studies have shown numerous challenges from various aspects (Kaur, 2013; Draffan & Rainger, 2006; Vaughan, 2007). Kaur (2013) highlighted that the time and budget allocated to design content

---

Y. H. Chan (✉) · W. S. Cheung
National Institute of Education, Nanyang Technological University, Singapore, Singapore
e-mail: yhuichan@yahoo.com

W. S. Cheung
e-mail: wingsum.cheung@nie.edu.sg



W. W. K. Ma et al. (eds.), *Learning Environment and Design*, Educational Communications and Technology Yearbook,
https://doi.org/10.1007/978-981-15-8167-0_7

are too little, as compared to technology implementation. Various studies on blended learning challenges have been presented from teachers' perspectives (Draffan & Rainger, 2006; Lameras, Levy, Paraskakis, & Webber, 2012), learners' perspectives (Draffan & Rainger, 2006; Bidder, Mogindol, Saibin, Andrew, & Naharu, 2016), and administrative (institution's) perspectives (Vaughan, 2007). In this study, we focused on instructional designers' perspectives as they are the subject matter expert in blended learning instructional design.

## 2 Problem Statement

Many organizations want to implement blended learning. Instructional designers often face challenges in designing blended learning. Studies concluded that blended learning is useful, when it is properly designed (Spanjers, Konings, Leppink, Verstegen, Jong, Czabanowska, & Merrienboer, 2015; Alten, Phielix, Janssen, & Kester, 2019). It is imperative to provide blended learning design guidelines to guide practitioners (Kim, Bonk, & Teng, 2009). To design blended learning properly, the problems faced by many novice instructional designers are under what situations face-to-face activities, asynchronous online activities, and synchronous online activities should be used, respectively. To guide blended learning design, various perspectives need to be sieved out from experienced instructional designers in blended learning.

## 3 Literature Review

Blended learning is a loosely defined term, which is widely used. There is ambiguity as to the definition of blended learning. There are several widely used definitions of blended learning. According to Graham (2006), "Blended learning systems combine face-to-face instruction with computer-mediated instruction" (p. 5). The definition by Garrison and Kanuka (2004) is that "blended learning is the thoughtful integration of classroom face-to-face learning experiences with online learning experiences" (p. 96). In this study, we adopted the definition of blended learning from the proposition by Horn and Staker (2011): "blended learning is any time a student learns at least in part at a supervised brick-and-mortar location away from home and at least in part through online delivery with some element of student control over time, place, path, and/or pace" (p. 4). A fully online course is not considered as blended learning (Hew & Cheung, 2014).

We went through literature review of blended learning design model from various studies. Pesavento, Klein, Macasaet, Shorter and Wagstaff (2015) identified blended learning models, which are based on learning objectives, including: (1) backward design model—design is based on learning objectives, (2) rapid prototype model—design process begins with lesson sequencing and is iterative, (3) multimodal design model—design content by incorporating multiple learning modalities (with various online and physical spaces), and (4) learner-driven design model—design is based on learning objectives in which learners play an active role in determining them. Blended learning lesson plan design procedure presented by Wikiversity (2016) is also based on learning objective, combining face-to-face components and computer-based components. Pappas (2014) presented blended e-learning strategy which is based on learning course's objectives.

Horn and Staker (2012) identified four broad blended learning models, including: (1) rotation model—design by rotating among learning modalities, with at least one online learning, (2) flex model—individually customized, design mainly for online learning and supported by face-to-face learning, (3) self-blend model—a course designed entirely online while taking other face-to-face courses, and (4) enriched virtual model—design as a whole-school experience in which within each course, students attend face-to-face learning and remote online learning. Rotation model can be subdivided into station rotation, lab rotation, flipped classroom, and individual rotation (Horn & Staker, 2012).

From the aforementioned literature review on blended learning design models, we realized that none of the studies addresses blended learning in the form of blended learning guidelines from instructional designers' point of view—under what situations that face-to-face activity, asynchronous online activity, and synchronous online activity should be used in a blended learning environment. Therefore, it is important to have blended learning guidelines to address the aforementioned gap from instructional designers' perspectives.

## 4 Research Questions

The research questions of this study are:

1. What are the appropriate situations that face-to-face activity, asynchronous online activity, and synchronous online activity should be used in a blended learning environment? Why?
2. What are the challenges of using face-to-face activity, asynchronous online activity, and synchronous online activity in a blended learning environment? Why?

## 5 Methodology

We approached this qualitative study by using interviews. The interview questions were designed in alignment with the two research questions of this study. There are three sections in the interview, namely face-to-face activities, asynchronous online activities, and synchronous online activities, respectively. Some of the interview questions were designed systematically based on in situ variables (e.g. nature of content and learning environment) and independent variables (e.g. pedagogical approach) of the blended learning research constructed by Hew and Cheung (2014) with the intention to obtain rich information of the various aspects leading to outcomes. The duration of each individual interview was approximately an hour.

## 5.1 Data Collection

In this study, convenient sampling was used. The seven instructional designers' age ranges from 26 to 60 years old, have at least three years blended learning design experience, have worked in blended learning instructional design to cater to the needs of learners from primary, secondary, and post-secondary education institutions were interviewed. The three years instructional design experience in blended learning was a reference to shortlist for the right candidates. All participants have teaching experiences, ranging from primary school learners, secondary school learners to adult learners. Detailed demographic information on participants are given in Table 1.

One of the authors took interview notes throughout the interview sessions. Member checking was conducted at the end of each interview session to ensure data validity—ensuring participants' practices and opinions had been accurately

**Table 1** Participants' demographic information

| Participant | Age group range | Teaching experience | Instructional design experience |
|---|---|---|---|
| A | 36–40 years old | 9 years in higher education institutions teaching young adults and adult learners | 6 years |
| B | 36–40 years old | 8 years in higher education institution | 6 years |
| C | 41–45 years old | 20 years, including teaching primary school students and adult learners | 13 years |
| D | 36–40 years old | 17 years in higher education institution teaching young adults and adult learners | 7 years |
| E | 56–60 years old | 26 years, including teaching primary school students and adult learners | 19 years |
| F | 26–30 years old | 7 years, including teaching secondary school students and adult learners | 4 years |
| G | 41–45 years old | 14 years in higher education institutions teaching adult learners | 14 years |

captured. Another round of interview would had been arranged, if necessary, during analysis stage of the study, to further confirm findings.

### 5.2 *Data Analysis*

The interview data were analyzed. Content analysis was done by identifying thematic patterns for each research question. To address the research questions, we performed content analysis for the entire interview data. The thematic unit as the unit of analysis was adopted in this study.

We used pedagogy categorization by Bower, Hedberg, and Kuswara (2010): (1) transmissive pedagogy, (2) dialogic pedagogy, (3) constructive pedagogy, and (4) co-constructive pedagogy. Transmissive pedagogy is the approach of broadcasting information or content to learners. Dialogic pedagogy involves communications or discussions pertaining to the learning. Constructive pedagogy is the approach of creating something pertaining to the learning. Co-constructive pedagogy involves learners working in groups in creating something pertaining to the learning. The pedagogy categorization by Bower et al. (2010) is based on the degree of negotiation and production, which is suitable to be used in different subject domains.

Even though Bower et al.'s pedagogy categorization was used a priori, we did not force feed any of the coding categories onto our data. We allowed new pedagogy category to emerge inductively, if any. Constant-comparative grounded approach by Lincoln and Guba (1985) was used in this study by going back and forth of data sets to confirm the coding categories.

We adopted the cognitive domain taxonomy by Anderson and Krathwohl (2001), with the extension of digital taxonomy by Churches (2008), because the cognitive domain taxonomy cuts across both face-to-face and online learning environment. Based on the interview data, the seven instructional designers provided insights into blended learning in cognitive domains and psychomotor domains.

## 6 Findings

Themes emerged showed that there are various situations and challenges of the three different learning modes—face-to-face, asynchronous online, and synchronous online.

### 6.1 The Use of Face-To-Face Activities, Asynchronous Online Activities, and Synchronous Online Activities in Blended Learning

In this section, we address the first research question: "What are the appropriate situations that face-to-face activity, asynchronous online activity, synchronous online activity should be used in a blended learning environment? Why?". Findings were arranged in Tables 2, 3, and 4, for the three learning modes, respectively.

For face-to-face activities, most participants shared about introductory session between instructors and learners to be carried out at the beginning of a course to build human relationships, perform administration on groupings, familiarize learners with online learning environment, and find out learners' prior knowledge. Most participants had used face-to-face activities for hands-on tasks and psychomotor tasks. Participants also offered insights on the connection between face-to-face activities and asynchronous online activities. It was also mentioned that during face-to-face, instructors access learners' learning and address the gaps from online learning.

Participant A provided an example, "At the beginning (of a term) let participants to have face-to-face learning so that they know how to login using credentials so that you don't have to spend time online to sort out their login credentials, for example". Six participants mentioned dialogic pedagogy to be used face-to-face, for learners to engage in small group discussion and/or instructor-led discussion, present their works, engage in role playing. Insights on transmissive pedagogy by Participant G, "More difficult concepts which requires more explanation from the lecturer because those basic information students can read online, and students can read it themselves".

Most participants had used face-to-face activities for hands-on tasks and psychomotor tasks. Participant F offered insights, "Learning objectives need demonstration because the learner can see clearly, the learner can interject for clarification; compared to demonstration through video online". Participant B shared, "Something requires equipment (e.g. special software, hardware equipment) because they need to be physically around to use the equipment. E.g. students are required to configure Sophos firewall. Students are to configure firewall on the hardware according to a job sheet. The hardware is too expensive for each student to purchase it and do it at home". Participant E emphasized, "When this is a critical content, the consequence is danger. Need to correct them the first time they learn face-to-face if they do it wrongly. Cannot afford to let them learn wrongly on their own online and then correct them face-to-face".

Participants also offered insights on the connection between face-to-face activities and asynchronous online activities. Instructors access learners' learning from online activities and address the gaps from online learning. Participant D shared, "Follow up to check students' understanding, to make sure they do their work and on the right track, not somebody doing the work for them".

In terms of asynchronous online activities, six participants shared that transmission of theory knowledge on remembering and understanding cognitive processes can be put online for asynchronous online learning. Conventionally, the transmission of

**Table 2** The use of face-to-face activities in blended learning

| Under What Situation | Why |
|---|---|
| Instructors introductory session with learners (beginning of course) | Familiarize learners on learning management system, provide briefing on technical aspect<br>Do groupings to prepare for future group work<br>Instructors build rapport with learners<br>Instructors find out learners' prior knowledge |
| Transmissive—learners do not have the capability to learn on their own | Do not have necessary foundational skills or knowledge to learn on their own |
| Transmissive—instructors present complex concepts/contents | Complex concepts require lecturer explaining because students cannot understand on their own through reading |
| Dialogic—learners engage in small group discussion | Want to have social interaction (learner–learner, learner–instructor) for group work to provide immediate feedback, clarification |
| Dialogic—learners engage in whole class instructor-led discussion | Learners learn from peers to wrap up online learning<br>Clarify learners' understanding through question and answer<br>Elicit learners' views on points of discussion from online activities |
| Dialogic—learners present their works | Instructors and other learners provide feedback |
| Dialogic—learners role play given scenario | Need role modeling soft skills, e.g. communication skills |
| Hands-on and psychomotor—instructors demonstrate tasks | Learners can interject for clarification face-to-face, provide learning environment which resembles or close to real-world situations |
| Hands-on and psychomotor—learners practice hands-on skills | Need to practice with equipment and/or specialized software in lab/workshop<br>Sensory—need to touch, smell, taste<br>Lecturer can clarify if students have any issue<br>Lecturer can see learners carry out certain task to ascertain they learn<br>For critical tasks, cannot afford to let them learn wrongly on their own online and then correct them face-to-face |
| Instructors access learners' learning | Check learners' understanding (follow-up from online learning—formative assessment)<br>Clarify their doubts, summarize learning and prepare for final assessment<br>Carry out summative assessment (end of course) |
| Instructors review contents with learners | Address gaps from online learning |

**Table 3** The use of asynchronous online activities in blended learning

| Under What Situation | Why |
|---|---|
| Transmissive—impart theory knowledge on remembering and understanding | Cut down face-to-face didactic teaching time to use face-to-face time for higher-order thinking, group work, problem-based learning, scenario-based activities, hands-on practice<br>Learners can learn at their own pace because learners can review, especially for slower learner<br>Learners can learn at their own pace. Flipped classroom. Subsequently, face-to-face hands-on practice, scenario, problem-based learning<br>For non-critical tasks, can let learners learn on their own at their own time |
| Dialogic—learners discuss on subjective questions from various perspectives | Peer learning through viewing other learners' posts<br>Check learners' understanding on online learning<br>Provide learners longer time (not constraint by face-to-face time) for online discussion |
| Dialogic—learners do self-reflection on their learning | Learners reflect and see each other's reflection on virtual wall |
| Dialogic—introverted/quiet learners expressing themselves in written form | Express themselves in online written activities |
| Dialogic, constructive—learners discuss on artifacts produced by individual learners from various perspectives | Improve learners' work (design project, creative writing) through other learners'/instructors' online comments<br>Improve learners' motivation and interest through positive reinforcement given in the form of comments |
| Dialogic, co-constructive—learners discuss on artifacts produced by groups of learners from various perspectives given role play scenarios | Improve learners' group work given scenarios through other learners' online comments |
| Learners not available at the same time | Adult learners with work and/or family commitment |

theory knowledge is being done as face-to-face didactic teaching. The reasons from participants were that by cutting down face-to-face didactic teaching, face-to-face time can be used for higher-order thinking activities, group works, problem-based learning, scenario-based activities, and hands-on practices. Participant F mentioned, "Comparing to spending time didactic teaching face-to-face, instructors' time is put to better use for meaningful learning". Meaningful learning equips learners with the cognitive processes needed for successful problem-solving (Anderson & Krathwohl, 2001). Participant D offered insights, "For content, learners need some time to understand. Can replay video and pause. Learners can think and review. Compare

**Table 4** The use of synchronous online activities in blended learning

| Under What Situation | Why |
|---|---|
| Learners have access to experts who are not located at the same location as learners | Guest lecturers from industry share practical experiences<br>Instructors at overseas<br>Need immediate answers from experts<br>Need real-time interactions among learners<br>Learners unable to travel due to physical condition<br>Learners are separated geographically |
| Learners save time traveling to class | Adult learners with work and/or family commitment can interact with one another |

to the lecture, if students miss it, it is unable to replay or hard to repeat. Easier for learners who are slower and need to catch up to replay".

Online discussion boards had been commonly used among participants to get learners discussing on subjective questions from various perspectives through dialogic pedagogy. Participant G shared an example, "How to make buildings consume less energy (green building concept)?". One participant shared that learners reflect on virtual wall and see each other's reflection. Participant C said, "For the quiet ones, they may express differently in online activities. Some students are afraid of making mistakes face-to-face".

Participants shared dialogic and constructive pedagogies used in situations where learners upload their individual writings and design projects onto online discussion board for others to comment. One participant shared dialogic and co-constructive pedagogies for learners to share their group works online for others to comment. Learners can then improve their works.

Three participants shared the notion that adult learners have work and family commitment, and often not available at the same time. Thus, asynchronous online activities are suitable for this type of learners.

For synchronous online activities, six participants shared that synchronous online activities are used under the situations where learners have access to experts (instructors) who are not located at the same location as learners. Participant G shared, "Guest lecturers (from industry to share real life experience) can through webinar because (students) get to know industry practices, which might open students to industry practices for their works next time". Participant E shared, "Webinar allows active interaction between instructor and learner. E.g. fairly complex content, if students (adult learners) were to learn on their own, the asynchronous response time may interfere their learning. If students have issue, they can clarify immediately".

Another theme emerged as one participant shared that synchronous online activities can be used so that learners can save time traveling to class for adult learners who have both work and family commitment. The advantage of working professional learners attending synchronous online sessions was the saving of traveling time (Jong, Savin-Baden, Cunningham, & Verstegen, 2014).

## 6.2 Challenges of Using of Face-To-Face Activities in Blended Learning

In this section, we address the second research question: "What are the challenges of using face-to-face activity, asynchronous online activity, and synchronous online activity in a blended learning environment? Why?". Findings were arranged in Tables 5, 6, and 7, for the three learning modes, respectively.

In the aspect of face-to-face activities, six participants shared the challenge of different levels of readiness from online learning prior to face-to-face lessons. Most participants raised the challenge over the fact that learners not learning online asynchronously prior to the face-to-face lesson, while face-to-face instructions are designed based on the assumption that learners had equipped themselves with prior learning. Participant D offered insights, "Sometimes students are at different level of readiness from the online learning. Some people don't understand, some never do, so they are at different levels of readiness. When we assume certain level of prior knowledge, but they did not achieve that".

One participant shared that some introverted learners do not express verbally face-to-face in class for fear of making mistakes. Another participant raised the challenge

**Table 5** The challenges of using face-to-face activities in blended learning

| Challenge | Why |
|---|---|
| Different levels of learners' readiness from online learning (including learners who did not learn) | Learners forget or do not read up the online material<br>Learners do not learn on their own. Instruction that are designed based on prior learning (flipped classroom approach) will not work<br>Some learners do not understand, some learners never do. Learners did not achieve certain level of prior knowledge<br>Learners come unprepared without doing online learning, with various reasons, e.g. no time, busy with part-time work, busy with co-curricular activities<br>Learners with mixed ability. Half the class have learnt; the other half have not learnt |
| Introverted learners do not express themselves | Some learners are afraid of making mistakes face-to-face |
| Some learners are unable to follow instructor-led pace | Learners who are unable to follow instruction |
| Time constraint on the conduct of learning activities | If the number of students is too big, the face-to-face activities would be ineffective<br>Due to time limitation, it may be a constraint for the number of role plays that can be carried out |

**Table 6** The challenges of using asynchronous online activities in blended learning

| Challenge | Why |
|---|---|
| Additional workload on instructors to manage online activities | Might be stressful for instructors to track (the number of) comments and (the number of) replies (if this is stated as criteria for grading)<br>Marking of attendance is challenge if it is required by institution<br>Manage a large class size<br>Instructors need to facilitate online discussions |
| Learners do not participate as intended | Learners do not post due to various reasons, including negative peer influence, when no accountability<br>Discussion becomes out-of-topic<br>Plagiarize others' answer/idea<br>Learners' writing ability not on par with the written task given<br>Learners lost interest after some time<br>High reliance on learners to be self-directed |
| Improper use of online tools for an intended learning activity | Learners feel lost because "improper tool is used for the task" |
| Technology infrastructure (compatibility of desktop and mobile versions) | Mobile platform problems. Content is not suitable for displaying on small device.<br>Discussion forum is difficult to be accessed using mobile device<br>Platform problems. Commenting feature of the tool is hard to use |
| Technology infrastructure (compatibility of web browsers) | Incompatibility of web browsers |
| Technology infrastructure (server downtime) | Server downtime during server upgrading affecting learning environment availability |

of slow learners being unable to follow instructor-led pace and the challenge of time constraint during face-to-face activities with a big class size.

## 6.3 Challenges of Using Asynchronous Online Activities in Blended Learning

In terms of asynchronous online activities, most participants shared the additional workload on instructors to manage asynchronous online activities. Three participants mentioned the workload that was added on instructors to facilitate online discussions. One participant offered the insights of stress added on instructors if instructors need to grade and track all learners' online posts.

**Table 7** The challenges of using synchronous online activities in blended learning

| Challenge | Why |
|---|---|
| Demand on instructor in managing synchronous session | Instructors face challenge to present content and response to both verbal and text questions<br>Instructors need to manage learners in terms of facilitation and manage technical aspect |
| Learners are not aware of webinar etiquette | Microphone causing disturbance to video conferencing |
| Difficulty in scheduling common time (learning environment scheduling of online activities) | Adult learner with family and work commitment, difficult to schedule common time |
| Technology infrastructure (learning environment bandwidth) | Unable to login due to bandwidth problems<br>Authenticity issue whether the learners still present<br>Technical problems when all learners attempt to upload files at the same time |
| Technology infrastructure (server downtime) | Server downtime affecting learning environment availability |

Most participants also shared challenges related to learners not participating in the asynchronous online activities as intended. Some of the challenges shared by participants include learners not posting due to negative peer influence as a result of "nobody started to post", students do not attempt when there is no accountability, and discussions go out-of-topic. One participant shared challenges of plagiarism, learners' ability in terms of writing not on par with the instructional design of the task, and learners lost interest after some time. Participant F highlighted the importance of learners' self-discipline, "Because asynchronous, learners can complete tasks at their own time. They can do it on-time, they can do it last minute, they might forget to do it. Assume they will do readings before posting to forums. Instructors has little control over how much preparatory work learners choose to do. Instructors can only advise".

Participant F shared the confusion caused when an online tool used does not match with the intended task, "The (weblog) doesn't have threads, only have commenting features but participants cannot reply to comment features. It can confuse the learners and instructors if improper tool is used".

Several participants brought up various technology infrastructure issues, including the incompatibility of desktop and mobile versions of the learning environment or learning tool, compatibility of different web browsers of the learning environment or learning tool and server scheduled downtime during server upgrading.

### 6.4 *Challenges of Using Synchronous Online Activities in Blended Learning*

For synchronous online activities, two participants shared the view that using synchronous online activities is demanding to instructors if they need to manage both verbal and text channels during the sessions. Participant D shared, "Need to make sure the technical aspect, familiarize with the platform, during the session, how to ask questions to see if learners understand. Some webinars have breakout groups, more challenging to manage learners in terms of facilitation and technical aspect".

Another participant shared the challenge related to learners being unaware of webinar etiquette. Participant E shared insights, "Learners are not aware the etiquette, e.g. mic not turned off while learning, need to indicate interest to talk in the system before starting to talk".

Three participants shared operational challenges, which are related to difficulty in scheduling common time for synchronous online activities to take place. All three participants elaborated that adult learners have family and work commitment and therefore, it is difficult to schedule common time for synchronous online activities.

Most participants shared about technology infrastructure problems, including learning environment bandwidth problems resulting in learners and instructors unable to login, and authenticity issues of whether learners are still in the session or disconnected due to bandwidth issue. One participant offered insights on server downtime disrupting synchronous online activities. The technology infrastructure challenge was consistent in both asynchronous online learning and synchronous online learning.

## 7 Discussion

From the findings, we postulated blended learning design guidelines: (1) incorporating the various situations where face-to-face activities, asynchronous and synchronous online activities should be used in blended learning, and (2) addressing the challenges of using face-to-face activities, asynchronous and synchronous online activities in blended learning. The five guidelines are:

1. Using transmissive pedagogy for face-to-face activities and asynchronous online activities.
2. Using dialogic pedagogy for face-to-face activities and asynchronous online activities.
3. Using affordance analysis to guide the design of online learning task.
4. Connecting face-to-face activities and asynchronous online activities.
5. Supporting instructors in managing online learning.

## 7.1 Using Transmissive Pedagogy for Face-To-Face Activities and Asynchronous Online Activities

We suggest using transmissive pedagogy for both face-to-face activities and asynchronous online activities. One participant provided insights on transmissive pedagogy for face-to-face activities which can be used to teach difficult concepts. Face-to-face activities can be used when learners do not have the prior knowledge or the capability to learn on their own. Asynchronous online activities can be used for learners to learn at their own pace and time, especially useful for slow learners who are often unable to keep up to the instructor-led pace in face-to-face sessions. Learners who are motivated and able to learn independently on their own are suitable for asynchronous online activities. Transmissive pedagogy is less suitable for synchronous online activities because it is demanding on instructors to manage both transmission of content to learners and technical aspects at the same time.

To addressing the challenge of learners do not do asynchronous online learning on their own, instructors can set expectations on learners' participation. Allocation of online participation marks and online quiz marks can be set to increase learners' extrinsic motivation. However, this strategy had shown mixed implementation results from various studies (Hew & Cheung, 2012). Instructors can communicate the connections between face-to-face activities and asynchronous online learning to learners. Communication on the purpose and value of online learning to learners can perhaps help learners to see the value of asynchronous online activities (Aloni & Harrington, 2018).

## 7.2 Using Dialogic Pedagogy for Face-To-Face Activities and Asynchronous Online Activities

It seems to us that dialogic pedagogy is suitable for both face-to-face activities and asynchronous online activities. Face-to-face activities can be used for learning tasks which require immediate feedback, where learners can engage in learner-learner and learner-instructor social interactions, for example through discussion, role play, and presentation. On the other hand, asynchronous online activities can be used for learning tasks which require time to formulate thinking and reflection. Through online discussion board, learners, including those who are introverted and do not express themselves during face-to-face activities, express themselves in written form. When using dialogic pedagogy through written activities, it shall match learners' writing ability for learning to take place.

From the findings, we realized that dialogic pedagogy can be used with constructive/co-constructive pedagogy for learning tasks, for example project works that require individual learners/groups of learners to work on over a duration beyond institutional curriculum hours. Through online discussion, learners can improve their works based on learner-learner and/or learner-instructor interactions. The study by Meyer (2003) reported that online discussion forums allowed learners to spend time related to learning objectives, and higher-order thinking occurred among graduate-level learners in the study.

Dialogic pedagogy is less suitable for synchronous online activities because it is demanding on instructors to manage both real-time discussion and technical aspects at the same time. Six out of seven participants indicated that synchronous online activities can be used where learners need to have access to experts who are geographically separated from learners. We suggest that geographically separation which requires long traveling hours and/or high traveling costs is the factor that can be considered for using dialogic pedagogy for synchronous online activities in view of the costs versus benefits in comparison.

## 7.3 Using Affordance Analysis to Guide the Design of Online Learning Task

One challenge faced by a participant was the confusion faced when an online learning task does not match with the tool used for an asynchronous online activity. To overcome the challenge to learners, affordances of a technological tool should match with the intended learning task. In this regard, instructional designers can make use of affordance analysis by Bower (2008), to match the affordance required for certain learning tasks to the affordance of online tools. Affordances of the technological tool should match with learning tasks (Lieser et al., 2018). This is crucial to prevent improper use of online tools for an intended learning activity, which will pose challenges to learners and instructors.

## 7.4 Connecting Face-To-Face Activities and Asynchronous Online Activities

Participants offered insights on the connection between face-to-face activities and asynchronous online activities. The findings suggested that it is appropriate to complement face-to-face activities with asynchronous online activities. Face-to-face activities and online activities are mutually supporting one another (Linder, 2016; Kenny & Newcombe, 2011). For example, after learning from face-to-face activities, learners go through asynchronous online activities, for example, upload their works or artifacts into an online discussion forum, followed by commenting one another's artifacts to improve on their works. According to Vaughan and Garrison (2005, p. 4), "the true benefit of blended learning is in integrating face-to-face verbal and online text-based exchanges and matching each to appropriate learning tasks". Flipped classroom approach is a form of blended learning where learners learn asynchronously before face-to-face learning (Lee, Lim & Kim, 2017; Galvis, 2018).

To address the challenge of learners' different levels of learning from online learning, differentiated instructions can be used during face-to-face learning. Another way to overcome the different levels of learning is by identifying common areas that students are facing difficulties in and designing the face-to-face learning at this level (Abeysekera & Dawson, 2014). In addition, another possible approach is that institutions may allocate additional instructor to facilitate learning with the group of learners who have not achieved certain level of learning from online learning; while the existing instructor proceed with the group of learners who have attained the level of learning from online learning (Lo & Hew, 2017).

### *7.5 Supporting Instructors in Managing Online Learning*

We suggest to providing instructors with technical support when implementing asynchronous and synchronous online activities. Technical support to instructors is crucial to overcome challenges which will otherwise pose a hindrance to the implementation of online learning, especially for synchronous online activities.

We also suggest that institutions shall consider allocating manpower resources for online learning. To address the challenge of instructors' additional workload for managing asynchronous online activities, institutions may include the number of man-hour instructors spend on managing online activities during instructors' resource planning. The man-hour spent by instructors on managing online activities may include: (1) the time instructors spend to facilitate online discussion, especially if the class size is big, and (2) the time used for grading online activities. During the synchronous online learning through webinar, instructors often fail to notice the text questions raised by learners. To address this, institutions may allocate additional instructor to manage text questions; alternatively, instructor may set a ground rule to learners that only verbal questions are accepted so that instructors can focus on verbal channel.

## 8 Limitations and Recommendations for Future Study

This study focused on qualitative data from the instructional designers with at least three years of blended learning experience, who were convenience samples based on accessibility or proximity to us. There was a possibility of biasness of the views from the convenience samples toward certain groups of learners that they had experienced with.

Reaching out to more instructional designers with relevant experience in blended learning for all levels of learners, from primary school learners to adult learners will provide a wide spectrum of experiences and opinions in the area for future study.

Though opinions from instructional designers were captured, they were limited as neither teachers nor learners were included in the study. Opinions from both teachers and learners will allow triangulation of data from instructional designers for data verification from two or more sources.

## 9 Conclusion

Findings from this study provide insights on the key situations that instructional designers consider or opine the usage of face-to-face activities, asynchronous and synchronous online activities in blended learning, alongside with the potential challenges. Generally, instructional designers consider the delivery medium by taking into consideration the types of content and pedagogical approaches. Other considerations include the complexity of the content, target learners, class size, and resources allocation by institutions, including instructor, technical support, and technology infrastructure.

The usage of face-to-face activities, asynchronous online activities, and synchronous online activities in blended learning shall complement or support one another, in the learning context for the target audience—with the profile of learners in mind. We postulated blended learning guidelines on various situations of using face-to-face activities, asynchronous, and synchronous online activities; and preventing some potential challenges.

## References


Abeysekera, L., & Dawson, P. (2014). Motivation and cognitive load in the flipped classroom: definition, rationale and a call for research. *Higher Education Research and Development, 34*(1), 1–14. https://doi.org/10.1080/07294360.2014.934336.

Aloni, M., & Harrington, C. (2018). Research based practices for improving the effectiveness of asynchronous online discussion boards. *Scholarship of Teaching and Learning in Psychology, 4*(4), 271–289.

Alten, D. C. D., Phielix, C., Janssen, J., & Kester, L. (2019). Effects of flipping the classroom on learning outcomes and satisfaction: a meta-analysis. *Educational Research Review, 28,* 1–18.

Anderson, L. W., & Krathwohl, D. R. (2001). *A taxonomy for learning, teaching, assessing: a revision of bloom's taxonomy of educational objectives.* New York: Longman.

Bidder, C., Mogindol, S. H., Saibin, T. C., Andrew, S. A., & Naharu, N. (2016). Students' perceptions of blended learning and achievement. In J. Luaran, J. Sardi, A. Aziz, N. Alias (Eds), *Envisioning the future of online learning.* Singapore: Springer. https://doi.org/10.1007/978-981-10-0954-9_19.

Bower, M. (2008). Affordance analysis—Matching learning tasks with learning technologies. *Educational Media International, 45*(1), 3–15.

Bower, M., Hedberg, J. G., & Kuswara, A. (2010, September). A framework for Web 2.0 learning design. *Educational Media International, 47*(3), 177–198.

Churches, A. (2008, July 2). *Bloom's digital taxonomy.* Retrieved March 17, 2019, from Burton's Life Learning: https://burtonslifelearning.pbworks.com/f/BloomDigitalTaxonomy2001.pdf.

Draffan, E. A., & Rainger, P. (2006). A model for the identification of challenges to blended learning. *ALT-J Research in Learning Technology, 14*(1), 55–67.

Galvis, A. H. (2018). Supporting decision-making processes on blended learning in higher education: Literature and good practices review. *International Journal of Educational Technology in Higher Education, 2018*(15), 1–38. https://doi.org/10.1186/s41239-018-0106-1.

Garrison, D. R., & Kanuka, H. (2004). Blended learning: Uncovering its transformative potential in higher education. *Internet and Higher Education, 7,* 95–105.

Graham, C. R. (2006). Blended learning systems: Definition, current trends and future directions. In C. J. Bonk & C. R. Graham (Eds.), *The handbook of blended learning: Global perspectives, local designs* (pp. 3–21). San Francisco: Pfeiffer.

Hew, K. F., & Cheung, W. S. (2012). *Student participation in online discussions: Challenges, solutions, and future research.* New York: Springer.

Hew, K. F., & Cheung, W. S. (2014). *Using blended learning: Evidence-based practices.* New York: Springer.

Horn, M. B., & Staker, H. (2011). *The rise of K-12 blended learning.* Innosight Institute.

Horn, M. B., & Staker, H. (2012). *Classifying K-12 blended learning.* Innosight Institute.

Jong, N. D., Savin-Baden, M., Cunningham, A. M., & Verstegen, D. M. (2014). Blended learning in health education: Three case studies. *Perspect Med Educ, 2014*(3), 278–288. https://doi.org/10.1007/s40037-014-0108-1.

Kaur, M. (2013). Blended learning—Its challenges and future. *Procedia—Social and Behavioral Sciences, 93,* 612–617.

Kenney, J., & Newcombe, E. (2011). Adopting a blended learning approach: Challenges encountered and lessons learned in an action research study. *Journal of Asynchronous Learning Networks, 15*(1), 45–57.

Kim, K., Bonk, C. J., & Teng, Y. (2009). The present state and future trends of blended learning in workplace learning settings across five countries. *Asia Pacific Education Review, 10,* 299–308. https://doi.org/10.1007/s12564-009-9031-2.

Lameras, P., Levy, P., Paraskakis, I., & Webber, S. (2012). Blended university teaching using virtual learning environments: Conceptions and approaches. *Instructional Science, 40,* 141–157. https://doi.org/10.1007/s11251-011-9170-9.

Lee, J., Lim, C., & Kim, H. (2017). Development of an instructional design model for flipped learning in higher education. *Educational Technology Research and Development, 2017*(65), 427–453. https://doi.org/10.1007/s11423-016-9502-1.

Lieser, P., Taff, S. D., & Murphy-Hagan, A. (2018). The webinar integration tool: A framework for promoting active learning in blended environments. *Journal of Interactive Media in Education, 2018*(1), 1–8. https://doi.org/10.5334/jime.453.

Lincoln, Y., & Guba, E. (1985). *Naturalistic inquiry.* Newbury Park, CA: Sage Publications.

Linder, K. E. (2016). *The blended course design workbook: A practical guide.* Stylus Publishing.

Lo, C. K., & Hew, K. F. (2017). A critical review of flipped classroom challenges in K-12 education: Possible solutions and recommendations for future research. *Research and Practice in Technology Enhanced Learning, 12*(4), 1–22. https://doi.org/10.1186/s41039-016-0044-2.

Meyer, K. A. (2003). Face-to-face versus threaded discussions: The role of time and higher-order thinking. *Journal of Asynchronous Learning Networks, 7*(3), 55–65.

Pappas, C. (2014). 7 tips to create an effective blended elearning strategy. Retrieved from https://elearningindustry.com/-tips-create-effective-blended-elearning-strategy.

Pesavento, T., Klein, J., Macasaet, D., Shorter, C., & Wagstaff, S. (2015). *Teaching with Technology.* UW-Madison: L&S Learning Support Services.

Spanjers, I. A. E., Konings, K. D., Leppink, J., Verstegen, D. M. L., Jong, N., Czabanowska, K., & Merrienboer, J. J. G. (2015). The promised land of blended learning: Quizzes as a moderator. *Educational Research Review, 15,* 59–74.

Vaughan, N., & Garrison, D. R. (2005). Creating cognitive presence in a blended faculty development community. *The Internet and Higher Education,* 1–12. https://doi.org/10.1016/j.iheduc.2004.11.001.

Vaughan, N. (2007). Perspectives on blended learning in higher education. *International Journal on E-Learning, 6*(1), 81–94.

Wikiversity. (2016). *Instructional design/blended learning lesson plans/designing a blended lesson*. Retrieved from https://en.wikiversity.org/wiki/Instructional_design/Blended_Learning_Lesson_Plans/Designing_a_Blended_Lesson.

**Ms. Ying Hui Chan** holds a Master of Arts in Instructional Design & Technology from the National Institute of Education, Nanyang Technological University, Singapore in 2019. She has been teaching at an Institute of Higher Learning for 10 years in Singapore. Her research interests include instructional design areas in blended learning and e-learning.

**Dr. Wing Sum Cheung** is an associate professor in the Learning Sciences and Assessment Academic Group at the National Institute of Education, Nanyang Technological University in Singapore. He has taught in USA, Macau, Australia, and Singapore. His research focuses are e-learning, blended learning, and asynchronous online discussion activities. He has published over 130 publications including journal articles, conference papers, and book chapters. He has also conducted workshops and courses for schools, universities, business sectors, and government sectors. He has published two books. They are *Student Participation in Online Discussions: Challenges, Solutions, and Future Research* and *Using Blended Learning: Evidence-Based Practices*.

# The Adoption of Free and Open Source Software in Teaching and Learning: An Empirical Study of General Education in Chu Hai College of Higher Education

**Chenggui Duan and Kwok Fong Lee**

**Abstract** The purpose of this study is to explore the integration of free and open source software (FOSS) with the daily teaching and curriculum development in order to improve teaching effectiveness. The methods of questionnaire survey, semi-structured interview and focus group discussion were used in this project. A total of 35 participants, including teaching faculty, research staff, teachers and students, from the general education programme in Chu Hai College of Higher Education were selected and actively involved in this project. The project started in September 2018 and a number of free and open source education software promotion activities and training workshops were provided. Both quantitative and qualitative data through questionnaire survey and interview were collected. The quantitative data were analyzed by using SPSS version 22, while qualitative data were used as supplement. It was found from this study that the majority of learners were positive about FOSS and agreed that FOSS was easy to learn, convenient to use and conducive to learn. It is hoped that the research results could bring some insights and inspiration to the implementation and application of FOSS in Chu Hai College of Higher Education.

**Keywords** FOSS · Learning efficiency · Curriculum development · Teaching quality

---

C. Duan
South China Normal University, Guangzhou, China
e-mail: cgduan@chuhai.edu.hk

Chu Hai College of Higher Education, Hong Kong, China

K. F. Lee (✉)
Department of Journalism and Communication, Chu Hai College of Higher Education, Hong Kong, China
e-mail: tracylee@chuhai.edu.hk



W. W. K. Ma et al. (eds.), *Learning Environment and Design*,
Educational Communications and Technology Yearbook,
https://doi.org/10.1007/978-981-15-8167-0_8

# 1 Introduction

The rapid development of information and communication technology (ICT) has transformed the ways of learning, understanding and accessibility, which has provided new opportunities for general learners (Scheuermann & Pedró, 2010). However, because of the proprietary of the software and other learning tools, many learners have to spend a large sum of money to access the technology development. Therefore, the cost of technology, including the purchase of software, has become a major barrier for technology implementation (Anderson & Dexter, 2005; Pelavin Research Institute, 1997). FOSS (Free and Open Source Software) aims to break these barriers and the software's users have freedom to use them, which encourages and enables free sharing of software and their distribution to serve the community (Gupta & Surbhi, 2018). The widespread and great potential of FOSS has attracted the attention of Chu Hai College of Higher Education (CHCHE) and lots of initiatives and projects have been carried out.

Chu Hai College of Higher Education (CHCHE) is a post-secondary college under the Post-secondary Colleges Ordinance in Hong Kong, which has provided 70 years of quality education and excellence to the Hong Kong community and remains committed to enhancing the educational opportunities for young people in Hong Kong by enlarging its array of programs. Its vision is to nurture future leaders, strengthen the competitiveness of Hong Kong and help bring about a stable, prosperous and harmonious society in a knowledge-based economic environment. Currently, Chu Hai College of Higher Education comprises three faculties: Faculty of Arts and Social Sciences, Faculty of Science and Engineering and Faculty of Business, and nine academic departments offering 14 Bachelor's degree programs and seven Master's degree programs accredited by the Hong Kong Council for Accreditation of Academic and Vocational Qualifications. In order to nurture students as individuals equipped with knowledge, professional skills, moral values and social engagement, Chu Hai College of Higher Education has always been emphasizing the potential benefits brought by the development of ICT and encouraging the teaching staff to explore the application of the abundant free resources and open source software available on the internet. FOSS project is one of the projects with the purpose of helping the teachers and students of Chu Hai College of Higher Education to better master and use free and open source education software in teaching, and promote the popularization, application and development of new teaching methods.

The FOSS project team of Chu Hai College of Higher Education (with the funding support from QESS of the Education Bureau of Hong Kong Special Administrative Region (HKSAR)) has organized a number of FOSS promotion activities and training seminars to carry out this project, with the purpose of promoting the daily teaching and curriculum development for teachers and students of Chu Hai College of Higher Education by using FOSS and more effectively integrating FOSS with teaching and improve teaching quality. What are the views and opinions of teachers and students on applying FOSS? This study aims to describe the project and find out whether the integration of FOSS can improve the efficiency and quality of learning and whether the application of FOSS can improve the digital literacy of teachers, bridge the digital divide, and achieve the ideal of equal education.

## 2 Literature Review

Free open source software (FOSS), sometimes simply called open source or free software, is software that is licensed to be free to use, modify, and distribute. In literature, the concept of FOSS has been used interchangeably as open source software (OSS), free software, and Libre software. Ghosh (2007) points out that many researchers in Europe use the term free/libre/open source software (FLOSS). The United Nations Educational, Scientific and Cultural Organization (UNESCO) uses the term FOSS. The present study also uses the term FOSS. FOSS provides more capability, security, and sustainability than commercial software at much less cost, and it runs much of the internet, and is increasingly the first software option for individuals and organizations alike. FOSS has been getting steadily better for several decades now and the original history can trace back to 1950s.

In the 1950s and early 1960s, computer software was mainly the concern of academics and corporate researchers, so it was treated more like scholarly work (in the public domain) than a proprietary asset. Computer manufacturers focused on selling hardware, so they threw in the software for free. Both IBM and DEC had active user groups that shared tips and software with other users (Miller, Voas & Costello, 2010). In the second half of the 1960s, as software costs increased for manufacturers, software was either bundled with hardware or priced separately. Licensing agreements and copyright protection became common ways to legally protect software. In 1966, the UK issued what was probably the first software patent (Haselden, 2001).

In 1984, Richard Stallman launched the GNU project (www.gnu.org). Stallman, who also founded the Free Software Foundation (FSF), has been an outspoken and controversial advocate for free software. GNU's General Public License (GPL), which has several different versions, has been widely influential. However, GNU's success didn't lead to a monolithic movement led by Stallman; rather, different groups have championed slightly different licensing agreements (Free Software Foundation, 2020).

Another important group, founded in 1998, is the open source initiative (OSI). Although both FSF and OSI promote alternatives to proprietary, closed-source software, relations haven't always been cordial between the two organizations. Stallman has accused the OSI of ignoring important freedoms in favor of corporations, while OSI proponents have criticized Stallman for his overly rigid "social activism". Despite their distinct philosophical views of FOSS, Stallman and others insist that FSF and OSI, as well as other independent producers of FOSS, are allies.

Nowadays, FOSS has been widely used in different areas. FOSS has become popular and has begun to get noticed by many governments around the world, and many initiatives have been launched to reap its benefits (Gupta & Surbhi, 2018). Hon,

Russell & Welch (2010) explore the benefits of FOSS for state and local governments. They present the three key benefits to such organizations: low initial acquisition costs, the ability to analyze source code for backdoors and other security risks, and system interoperability. Gangadharan, Dandrea & Weiss (2007) propose applying the concepts behind FOSS to advance the field of service-oriented computing. They argue that by having access to the source code of the interfaces, novel composite and derivative services can be created. Wang, Blue & Plourde (2010) put forward the increased use of FOSS in higher education and show how the University of Delaware put the principles of FOSS into practice.

The popularity of FOSS stems from the benefits and advantages it brings. Firstly, FOSS provides interesting tools and processes with which women and men can create, exchange, share and exploit software and knowledge efficiently and effectively (UNESCO, 2020; Hoe, 2006). Besides, from the perspective of their proponents, FOSS are safer, more efficient, and work more reliably than their proprietary counterparts. They encourage users to access, view, and modify software whenever they like, as long as they let others do the same when they share their work. Furthermore, people prefer FOSS to proprietary software for a number of other reasons, including: more control, easy training, more security, better stability. Among the benefits, cost is one of the major driving forces behind the FOSS projects. Studies have shown that FOSS integration reduces costs, and this is the main motivation of the deployment of FOSS (Ajila & Wu, 2007; AlMarzouq et al., 2005; Feller et al., 2007). UNESCO claims that FOSS can play an important role as a practical instrument for development as its free and open aspirations make it a natural component of development efforts in the context of the Sustainable Development Goals (UNESCO, 2020).

However, there are still some viewpoints concerning the disadvantages of FOSS. For example, some researchers discuss the generic complaints about FOSS—such as the lack of technical support—but also mentions complaints specific to law enforcement, such as the lack of relevant applications (Hon, Russell & Welch, 2010). Other researchers claim that despite the benefit in the initial purchase price of software, deploying software requires total cost that goes beyond the initial purchase price. Total cost is a silent issue of FOSS and can only be evaluated in the particular environment in which it is adopted (Thankachan & Moore, 2017). Even though FOSS has the potential to enhance the education sector, the prediction of Roger's diffusion of innovation is often substantiated: "Getting a new idea adopted, even when it has obvious advantages, is difficult" (Rogers, 2003).

Despite a good deal of controversy between the different organizations involved in promoting FOSS, there's a long list of thriving FOSS projects. If you or your organization uses GNU/Linux, TeX, Mozilla Firefox, the Apache server, the Open Office suite, or Java, then FOSS is part of your technical world.

FOSS has also inspired many academic researchers. Economists are interested in the business model of entities that often gives away products. Ethicists study the motivations of programmers involved in FOSS projects and the philosophical impact of a viable alternative to proprietary software. Computer scientists are examining the technical effectiveness of a development model that employs "many hands," only a few of whom make their living developing or maintaining FOSS. And never forget that FOSS and commercial software need each other; the competition between the two keeps the software marketplace in check with added diversity and innovation (Miller, Voas & Costello, 2010).

Obviously, FOSS is popular in lots of areas in the whole world, and opens a new path for the improvement of teaching and learning efficiency. However, there are still little evidence and data concerning the application of FOSS in Hong Kong. Quantitative and qualitative researches are essential and efficient methods for the policymakers and practitioners to have a broad view of FOSS. This study employs the method of questionnaire survey, semi-structured interview and focus group discussion with purpose of finding out the final practical consequences in order to further promote the popularization, application and development of FOSS.

### *2.1 Research Questions*

There are two major research questions in this project:

1. What are the results and efficiency of integrating FOSS in teaching and learning?
2. What are the barriers and challenges of implementing FOSS-based education in Chu Hai College of Higher Education?

## 3 Methods

The study was carried out from September 2018 to May 2020 at Chu Hai College of Higher Education. Training workshops and seminars on the technical operation, pedagogical relevance and application of FOSS tools were implemented sequentially, in which 35 persons actively participated in the study, including teaching faculty, research staff, teachers and students. In order to enhance the efficiency of FOSS application, the project team reviewed the FOSS initiatives and websites around the world and categorized the FOSS into the following categories: media editing, classroom interaction, courseware production, communication and collaboration, learning tools and OERs. In addition, the teachers' FOSS handbook, the repository of FOSS and online virtual community of practice platform were developed and launched.

With the purpose of exploring the efficiency and reflection of FOSS application, the methods of questionnaire survey, semi-structured interview and focus group discussion were employed among our teachers and students. The 35 valid questionnaires were received and 30 persons' workshop feedback materials were collected. The participants were from different departments including Chinese Department, Architecture Department, Civil Engineering Department, English Department and Department of Journalism and Communication. Of the participants, 17.65% are male

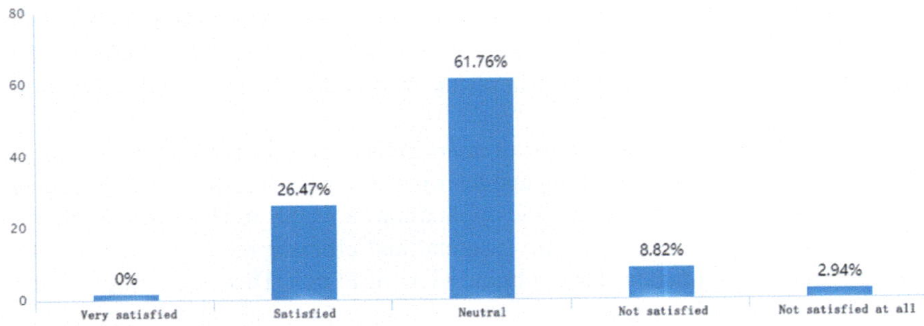


**Fig. 1** Degree of satisfaction on information technology application ability

and 82.35% are female; 8.82% are from (Faculty of Science) school, 2.94% from Faculty of Business and 88.24% from (Faculty of Arts and Social Science).

The statistical package, SPSS version 24.0, was used. For statistical analysis, we looked into the frequency, mean and standard deviation (M&SD) of the data. NVIVO version 11 was used to analyze the text.

# 4 Results

## *4.1 Degree of Satisfaction on ICT Application Ability*

In digital era, ICT application ability is crucial for teaching staffs and learners. Concerning current situation, the survey on the degree of satisfaction was carried out.

It can be seen from Fig. 1 that only 26.74% of respondents feel satisfied with their current ICT application ability while about 12% feel unsatisfied; 61.76% of respondents show a neutral attitude.

## *4.2 Understanding and Application of FOSS*

FOSS is widely used around the world, but it is still rather strange toward some users. About the current application and knowledge of FOSS, we designed four items to have a full understanding.

Figure 2 shows that only 17.65% of respondents have heard of FOSS before. However, after the explanation to the above respondents, 66.67% of them found that they have used FOSS during their studies, as displayed in Figs. 3, 4, 5.

About the freeware, it was found that about 32.35% of respondents have heard of freeware before, but 63.64% of them have never used freeware.

Fig. 2 Knowledge of FOSS

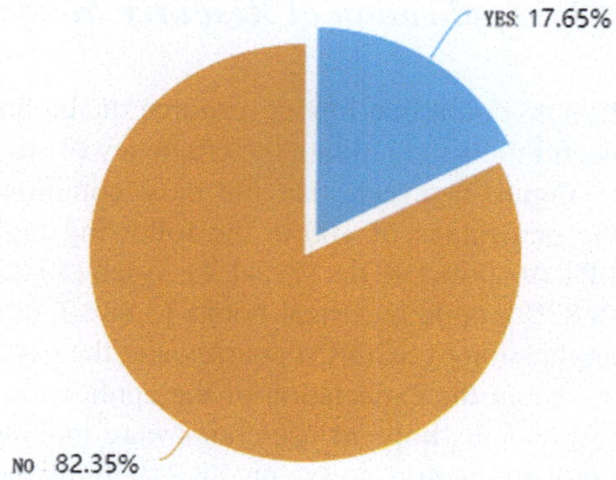


Fig. 3 Application of FOSS

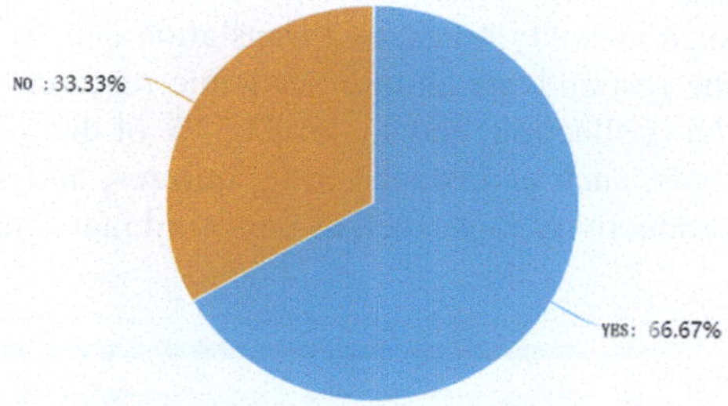


Fig. 4 Knowledge of freeware

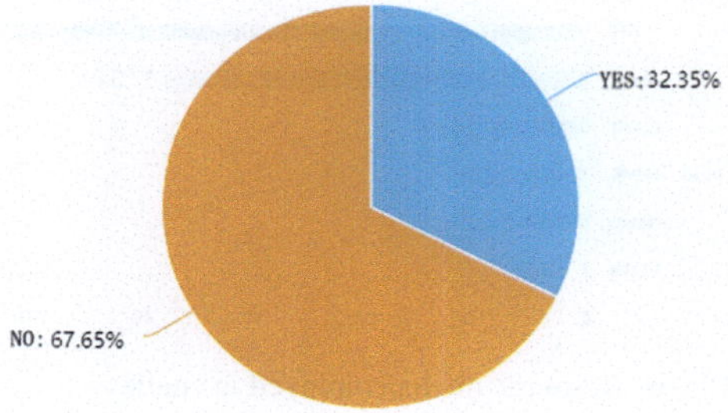


Fig. 5 Application of freeware

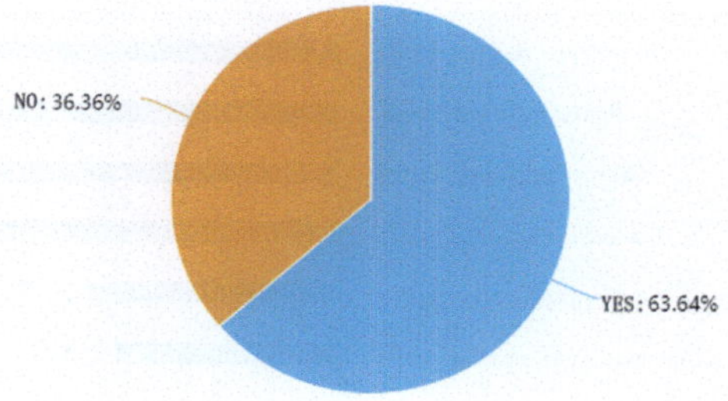

### 4.3 Application of Resource Media and FOSS in Studies

It is no doubt that lots of resource media have been used in our studies, which are useful tools to facilitate the efficiency of study.

Figure 6 reveals that the most commonly used resource media is Word with the percentage of 100%, the following high coverage of resource media include PPT (97.06%), PDFs (79.41%), video (59.82%), audios (38.24%), flash courseware (8.82%), epub electrical books (5.88%), others (5.88%). However, concerning the application of SCORM courseware, the percentage is zero (Fig. 7).

About the expectation of the application of FOSS in daily learning, 88.24% of respondents hope to use courseware making tools, such as PPT beautification, e-book production, and so on; 88.24% think media editing tools, such as video editing, picture editing, and so on are their favorite; 52.94% think online learning tools, such as study notes, text translation and so on, are important for them; 41.18% of the respondents choose academic research tools, such as document management, data collection, and so on; 23.53% of the respondents choose classroom interactive tools, such as instant voting, quizzes, and so on; 26.47% of the learners insist on productivity tools such as password management, text recognition, and so on.

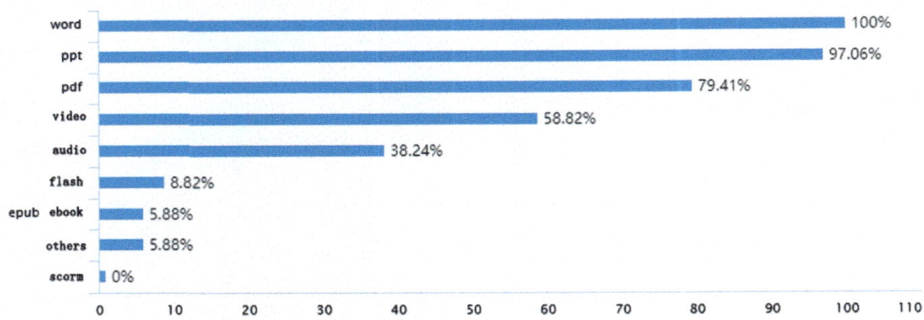


**Fig. 6** Resource media employed in studies

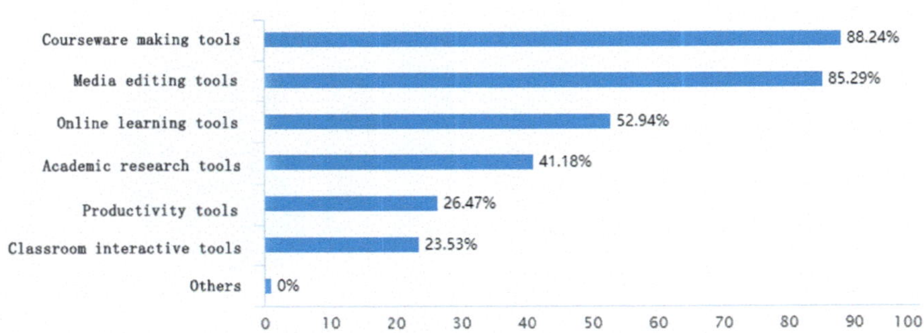


**Fig. 7** Expectation of FOSS application in daily learning

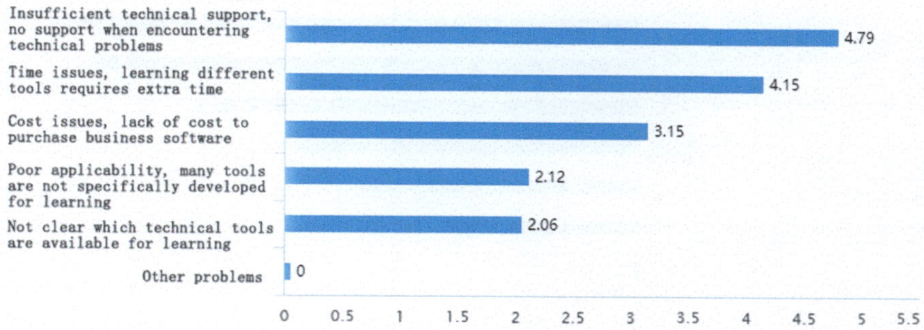


**Fig. 8** Difficulties in the Application of FOSS

## 4.4 *Difficulties in the Application of FOSS*

During the process of the application, different learners may encounter different problems. We have used six points to present the difficulties from the most difficult (1) to the least difficult (6).

Figure 8 shows that 'insufficient technical support, no support when encountering technical problems' ranks highest with 4.79 points, and the second is 'time issues, as learning different tools requires extra time'. 'Cost issues, lack of cost to purchase business software' ranks the next with 3.15 points. 'Poor applicability, many tools are not specifically developed for learning' and 'it is not clear which technical tools are available for learning' are ranked almost the same, with 2.12 and 2.06, respectively.

## 4.5 *Considerations for Recommending FOSS*

About the considerations for recommending free and open tools, different respondents have different recommendations. We also used six points to present the degree of recommendation from the most important (1) to the least important (6).

Figure 9 shows that 'free to use, that is not chargeable' ranks highest with 5.68 and the second rank is 'easy to use, who do not want to spend too much time studying how to use'. 'Stable and reliable, which is the classic software in related fields' ranks the third with 4.12 points. The next one with 3.35 is 'learning related, especially closely related to daily learning'. Furthermore, the respondents also think 'novel and easy to use' is very important and holds the view that they hope to see the latest tools. Surprisingly, the lowest one is free and open source, including the software's original code, and so on.

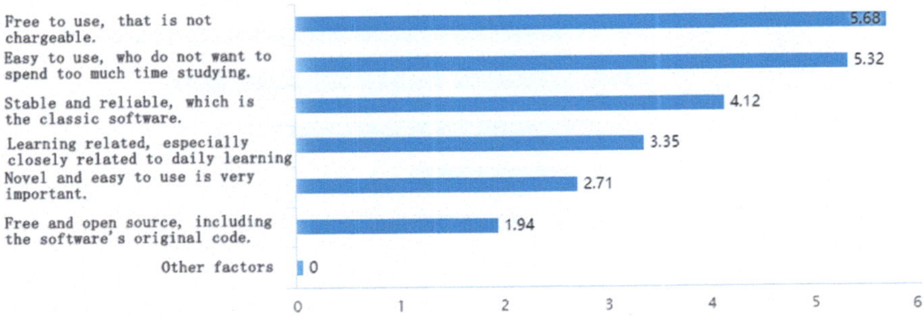


**Fig. 9** Considerations for recommending FOSS

# 5 Views on the Application of FOSS

In order to understand user's experience and perceptions after using FOSS, the students were invited to be interviewed. The results could be classified into three types.

Most students hold the positive views. It is believed that FOSS is more popular and easier to learn. It is also very convenient to use and can easily edit videos and pictures. FOSS is conducive to completing homework quickly and can also be used in life. They expressed their plan and expectation to continue using FOSS in the future. Below are a few examples of typical comments.

> Using this software, I can quickly edit videos using my mobile phone, and then I can do my homework and even do it on the bus.

> It's very convenient for editing. You don't have to upload videos to your computer for editing. Inshot is really convenient. Sometimes I use video editing tools for homework editing.

> Our last silhouette film was a news film, so we needed a lot of scenery to go too far. For example, when other people are talking, they are not just shooting them, they may need other scenes of Hong Kong. The website where you can find pictures and videos for free has a lot of pictures in it. I think this website is very convenient and we can continue to use this website in the future.

> FOSS software is relatively popular, whether it is a person who has taken relevant courses, or the Chinese Department, or the Civil Engineering Department. Everyone has a chance to get started, even in mastery of several different programs in a class. In addition, I think there is one that is really useful, called Canva, which is the one who made the poster, and that one is really available no matter which classmate you are.

> I Usually use it, for example, there are some movies in life. I also use this to cut it, or add some background music, so using the software is not just limited to homework.

The second type is negative evaluation. FOSS is considered to be used very little. After learning, it is easy to forget. Some learners prefer the traditional learning method, which contradicts the use of FOSS, which reflects the strong professional characteristics of different discipline.

> I study architecture, and I rarely use the tools of silhouette films.

> I don't need to use those that I read, so there is no particular opinion.

> Our Chinese homework assignments are all hand write, So they are rarely used.
>
> I am relatively primitive, but I always feel that these are a bit Complicated, and I usually prefer to hand write my homework.
>
> Because I can't use it in my own discipline, I don't mean that I can't use it in daily life, but there are not many opportunities. It's really useful, but one thing is not commonly used, and I don't remember any more.
>
> It's always necessary to use cuts, but there are really few opportunities to use them. it is sometimes used when you suddenly want to record your life.
>
> It's very easy to use, but we have not applied. It's not necessary to use the app in our course, such as the clippers, which will not be used for our Architecture Department students.

The third type of evaluation is neutral. Two students believe that FOSS and professional paid software each have their own strengths. If you choose to use FOSS software, you cannot deny other software. For difficult tasks, APP does not have the function to replace professional computer software, but FOSS software's convenience should be supported as well.

> For example, Inshot, there are always restrictions on silhouettes in mobile phones, otherwise not so many people use computer silhouettes. But if it is only for homework, or usually, sometimes my computer is not easy to use, these apps are also easy to operate. But I know that if you want to separate the sound in the movie from the movie, then separate the two with sound. If all the videos are integrated, the App can't accomplish the task, and it must be done with a computer, so in these cases, there will be this limitation. But in general, using the App silhouette film is already very convenient.
>
> Our department is actually written by the teacher on the blackboard. I think the action written on the blackboard is an element needed, so I find it more difficult to get it, and we don't use any apps for class, so I think they have their own places.

## 6 Discussion

### *6.1 Application of FOSS Can Bring Learning Efficiency*

It was found from this study that FOSS is very convenient to use, which help the learners to improve their learning efficiency and accomplish the learning tasks. The learners can use FOSS to edit videos, pictures, PDF files and so on, which can be achieved without space and time limitations. Most of the teachers and students show their confidence. These facts can be specifically proved by the lecturer of this course. She has been using Tianruo OCR, as she finds it is easy for her on text copying and editing. For some teaching materials she finds online, she needed to type it again if the original text is on PDF format before learning the tool. By adapting Tianruo OCR, she finds it more convenient and time-saving to copy and edit text now. Among the eight tools, the majority of students in her class found Headliners, Inshot more useful and they have applied the tools in this course. For Headliners, students suggest that it is easy for them to transcribe the audio into subtitles, adding captions to the video clips that they produce for this course, without typing the subtitles manually word by word. And for Inshot, the tool helps them to edit the video clips and soundbites by using their mobile phones, without paying for adobe premiere or final cut pro, two licenses that are needed for video editing. They can also work on their video production anytime and anywhere, without the restrictions of sitting in front of a computer or mac computer.

### 6.2 Major Background Influences Learners' Feedback

FOSS is useful and convenient to the learners' study, but major background has a strong influence on their evaluation and reflection. Some majors, such as Journalism and Communication, show very active attitude toward the application of FOSS. The main reason is that they need FOSS to edit videos and pictures in a limited period. However, for some students, such as from Chinese department and Architecture department, they use FOSS less and prefer professional software or even traditional methods. Life habit is also a factor to affect the application of FOSS. Some learners prefer their own learning style and resist new methods.

### 6.3 Specific Strategies Should Be Used Concerning the FOSS Application

Most learners hold the view that FOSS is beneficial for their study and even daily life, but different learners may have different needs and requirements. Therefore, concerning different disciplines and professions, it will be more appropriate to introduce different FOSS tools to different learners, which will be more efficient and convenient. In this study, we prepared 100 teaching tools, and in the future we would encourage the learners to know them and guide them the use effectively and efficiently.

## 7 Conclusion

The two-year FOSS project has been carried out smoothly in Chu Hai College of Higher Education, during which a number of FOSS promotion activities and training seminars have been organized with the purpose of promoting the daily teaching and curriculum development and improve teaching quality. It was found that only a small percentage of respondents have a good knowledge of FOSS before, but after the project, lots of them show positive responses, which indicates that FOSS is more popular to learn, very convenient to use, easy to edit videos and pictures, conducive to completing homework quickly and suitable to be used in daily life. However, there are still some difficulties concerning the application of FOSS, such as insufficient technical support, no support when encountering technical problems, and so on. For the considerations for recommending free and open tools, it is suggested that free to use and easy to use be crucial. During the project, teaching staff, researchers and learners cooperated closely to find out the direct evidence in the application of FOSS. The experiences have been accumulated, which are crucial and beneficial for further research and practice. However, there are still some limitations such as short time and small samples, which call for broader and larger scale applied research.

**Acknowledgements** This work was supported by the Quality Enhancement Support Scheme from Education Bureau, Hong Kong SAR Government [grant number: 01/QESS/2018]. The authors are immensely grateful to Professor Zhang Weiyuan of Beijing Normal University for his comments on an earlier version of the manuscript. Also, thanks to Mr. Lee Koon Lam, Research Assistant of this Project, for assistance in conducting the semi-structure focus group discussions.

## References

Anderson, R. E., & Dexter, S. (2005). School technology leadership: An empirical investigation of prevalence and effect. *Educational Administration Quarterly, 41*(1), 49–82. https://doi.org/10.1177/0013161X04269517.

Ajila, S., & Wu, D. (2007). Empirical study of the effects of open source adoption on software development economics. *The Journal of Systems & Software, 80*(9), 1517–1529.

AlMarzouq, M., Zheng, L., Rong, G., & Grover, V. (2005). Open source: Concepts, benefits, and challenges. *Communications of the Association for Information Systems, 16*(1), 37.

Feller, J., Fitzgerald, B., Hissam, S. A., & Lakhani, K. (Eds.). (2007). Perspectives on free and open source software. Cambridge, MA: The MIT Press.

Free Software Foundation (FSF). (2020).*GNU Operating System*. Retrieved March 20, 2020, from https://www.gnu.org/home.en.html#mission-statement.

Gangadharan, G. R., Dandrea, V., & Weiss, M. (2007).*Free/Open Services: Conceptualization, Classification, and Commercialization*. Open source systems.

Ghosh, R. A. (2007). Understanding free software developers. In J. Feller, B. Fitzgerald, S. A. Hissam, & K. Lakhani (Eds.), *Perspectives on free and open source software*. Cambridge, MA: The MIT Press.

Gupta, D., & Surbhi (2018). Adopting Free and Open Source Software (FOSS) in Education. *Journal on Educational Technology, 14*(4).

Hoe, N. S. (2006). *Breaking barriers: The potential of free and open source software for sustainable human development—A compilation of case studies from across the world*. Bangkok: NUDP PDIP, Elsevier. Retrieved from https://www.unapcict.org/ecohub/resources/breaking-barriers.

Hon, M., Russell, G., & Welch, M. (2010). Open source software considerations for law enforcement. *IT Professional, 12*(6), 18–23.

Haselden, D. M. (2001). Patenting software under the European patent convention, keith beresford, sweet and maxwell, London, March 2000, *World Patent Information, 23*(1), 101–102.

Miller, K., Voas, J., & Costello, T. (2010). Free and open source software. *IT Professional, 12*(6), 14–16.

Pelavin Research Institute. (1997). *Investing in school technology: Strategies to meet the funding challenge*. Washington, D.C.: U.S. Dept. of Education, Office of Educational Technology.

Rogers, E. M. (2003). *Diffusion of innovations*. New York: The Free Press.

Scheuermann, F., & Pedró, F. (Eds.). (2010). *Assessing the effects of ICT in education indicators, criteria and benchmarks for international comparisons*. Ispra (VA): Joint Research Centre- European Commission. Retrieved from https://libproxy.cbu.ca:2048/login?url=https://dx.doi.org/10.1787/9789264079786-en.
Thankachan, B., & Moore, D. R. (2017). Challenges of Implementing Free and Open Source Software (FOSS): Evidence from the Indian Educational Setting. *The International Review of Research in Open and Distributed Learning, 18*(6), 186–199.
Wang, H. J., Blue, J., & Plourde, M. (2010). Community source software in higher education. *IEEE Computer Society, 10*(6), 31–37.
UNESCO. (2020). *Free and Open Source Software (FOSS)*. Retrieved March 20, 2020, from https://en.unesco.org/freeandopensourcesoftware.

**Chenggui Duan** candidate at South China Normal University; Project Manager at Centre of Buddhist Studies, Chu Hai College of Higher Education, Hong Kong. Email: cgduan@chuhai.edu.hk. He was Research Associate at the Research and e-Learning Unit, School of Professional and Continuing Education at the University of Hong Kong (2007–2016) and Project Manager of the Open University United Kingdom's Marketing Representative Office in Beijing (2002–2007). His professional and research interests are mainly in free and open source software, online education, open educational resources, instructional design, online course development, especially based upon the Moodle platform.

**Kwok Fong Lee** is an associate professor of the Department of Journalism and Communication at Chu Hai College of Higher Education, Hong Kong, email: tracylee@chuhai.edu.hk. Her research interest includes media studies and gender studies. Her interest also includes how technology can upgrade teaching and learning.

# Pre-class Learning Strategies of a Flipped Lesson: A Randomized Comparative Study of Student Achievement and Cognitive Load

**Ziyan Che, Khe Foon Hew, and Chung Kwan Lo**

**Abstract** In this study, we examined the effects of different pre-class learning strategies of flipped learning on students' learning achievement and perceived cognitive load. Based on cognitive load theory and cognitive theory of multimedia learning, three strategies were designed and testified in a secondary school ICT (information and communication technology) lesson in Wuhan, China. The three strategies were: (1) segmenting video lecturing (watching a series of six bite-sized videos), (2) conventional video lecturing (watching one 10-min instructional video), and (3) interactive reading (studying through an interactive web page). Eighty-four Senior Secondary 3 students were randomly assigned to learn under these three instructional approaches. The results show that the segmenting video class ($n = 28$) and conventional video class ($n = 28$) had a significantly better performance in their pre-class test compared to the interactive reading class ($n = 28$). In their pre-class cognitive load survey, students in the segmenting video class was rated significantly lower than those in the other two classes. Therefore, segmenting video lecturing was better than the other two pre-class learning strategies when flipping a secondary school ICT lesson. Future research can further testify the use of different pre-class learning strategies in other subject disciplines.

**Keywords** Flipped learning · Flipped classroom · Class preparation · Student learning · Cognitive load · Experimental study

Z. Che · K. F. Hew
Faculty of Education, The University of Hong Kong, Hong Kong, China
e-mail: czy0312@connect.hku.hk

K. F. Hew
e-mail: kfhew@hku.hk

C. K. Lo (✉)
Department of Mathematics and Information Technology, The Education University of Hong Kong, Hong Kong, China
e-mail: chungkwanlo@eduhk.hk



W. W. K. Ma et al. (eds.), *Learning Environment and Design*, Educational Communications and Technology Yearbook,
https://doi.org/10.1007/978-981-15-8167-0_9

# 1 Introduction

Flipped learning has become increasingly widespread in the education sector. This is a technology-enhanced pedagogy, aiming to free up class time for more interactive learning activities, such as group discussions and teacher's one-to-one assistance (Bishop & Verleger, 2013). Different from the traditional pattern of teaching, students can access course content and learning tasks through instructional videos and text-based materials prior to class at their own pace. They can further prepare themselves for in-class collaborative problem-solving and discussions that clarify concepts and contextualized knowledge through application and discourse analysis (Akçayır & Akçayır, 2018; Bishop & Verleger, 2013; Lo et al., 2017).

Multiple articles have been published on flipped learning. Searching only in the *Web of Science* database, Lo and Hwang (2018) found more than 1,800 documents about flipped learning from 2013 to 2017 (5 years). With a growing body of the literature, researchers currently have a clear understanding of its benefits and challenges. For example, students in a flipped learning environment have the flexibility to study or review pre-class materials according to their schedule and needs; but their unpreparedness for pre-class learning tasks becomes a major challenge of using this instructional approach (see Akçayır & Akçayır, 2018; Lo et al., 2017 for a review). As for its effects on students' learning achievement, van Alten et al. (2019) are able to retrieve more than 100 comparative studies of flipped learning for a meta-analysis. Their study has provided solid evidence of a small but significantly positive effect in favor of flipped learning over traditional lecture-based instructions.

Despite our increasing familiarity of flipped learning, it is not necessary that we are flipping our classroom with the best practice; and that we have reached a ceiling of the power of this instructional approach. Gaughan (2014), for example, found room for improvement about the use of pre-class videos in her flipped history course. Although her students generally had a positive attitude toward flipped learning, she found that almost all of her students did not watch the longest pre-class video (40 minutes). Clearly, instructional videos with a long duration impair student learning because these videos are not only disengaging (Guo et al., 2014) but also increase students' cognitive load (Mayer, 2014). Hew and Lo (2018) examined the effects of using different video-styles and pre-class testing strategies. They suggested Khan-style videos with short-answer quizzes be used in future practice of flipped secondary school mathematics lessons. Apart from videos, how about the use of pre-class readings? In the literature, there are several examples of flipped learning that assigned text-based materials without pre-class videos (Moravec et al., 2010; Talbert & Bergmann, 2017).

The overarching goal of this study is to identify some effective pre-class learning strategies that can enhance students' learning achievement and reduce their perceived cognitive load. The following section first explicates our theoretical framework and instructional design. We then present the method and results of our intervention in an ICT (information and communication technology) lesson of a secondary school in Wuhan, China. After that, the results and the agenda of future research are discussed.

## 2 Theoretical Framework and Instructional Design

Lo (2018) establishes the theoretical grounding of flipped learning, where two theories are drawn with a specific purpose of reducing students' cognitive load. In this study, we apply these two theories—cognitive load theory (Sweller, 2019) and cognitive theory of multimedia learning (Mayer, 2014)—to inform the design of our pre-class learning strategies.

### *2.1 Cognitive Load Theory*

Sweller's (2019) cognitive load theory describes human cognitive architecture in which learning, thinking, and problem-solving take place. Given that human beings have a limited cognitive processing capacity, cognitive load imposed on us during a learning task determines our performance (Sweller, 1988). For example, balancing a chemical equation usually requires high cognitive load because altering the amount of substances involved has consequences for the entire equation, whereas learning the nouns of a foreign language demands a relatively lower level of cognitive load (see Sweller, 2019 for a review).

In addition to the nature of a learning task, its instructional design determines how much mental effort the task requires. With reference to the modality effect of cognitive load theory, Castro-Alonso and Sweller (2020) predict that "learning from visualizations supplemented with written text should be less effective than learning from the same visualizations supplemented with comparable spoken text" (p. 75). As they argue, human beings rely on two processors to handle information: (1) the visuospatial processor which handles visuospatial information (i.e., visualizations and visual text), and (2) the auditory processor which handles auditory information (i.e., visualizations and auditory text). They thus suggest that learning from visuospatial information only tends to overload the visuospatial processor. In contrast, if the information is presented using both visuospatial and auditory formats, the two processors can share the cognitive load of handling the materials (Castro-Alonso & Sweller, 2020).

### *2.2 Cognitive Theory in Multimedia Learning*

Mayer's (2014) cognitive theory of multimedia learning explicates how human mind processes instructional media. He establishes 12 design principles of multimedia instruction, aiming to decrease cognitive demand on learners. More specifically, these principles can (1) reduce extraneous processing, (2) manage essential processing, and (3) foster generative processing (see Mayer, 2014 for a review).

In the context of flipped learning, Lo (2018) highlights three principles of Mayer (2014) that have been adopted by practitioners. First, the personalization principle suggests putting words in conversational style rather than formal style. In fact, if teachers speak in a monotone, students may be disengaged from listening. As one student of Snyder et al. (2014) commented on their instructional videos, "I feel like I'm just reading and listening to facts, rather than you talking to us in person" (p. 314). Second, the signaling principle suggests highlighting essential materials during presentation. With this principle, the teacher can also direct students' note taking during video lecturing (Lo, 2018). Third, the segmenting principle suggests breaking a long instructional video into learner-paced segments. In the flipped engineering course of Velegol et al. (2015), for example, a majority of students recommended an online lecture be divided into a series of 10-min videos.

## 2.3 Instructional Design of Pre-Class Learning Strategies

In this study, we created a 10-minute instructional video and course materials for our flipped learning intervention. The video was a PowerPoint presentation with our teacher's talking head. We chose this video-style because pre-typed texts in a PowerPoint format enabled a smooth presentation while recording and its production cost was reasonable (Hew & Lo, 2018). As informed by the modality effect of cognitive load theory (Castro-Alonso & Sweller, 2020), learning materials were presented simultaneously in both visuospatial (i.e., text on slides) and auditory (i.e., teacher's explanation of problem-solving steps) formats. Furthermore, the video was produced using the personalization principle (i.e., speaking in a conversational style) and signaling principle (i.e., highlighting the key content on screen) of Mayer's (2014) cognitive theory of multimedia learning. In Appendix, we provide an overview of the three pre-class learning strategies in this study: (1) segmenting video lecturing, (2) conventional video lecturing, and (3) interactive reading.

For segmenting video lecturing, we divided the 10-minute instructional video into a series of six bite-sized videos (from 0:20 to 4:02) by concept. Such an attempt was supported by the segmentation principle of Mayer (2014). At the end of each video, students had to click an "Answer" button to view answers and to proceed.

For conventional video lecturing, we did not segment the 10-minute instructional video into some shorter ones. This followed the empirical recommendation of multiple flipped learning researchers (e.g., Mok, 2014; Velegol et al., 2015). As no segmentation had been done, all problem-solving steps and answers were showed in sequence.

For interactive reading, we provided students with text-based reading materials in a web page format. The web page was developed using the Wick Editor (https://www.wickeditor.com). This application enabled us to create an interactive web page on which students could navigate learning materials and click to check out the full problem-solving steps and answers. This kind of interactive learning materials is an alternative to pre-class instructional videos (Talbert & Bergmann, 2017).

# 3 Method

## 3.1 Objectives and Research Questions

In this study, we used a randomized experimental approach to compare the effect of the above three different pre-class learning strategies. We aim to examine students' learning achievement and perceived cognitive load under these three instructional environments. The following three research questions (RQ1 to RQ3) are posed to guide this study:

RQ1: How does segmenting video lecturing influence students' learning achievement compared to conventional video lecturing and interactive reading?
RQ2: How does segmenting video lecturing influence students' perceived cognitive load compared to conventional video lecturing and interactive reading?
RQ3: What is the relationship between students' learning achievement and their perceived cognitive load?

## 3.2 Research Participants and Research Design

This study was conducted in a secondary school in Wuhan, China. At the time of study, there was no systematic implementation of flipped learning in the school. Therefore, student experience of flipped learning was minimal. Our flipped lesson was offered in July 2019, introducing an ICT topic—social network analysis.

Students were first invited to participate in the research study. Those who consented to participate were selected as our research participants. A total of 84 Senior Secondary 3 students (female = 41, male = 43) participated in the study. Each student was randomly assigned into one of the three pre-class learning conditions: (1) segmenting video class ($n = 28$; female = 13, male = 15), (2) conventional video class ($n = 28$; female = 15, male = 13), and (3) interactive reading class ($n = 28$; female = 13, male = 15). This random assignment procedure was performed using computer. At the start of our intervention, all student participants attended an orientation session as suggested by Lo and Hew (2017), aiming to help them get accustomed to flipped learning. During the session, we explained our course requirements (e.g., studying online before class and completing a pre-class test) and logistics (e.g., the way to access learning materials).

Our flipped lesson comprised with pre-class and in-class components (Bishop & Verleger, 2013). In the pre-class component, students visited an online lecture using the computer facilities on campus or at home. Although the pre-class learning strategies in the three classes were different, the instructional materials were identical. They learned some new concepts and terms (e.g., directed and undirected network) and computations with examples (e.g., calculating closeness centrality of a node) of social network analysis. After that, they were required to complete a pre-class test and a cognitive load survey (see the Data Collection section). We estimated that all pre-class learning tasks could be completed within 30 minute.

In the in-class component, students attended a 90-minute face-to-face lesson. The same lesson plan was run in the three classes. The teacher first answered students' questions regarding the pre-class learning materials. The lecture then continued with a short lecture of advanced application problems (e.g., the application of network density and individual indicators). After that, students discussed in groups to complete various problem-solving exercises. Toward the end of the lesson, the teacher offered a Q&A section to address students' questions. Upon the completion of our intervention, students took a 30-minute post-class test to evaluate their learning achievement (see the Data Collection section).

## 3.3 *Data Collection*

This study focused on three data sources, including a pre-class test, a post-class test, and a pre-class cognitive load survey. The effect of different pre-class learning strategies on students' learning achievement was evaluated by comparing their test scores. Toward the end of pre-class learning, students were required to complete an online pre-class test. The test consisted of two sections: (1) 12 recall questions to assess students' ability to recall information as delivered in the pre-class component, and (2) 8 application questions to assess students' ability to apply what they have learned to solve related problems. Each question carried 1 mark. Thus, the possible score of the recall and application sections ranged from 0 to 12 and from 0 to 8, respectively. Figure 1 shows a sample question in each of the two sections.

In addition to the pre-class test, a written post-class test was conducted to evaluate students' learning achievement at the end of our intervention. There were 20 application questions. The difficulty level of the questions increased progressively. These questions assessed students' ability to apply what they have learned to solve simple and advanced problems (Fig. 2). Each question carried 1 mark. Thus, the possible score of the post-class test ranged from 0 to 20.

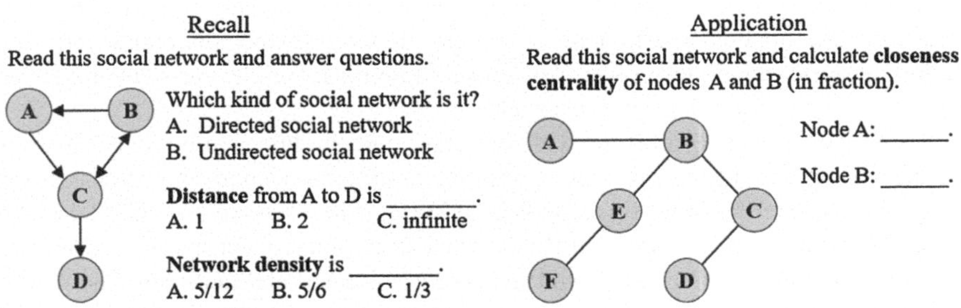


**Fig. 1** Sample questions in the pre-class test

**Simple problem**

Read this social network.

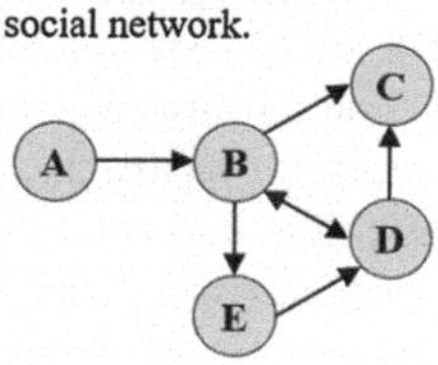


This is (a directed/an undirected) ________ network.
Distance from E to A is ________.
Distance from E to B is ________.
Indegree centrality of node B is ________.
Outdegree centrality of node B is ________.
Network density is ________.

**Advanced problem**

Draw the graphical representation of the social network which meets following requirements:
- it is an undirected social network;
- consists of nodes A, B, C, D, E, and F;
- degree centrality of node A is 5;
- degree centrality of nodes B, C, D, E, and F

Its network density is ________.

**Fig. 2** Sample questions in the post-class test

The effect of different pre-class learning strategies on students' perceived cognitive load was evaluated by comparing their ratings of an online survey. The survey was distributed at the end of students' class preparation. Following a highly cited study of Brünken et al. (2003), we examined students' perceived cognitive load using indirect and direct subjective measures: (1) the amount of mental effort invested (Paas et al., 1994), and (2) the perceived difficulty of materials (Kalyuga et al., 1999). First, we developed five items (M1 to M5 in Table 1) to examine students' mental effort invested in understanding the pre-class learning materials, such as terms and concepts. Following Paas et al. (1994), a 9-point Likert scale was used ranging from 1 (the lowest mental effort) to 9 (the highest mental effort). Second, we developed four items (D1 to D4 in Table 1) to examine students' perceived difficulty of the pre-class

**Table 1** Cognitive load survey items and Cronbach's alpha

| Constructs and survey items | Cronbach's alpha |
|---|---|
| *Mental effort* | |
| In pre-class learning, how much mental effort did you invest in understanding | |
| (M1) the topics? | 0.88 |
| (M2) the terms and concepts? | |
| (M3) the formulas? | |
| (M4) the explanations of exampled calculations? | |
| (M5) the key content? | |
| *Difficulty of materials* | |
| In pre-class learning, how do you rate the difficulty of | |
| (D1) the topics? | 0.91 |
| (D2) the terms and concepts? | |
| (D3) the formulas? | |
| (D4) the calculation? | |

learning materials. Similar to the mental effort items, a 9-point Likert scale was used ranging from 1 (extremely easy) to 9 (extremely difficult). Table 1 shows the results of our preliminary exploration of survey data ($n = 84$). The Cronbach's reliability coefficient alpha of the mental effort and difficulty of materials constructs was 0.88 and 0.91, respectively. Therefore, the items in both constructs had relatively high internal consistency (Field, 2009). The overall rating of each construct was obtained by averaging out the ratings of the items under a construct.

## 3.4 Data Analysis

To determine an appropriate statistical test for data analysis, we first checked the normality of the data (Field, 2009). Our preliminary exploration of data suggests that it was not appropriate to assume the normality of both the tests and survey data. According to the Kolmogorov–Smirnov test, there was a significant deviation form normality in the pre-class application test in the conventional video class ($p = 0.011$), the post-class test in the segmenting video class ($p = 0.036$), and the mental effort construct of the pre-class survey in the interactive reading class ($p = 0.031$), among others. Therefore, non-parametric tests were more appropriate for the data analysis in this study (Field, 2009).

To analyze the tests and survey data, the Kruskal–Wallis test was applied at a significance level of 0.05. This is a non-parametric test for comparing multiple groups (Field, 2009)—the segmenting video, conventional video, and interactive reading classes in this study. Following Field (2009), multiple Mann–Whitney tests were run as non-parametric post-hoc procedures for pairwise comparisons when significant differences were found. It is worth noting that running a series of the Mann–Whitney tests will increase the risk of making Type I errors if the significance level is kept at 0.05. Therefore, Bonferroni correction was applied, and the alpha value became $0.05/3 = 0.0167$ during post-hoc analysis.

Correlational analyses were conducted to examine the correlation between students' learning achievement and their perceived cognitive load. As Field (2009) suggests, Spearman's correlation coefficients ($r_s$)—non-parametric statistics—were computed between students' test scores (i.e., pre-class recall, pre-class application, and post-class application) and their ratings of cognitive load (i.e., mental effort and difficulty of materials). As there were two constructs of cognitive load, we further considered the value of $r_s$ square (i.e., $R_s^2$) which is the explaining power of a variable (Field, 2009).

# 4 Results

## 4.1 RQ1: How Does Segmenting Video Lecturing Influence Students' Learning Achievement Compared to Conventional Video Lecturing and Interactive Reading?

Table 2 shows the pre-class test results across different classes. The Kruskal–Wallis test of the recall ($H(2) = 17.44, p < 0.001$) and application ($H(2) = 8.77, p = 0.012$) sections indicates significant differences between the three classes. For the recall section, the results of the Mann–Whitney test show that the interactive reading class ($Mdn$ = 8.00) scored significantly lower than the segmenting video class ($Mdn$ = 11.00), $U$ = 152.00, $z$ = –3.97, $p$ < 0.001, and the conventional video class ($Mdn$ = 10.00), $U$ = 207.50, $z$ = –3.05, $p$ = 0.002. However, the difference between the segmenting video and conventional video classes was not significant, $U$ = 333.00, $z$ = –0.98, $p$ = 0.327. For the application section, the differences between the three classes were not significant in general, except that the segmenting video class ($Mdn$ = 4.00) significantly outperformed the interactive reading class ($Mdn$ = 2.00), $U$ = 202.50, $z$ = –3.15, $p$ = 0.002.

Table 3 shows the post-class application test results across different classes. The Kruskal–Wallis test indicates significant differences between the three classes, $H(2)$ = 6.28, $p$ = 0.043. The results of the Mann–Whitney test show that the interactive reading class ($Mdn$ = 12.50) scored lower than the segmenting video class ($Mdn$ = 14.00), $U$ = 262.50, $z$ = –2.15, $p$ = 0.031, and the conventional video class ($Mdn$ = 13.50), $U$ = 261.50, $z$ = –2.17, $p$ = 0.030. However, Bonferroni correction should be applied to avoid making Type I errors in the stage of post-hoc analysis. Therefore, we would not claim that the segmenting video and conventional video classes significantly outperformed the interactive reading class because none of these results reached the significant level of 0.0167. Also, the difference between the

**Table 2** Pre-class test results by class ($n$ = 28 for each class)

| Class | $M$ ($SD$) | $Mdn$ | Mean rank | Pairwise comparison |
|---|---|---|---|---|
| *Recall (full mark = 12)* | | | | |
| Segmenting video (SV) | 10.50 (1.60) | 11.00 | 56.18 | SV > IR***; CV > IR** |
| Conventional video (CV) | 10.00 (1.96) | 10.00 | 46.98 | |
| Interactive reading (IR) | 7.86 (2.43) | 8.00 | 27.34 | |
| *Application (full mark = 8)* | | | | |
| Segmenting video (SV) | 3.46 (1.71) | 4.00 | 51.05 | SV > IR** |
| Conventional video (CV) | 3.00 (2.07) | 2.50 | 44.21 | |
| Interactive reading (IR) | 1.96 (1.45) | 2.00 | 32.23 | |

*** $p < 0.001$; ** $p < 0.01$

**Table 3** Post-class application test results by class ($n = 28$ for each class)

| Class | *M (SD)* | *Mdn* | Mean rank | Pairwise comparison |
|---|---|---|---|---|
| *Application (full mark = 20)* | | | | |
| Segmenting video (SV) | 13.46 (1.55) | 14.00 | 47.11 | SV > IR^; CV > IR^ |
| Conventional video (CV) | 13.46 (1.48) | 13.50 | 47.18 | |
| Interactive reading (IR) | 12.43 (2.03) | 12.50 | 33.21 | |

^$p < 0.05$ but $p > 0.0167$ (Bonferroni correction).

segmenting video and conventional video classes was not significant, $U = 333.00$, $z = -0.98$, $p = 0.327$.

## 4.2 RQ2: How Does Segmenting Video Lecturing Influence Students' Perceived Cognitive Load Compared to Conventional Video Lecturing and Interactive Reading?

Table 4 shows the cognitive load survey results across different classes. The Kruskal–Wallis test of the mental effort ($H(2) = 25.98$, $p < 0.001$) and difficulty of materials ($H(2) = 13.10$, $p = 0.001$) constructs indicates significant differences between the three classes. For the mental effort construct, the results of the Mann–Whitney test show that the interactive reading class (*Mdn* = 4.70) rated significantly higher than the segmenting video class (*Mdn* = 2.60), $U = 107.00$, $z = -4.68$, $p < 0.001$, and the conventional video class (*Mdn* = 3.50), $U = 174.50$, $z = -3.57$, $p < 0.001$. In other words, students in the segmenting video and conventional video classes perceived a lower level of cognitive load in terms of their mental effort. However, at a significance

**Table 4** Cognitive load survey results by class ($n = 28$ for each class)

| Class | *M (SD)* | *Mdn* | Mean rank | Pairwise comparison |
|---|---|---|---|---|
| *Mental effort (rating = 1 to 9)* | | | | |
| Segmenting video (SV) | 2.76 (1.12) | 2.60 | 27.73 | SV < CV^; SV < IR***; CV < IR*** |
| Conventional video (CV) | 3.43 (0.98) | 3.50 | 39.32 | |
| Interactive reading (IR) | 4.85 (1.81) | 4.70 | 60.45 | |
| *Difficulty of materials (rating = 1 to 9)* | | | | |
| Segmenting video (SV) | 3.06 (1.58) | 2.63 | 29.71 | SV < CV*; SV < IR** |
| Conventional video (CV) | 4.12 (1.46) | 4.50 | 44.86 | |
| Interactive reading (IR) | 5.02 (2.08) | 4.50 | 52.93 | |

Remark: For the ratings of cognitive load, the lower the better. *** $p < 0.001$; ** $p < 0.01$; * $p < 0.0167$; ^ $p < 0.05$ but $p > 0.0167$ (Bonferroni correction)

level of 0.0167, the difference between these two classes was not significant, $U = 263.50$, $z = -2.11$, $p = 0.035$. For the difficulty of materials construct, the segmenting video class ($Mdn = 2.63$) was rated significantly lower than the conventional video class ($Mdn = 4.50$), $U = 241.00$, $z = -2.48$, $p = 0.013$, and the interactive reading class ($Mdn = 4.50$), $U = 185.00$, $z = -3.40$, $p = 0.001$. However, the difference between the conventional video and interactive reading classes was not significant, $U = 307.00$, $z = -1.40$, $p = 0.163$.

### 4.3 RQ3: What is the Relationship Between Students' Learning Achievement and Their Perceived Cognitive Load?

As Table 5 shows, the results of the Spearman's correlation coefficient ($r_s$) analysis indicate that the three test scores (i.e., pre-class recall, pre-class application, and post-class application) had a significantly negative correlation to the two constructs of cognitive load (i.e., mental effort and difficulty of materials) for all students ($n = 84$). For the recall and application sections in the pre-class test, the correlation coefficients ranged from –0.56 to –0.44, $p < 0.001$. As for the post-class application test, the correlation coefficient of mental effort and difficulty of materials had a smaller value of –0.39 ($p < 0.001$) and –0.27 ($p = 0.013$), respectively.

Considering the explaining power of mental effort and difficulty of materials, around 20–30% of students' pre-class test performance in the recall and application sections were accounted by these two constructs of cognitive load (Table 5). Even after the 90-min face-to-face lesson, a notable small portion of students' post-class test performance was still accounted by mental effort (15.2%) and difficulty of materials (7.2%) in their pre-class learning.

**Table 5** The correlation ($r_s$) and explaining power ($R_s^2$) of students' learning achievement and perceived cognitive load

| Variables | Pre-class recall | Pre-class application | Post-class application |
|---|---|---|---|
| Mental effort | – 0.56*** ($R_s^2 =$ 0.308) | – 0.50*** ($R_s^2 =$ 0.252) | – 0.39*** ($R_s^2 =$ 0.152) |
| Difficulty of materials | – 0.46*** ($R_s^2 =$ 0.215) | – 0.44*** ($R_s^2 =$ 0.196) | – 0.27* ($R_s^2 = 0.072$) |

*** $p < 0.001$; * $p < 0.05$

# 5 Discussion and Conclusions

## *5.1 Effects of Different Pre-Class Learning Strategies*

Among the three pre-class learning strategies, interactive reading generally had an inferior effect on students' pre-class learning compared to segmenting video lecturing and conventional video lecturing. There is thus a notable contrary to those studies in higher education (e.g., Moravec et al., 2010; Talbert & Bergmann, 2017). However, our result resonates with the studies of flipped learning in secondary education. In their high school calculus class, Grypp and Luebeck (2015) created instructional videos for students' class preparation. Their intervention of flipped learning ran smoothly in the first few lessons. In the last lesson, however, they could only provide a reading for their students to study before class. As a result, their students failed to handle the pre-class materials. Additional explanation was required to resolve their problems during class time.

Besides, students in the interactive reading class generally reported a higher level of cognitive load in pre-class learning. As the modality effect of cognitive load theory predicts, visualizations with visual text only are more likely to overload students compared to visualizations with both visual and auditory text (Castro-Alonso & Sweller, 2020). Our result further echoes with the qualitative findings of Muir and Geiger (2016). In their Grade 10 mathematics class, the students expressed strong disapproval of the sole use of pre-class readings. In the words of one student, "A book doesn't really walk through the steps on how to do something as well" (p. 164). Another student continued to explain the benefit of watching instructional videos: "it is easier to understand as the teacher is working through the question with you on the video and you can follow on" (p. 164). In other words, viewing instructional videos demanded lower mental effort and made it easier for the students to understand the pre-class materials.

As for the post-class application test, the differences between the three classes were not significant. Our teacher's remedial teaching inside the classroom might have resolved students' problems encountered in their per-class learning. In fact, the meta-analysis of Lo and Hew (2019) showed a significantly positive effect of offering a brief review of pre-class materials at the start of face-to-face lessons. Meanwhile, we further provided an end-of-class Q&A session for students to seek help. Students' learning achievement in the interactive reading class thus became statistically similar to those in the segmenting video and conventional video classes. However, remedial teaching should not be the focus of the in-class component of flipped learning. Moreover, students' pre-class cognitive load still accounted for a portion of in-class learning effectiveness as reflected in their post-class test performance. Given that the interactive reading strategy demanded a higher level of their perceived cognitive load, we suggest pre-class instructional videos be used in future practice of flipped learning in secondary education contexts.

In this study, the differences between segmenting video and conventional video lecturing were not significant in terms of students' learning achievement. However, there was evidence that the students in the segmenting video class had a lower level of their perceived cognitive load. In this class, we divided our 10-minute instructional video (as used in the conventional class) into six bite-sized videos by concept. Consistent with Mayer's (2014) cognitive theory of multimedia learning, dividing a long video into a series of short videos helped manage students' perceived cognitive load. When creating flipped learning videos, educators can further consider keeping each video within 6 minutes as this is the median engagement time for watching instructional videos (Guo et al., 2014; Lo, 2018).

### *5.2 Limitations and Future Work*

Although this study has laid the groundwork for designing effective pre-class learning strategies, several limitations must be acknowledged. First, the duration of the study was short. Our student participants might not get fully accustomed to these strategies. Therefore, the efficacy of their class preparation might have been impaired. Further studies with a longer duration (e.g., several lessons) are necessary.

Second, we examined the application of three different pre-class learning strategies in one single context (i.e., the flipped ICT lesson in the research school). Overall, the results show that the segmenting video class and conventional video class performed significantly better than the interactive reading class. Moreover, students in the segmenting video class scored significantly lower cognitive load than those in the other two classes. Therefore, our overall results suggest that segmenting video lecturing was better than the other two pre-class learning strategies when flipping a secondary school ICT lesson. However, we acknowledge that the topic of the lesson and the culture of the research school might have influenced students' preferred formatting and perceived cognitive load. Hence, we are hesitant to generalize the findings of our learning strategies to other educational contexts. We recommend further research be conducted to testify the use of these strategies in different subject disciplines (e.g., mathematics, science, and language). In this way, researchers can compare the results of different contexts to determine whether the effects of the learning strategies would apply across curriculum.

Third, this paper focuses on our quantitative analysis of students' learning achievement and perceived cognitive load. In addition to quantitative methods, future studies can incorporate qualitative approaches (e.g., learning journals and interviews) to explore students' learning experience and perceptions of different pre-class learning strategies.

## Appendix

Screenshots and descriptions of pre-class learning strategies.

## Segmenting Video Lecturing

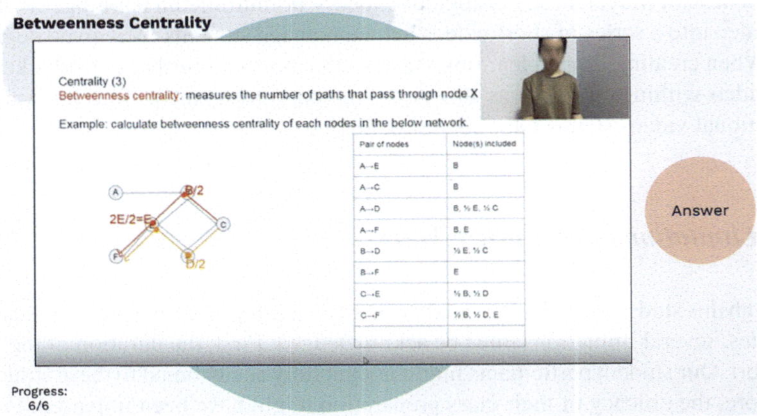


Six short instructional videos (a total of 10 min)
Showing problem-solving steps; students have to click an "Answer" button to view answers

## Conventional Video Lecturing

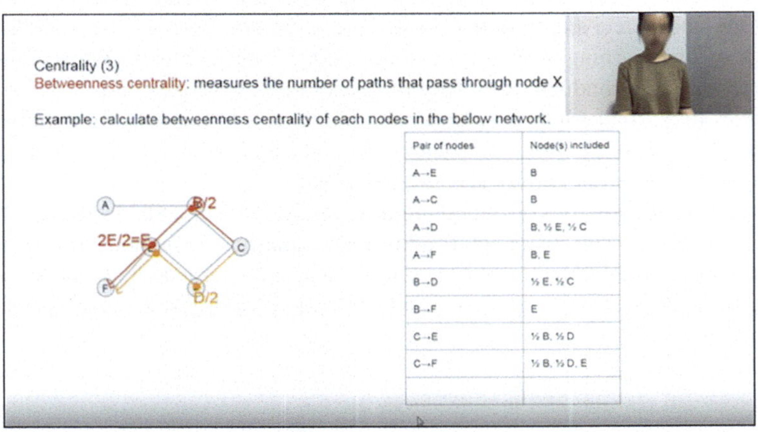


One 10-minute instructional video
Showing problem-solving steps and answers in sequence.

## Interactive Reading

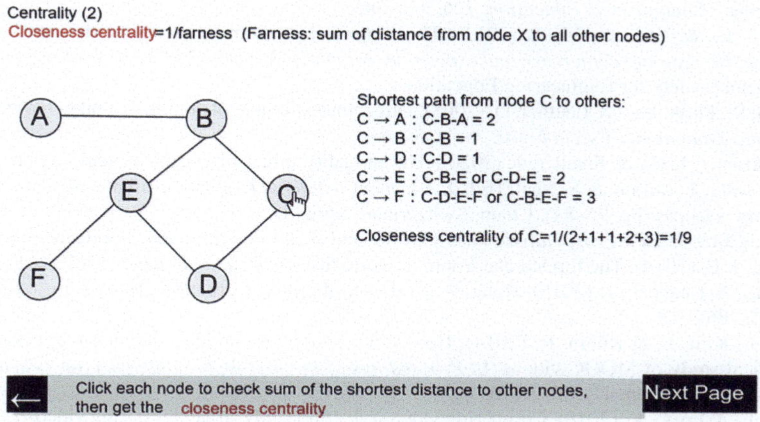


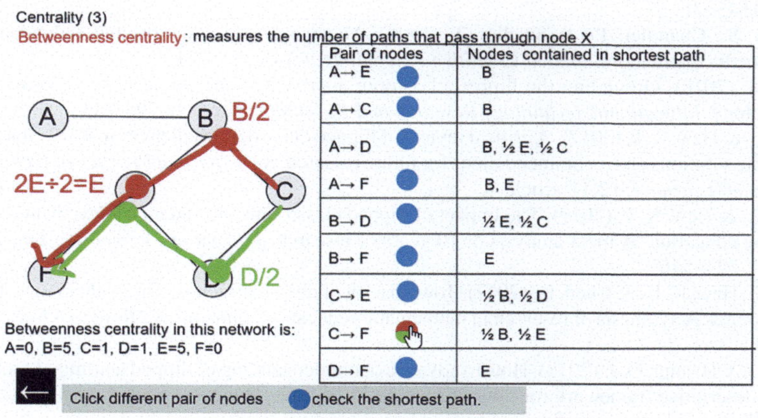


Text-based reading materials
Students click "Forward" or "Backward" buttons to navigate
Students have to click hyperlinks to view problem-solving steps and answers

## References

Akçayır, G., & Akçayır, M. (2018). The flipped classroom: A review of its advantages and challenges. *Computers & Education, 126,* 334–345.

Bishop, J. L., & Verleger, M. A. (2013). The flipped classroom: A survey of the research. In *120th ASEE national conference and exposition, Atlanta, GA* (paper ID 6219). Washington, DC: American Society for Engineering Education.

Brünken, R., Plass, J. L., & Leutner, D. (2003). Direct measurement of cognitive load in multimedia learning. *Educational Psychologist, 38*(1), 53–61.

Castro-Alonso, J. C., & Sweller, J. (2020). The modality effect of cognitive load theory. In W. Karwowski, T. Ahram, & S. Nazir (Eds.), *Advances in human factors in training, education, and learning sciences* (pp. 75–84). Cham, Switzerland: Springer.

Field, A. (2009). *Discovering statistics using SPSS* (3rd ed.). London: SAGE Publications Ltd.

Gaughan, J. E. (2014). The flipped classroom in world history. *History Teacher, 47*(2), 221–244.

Grypp, L., & Luebeck, J. (2015). Rotating solids and flipping instruction. *Mathematics Teacher, 109*(3), 186–193.

Guo, P. J., Kim, J., & Rubin, R. (2014). How video production affects student engagement: An empirical study of MOOC videos. In *Proceedings of the first ACM conference on Learning@ scale conference* (pp. 41–50). ACM.

Hew, K. F., & Lo, C. K. (2018). Comparing video styles and study strategies during video-recorded lectures: Effects on secondary school mathematics students' preference and learning. *Interactive Learning Environments.* Advance online publication. doi:https://doi.org/10.1080/10494820.2018.1545671

Kalyuga, S., Chandler, P., & Sweller, J. (1999). Managing split-attention and redundancy in multimedia instruction. *Applied Cognitive Psychology, 13*(4), 351–371.

Lo, C. K. (2018). Grounding the flipped classroom approach in the foundations of educational technology. *Educational Technology Research and Development, 66*(3), 793–811.

Lo, C. K., & Hew, K. F. (2017). A critical review of flipped classroom challenges in K-12 education: Possible solutions and recommendations for future research. *Research and Practice in Technology Enhanced Learning, 12*(1), article 4.

Lo, C. K., & Hew, K. F. (2019). The impact of flipped classrooms on student achievement in engineering education: A meta-analysis of 10 years of research. *Journal of Engineering Education, 108*(4), 523–546.

Lo, C. K., Hew, K. F., & Chen, G. (2017). Toward a set of design principles for mathematics flipped classrooms: A synthesis of research in mathematics education. *Educational Research Review, 22,* 50–73.

Lo, C. K., & Hwang, G. J. (2018). How to advance our understanding of flipped learning: Directions and a descriptive framework for future research. *Knowledge Management & E-Learning, 10*(4), 441–454.

Mayer, R. E. (2014). *The Cambridge handbook of multimedia learning.* New York, NY: Cambridge University Press.

Mok, H. N. (2014). Teaching tip: The flipped classroom. *Journal of Information Systems Education, 25*(1), 7–11.

Moravec, M., Williams, A., Aguilar-Roca, N., & O'Dowd, D. K. (2010). Learn before lecture: A strategy that improves learning outcomes in a large introductory biology class. *CBE–Life Sciences Education, 9*(4), 473–481.

Muir, T., & Geiger, V. (2016). The affordances of using a flipped classroom approach in the teaching of mathematics: A case study of a grade 10 mathematics class. *Mathematics Education Research Journal, 28*(1), 149–171.

Paas, F. G. W. C., van Merriënboer, J. J. G., & Adam, J. J. (1994). Measurement of cognitive load in instructional research. *Perceptual and Motor Skills, 79,* 419–430.

Snyder, C., Paska, L. M., & Besozzi, D. (2014). Cast from the past: Using screencasting in the social studies classroom. *The Social Studies, 105*(6), 310–314.

Sweller, J. (1988). Cognitive load during problem solving: Effects on learning. *Cognitive Science, 12*(2), 257–285.
Sweller, J. (2019). Cognitive load theory. In S. Tindall-Ford, S. Agostinho, & J. Sweller (Eds.), *Advances in cognitive load theory: Rethinking teaching* (pp. 1–11). New York, NY: Routledge.
Talbert, R., & Bergmann, J. (2017). *Flipped learning: A guide for higher education faculty*. Sterling, VA: Stylus Publishing LLC.
van Alten, D. C. D., Phielix, C., Janssen, J., & Kester, L. (2019). Effects of flipping the classroom on learning outcomes and satisfaction: A meta-analysis. *Educational Research Review, 28*, article 100281.
Velegol, S. B., Zappe, S. E., & Mahoney, E. (2015). The evolution of a flipped classroom: Evidence-based recommendations. *Advances in Engineering Education, 4*(3), 1–37.

**Ziyan Che** is a graduate of Master of Science Information Technology in Education Program at the Faculty of Education, the University of Hong Kong, Hong Kong. She is interested in the use of technology-enhanced learning, including flipped learning.

**Khe Foon Hew** is an Associate Professor at the Faculty of Education, the University of Hong Kong, Hong Kong. Trained as a Systems Engineer with a bachelor's degree in Computer Technology, Dr. Hew had worked for 4 years in Sony before moving on the educational field. He has a master's degree in Instructional Design and Technology, and a PhD in Instructional Systems Technology from Singapore and USA, respectively. His research interests focus on investigating knowledge sharing in social media-supported environments, and examining user engagement in online/blended learning contexts.

**Chung Kwan Lo** is an Assistant Professor in the Department of Mathematics and Information Technology at the Education University of Hong Kong, Hong Kong. He received his EdD from the University of Hong Kong. His main research interests are mathematics education, technology-enhanced learning (e.g., flipped learning and gamification), student engagement, and teacher professional development.

# Task Design to Enhance Learners' Collaboration and Engagement in an Online Chinese Learning Environment

**Sijia Guo**

**Abstract** Empirical research has shown that applying authentic tasks can create a collaborative learning environment for learners, which enables them to engage in meaningful language exchanges that promote their second language acquisition. Although new technologies make collaboration among learners possible in virtual classrooms, few studies have investigated task design in multimodal learning environments or its influence on learners' interaction and collaborative language learning. To fill the gap, this study was carried out to explore the best practice by applying a task-based language teaching approach via a web conferencing-based online Chinese class. The purpose of the current research was to answer the following questions: (1) How do communicative tasks stimulate learners' collaboration and interaction in an online Chinese language class?, (2) What are learners' perceptions of the task design in the web-conferencing-based multimodal learning environment?, and (3) What are the impacts of technology-mediated task-based language teaching on learners' learning experience? In this study, 16 university beginner learners of Chinese participated in this two-stage project. Data were collected through post-session surveys, in-depth interviews, and web conferencing archive recordings. The results confirmed that the designed communicative tasks showed great pedagogical value in facilitating learners' collaboration and interaction in an online learning environment.

**Keywords** Computer-assisted language learning (CALL) · Collaborative learning · Learners' interaction · Multimodal learning

## 1 Introduction

Derived from both the interaction approach to second language acquisition (SLA) and sociocultural theory, task-based language teaching (TBLT) has been recognised

S. Guo (✉)
Macquarie University, Sydney, Australia
e-mail: sijia.guo@mq.edu.au





W. W. K. Ma et al. (eds.), *Learning Environment and Design*, Educational Communications and Technology Yearbook,
https://doi.org/10.1007/978-981-15-8167-0_10

in abundant literature as a very effective language teaching approach in face-to-face environments (Ellis, 2003; Nunan, 2004; Samuda & Bygate, 2008; Van den Branden, 2006). The fundamental driving force behind the task-based language teaching approach is based on the rationale that tasks can stimulate learners' interactions, which in turn can facilitate second language acquisition (Pica, 1994; Pica & Doughty, 1985; Pica et al., 1996). Moreover, it can provide opportunities for learners to collaboratively work with others on tasks that they are not able to complete by themselves (Vygotsky, 1978).

In the last two decades, researchers' attention has been drawn to technology-mediated task-based language teaching (Chapelle, 2001; Ortega & González-Lloret, 2014; Stockwell, 2010). González-Lloret and Ortega (2014) argue that "[t]he imperative of integrating technology in education is undisputed today" (p. 1). In the field of computer-assisted language learning, a substantial body of research has focused on collaboration and interaction in multimodal learning environments and its influence on learners' second language acquisition (Abrams, 2016; Hampel & Stickler, 2012; Lin, 2015; Rouhshad et al., 2016; Stickler & Shi, 2013). Multimodal technologies, such as web conferencing, provide learners with efficient and diverse modes of communication. Although implementation of those tools in foreign language classes has become more pervasive now, the question of how multimodal software and language tasks should be integrated into a mutually informative environment remains largely under-researched. Salaberry (2000) urges language instructors to analyse how pedagogical goals can be achieved through activity design and implementation in computer-mediated communication (CMC) environments.

The current study was carried out to bridge this gap, by implementing a task-based approach in a web conferencing-based online Chinese class to explore technology-mediated task design principles and its influence on learners' collaboration and engagement.

## 2 Background of the Study

### *2.1 TBLT and Learners' Interaction*

Within the field of second language acquisition (SLA), numerous studies have been conducted on task-based language teaching. Empirical evidence has been found to support the Interaction Hypothesis in the context of face-to-face TBLT teaching practice (Ellis et al., 1994; Gass & Varonis, 1985; Long, 1985; Pica, 1994; Pica et al., 1991). In Ellis's (2009) review of TBLT literature, he provides a set of practical definitions, using language holistically to fulfil non-linguistic goals in order to achieve meaning-based communication: a) the primary focus should be on 'meaning'; b) there should be some kind of 'gap'; c) learners should largely have to rely on their own resources (linguistic and non-linguistic) in order to complete the activity; and d) there is a clearly defined outcome other than the use of language.

In his study, Long (1983) proposed a model that explained the relationship between conversational task, interaction, and second language acquisition. Further,

TABLE 2 *Communication task types for L2 research and pedagogy analysis based on: Interactant (X/Y) relationships and requirements in communicating information (INF) to achieve task goals*

| | *INF holder* | *INF requester* | *INF supplier* | *INF requester–supplier relationship* | *Interaction requirement* | *Goal orientation* | *Outcome options* |
|---|---|---|---|---|---|---|---|
| *Task Type:* | | | | | | | |
| Jigsaw | X & Y | X & Y | X & Y | 2 way (X to Y & Y to X) | + required | + convergent | 1 |
| Information gap | X or Y | Y or X | X or Y | 1 way > 2 way (X to Y/Y to X) | + required | + convergent | 1 |
| Problem-solving | X = Y | X = Y | X = Y | 2 way > 1 way (X to Y & Y to X) | − required | + convergent | 1 |
| Decision-making | X = Y | X = Y | X = Y | 2 way > 1 way (X to Y & Y to X) | − required | + convergent | 1+ |
| Opinion exchange | X = Y | X = Y | X = Y | 2 way > 1 way (X to Y & Y to X) | − required | − convergent | 1+/− |

**Fig. 1** Task typology proposed by Pica, Kanagy, and Falodun (1993)

he predicted that more negotiation may be generated in two-way tasks. Following that, Doughty and Pica (1986) examined the effects of task type and its influence on interaction pattern. The results suggest that tasks requiring information exchange between interlocutors may generate more interaction. Moreover, they noticed that task type played a key role in stimulating learners' conversational modifications. A communication task typology was later developed by Pica, Kanagy, and Falodun (1993), which covers interactant relationship, communication goals, and outcome options. It has since been used to locate, describe, and compare five task types—jigsaw, information gap, problem-solving, decision-making, and opinion exchange—and to assess their contributions in terms of stimulating comprehension, production, and providing feedback opportunities to L2 learners (see Fig. 1).

They note that the tasks in which the information is split into two-way flows (e.g., jigsaw) and the tasks that subsequently require interactants to exchange information (e.g., information gap and jigsaw) show higher potential in stimulating interaction and communication than other task stimuli. Tasks that require information exchange, such as jigsaw and information gap tasks, are considered to have great potential for eliciting negotiation. In jigsaw tasks, each participant holds part of the information and is required to exchange the information in order to achieve the task goal. Similar to jigsaw tasks, in information gap tasks, split information is provided, and interlocutors are asked to do one-way or two-way information exchange.

## 2.2 *TBLT and Computer-Assisted Language Learning*

In the current literature, the body of research examining technology-mediated TBLT has grown (Lai & Li, 2011; Ortega & González-Lloret, 2014; Thomas, 2013; Ziegler,

2016). TBLT not only provides a useful framework for designing and implementing instructional activities in computer-assisted language learning (CALL) contexts (Duran & Ramaut, 2006; 2005), but has also received positive reactions from learners and teachers (Hampel & Hauck, 2004; Lai et al., 2011).

According to González-Lloret and Ortega (2014), technology-mediated TBLT has long noted the importance of task design and its benefits on learners' interaction and collaboration (Chapelle, 2003; Doughty & Long, 2003). Similar to the findings in face-to-face settings, research has shown that tasks requiring information exchange have great potential in stimulating learners' interaction in technology-mediated TBLT (Blake, 2000; 2006; Smith, 2003).

In Blake's (2000) study, 50 intermediate-level Spanish language learners were asked to conduct jigsaw, information gap (one- and two-way), and decision-making tasks in a synchronous chat programme in order to compare learners' interaction and how learners' interaction was affected by different task types. Students' chat transcripts were analysed in terms of negotiation types, negotiation of miscommunication, and turn taking. The results showed that jigsaw tasks stimulated the most negotiations, whereas information gap tasks were not nearly as productive as a stimulus. In Smith's (2003) study, 14 non-native speaker dyads collaboratively conducted two jigsaw tasks and two decision-making tasks in a synchronous text chat. He found that learners' collaboration on decision-making tasks outperformed negotiation turns in jigsaw tasks. Keller-Lally 2006 investigated the impact of task type (jigsaw, decision-making, and opinion exchange) and group size (dyads and small group) on learners' frequency of negotiations and language production. In her study, 62 intermediate German language learners' online discussions were transcribed and coded in terms of L1/L2 language use and off-task communication units. The statistical results illustrated that learners' language output in decision-making tasks and opinion exchange tasks outweighed that in jigsaw tasks. In addition, no significant difference in the amount of negotiations between decision-making and jigsaw tasks was noticed in her study, which was contrary to Blake's (2000) and Smith's (2003) findings.

## 2.3 *Learners' Interaction and Engagement in the Multimodal Learning Environment*

The number of studies on multimodality learning environments is growing in the field of CALL (Guichon & McLornan, 2008; Levy & Stockwell, 2013). In multimodal environments, particularly in conferencing-based settings, how to design tasks to facilitate collaborative learning has become the interest of recent research (Hampel, 2006, 2010; Stockwell, 2010). Ciekanski and Chanier (2008), who focused on the impact of integrating audio and text on learners' collaborative writing competence, observed learners using multiple modes to make meaning and therefore facilitate collaboration. They maintained that, in multimodal environments, learners' focus and engagement with the learning activity are influenced during the process of implementing telecollaborative tasks in which fluency may be prioritised over accuracy. Wang (2004, 2006, 2008), who examined learners' interaction via desktop video conferencing, noted that oral–visual interaction, which is facilitated by the multimodal environment, enabled real-time facial expression and gestures between learners and teachers. It provided semiotic cues for meaning making, which was conducive to task completion. Guichon and Cohen (2014), who also observed learners' interaction in video conferencing and audio conferencing, noticed that there was more overlapping interaction in the former mode and more student silences in the latter. They suggested that video conferencing could facilitate a fast and seamless interaction with paralinguistic cues for conversation.

However, some researchers point out the limitations of using conferencing tools in teaching practice. For instance, Berglund (2009), who investigated participant rates and conversational feedback strategies in a video conferencing-based class, found that, without teachers' presence, students' engagement resembled that of instructed discussion. However, long monological turns were identified in learners' contribution as well. As Salaberry (2000) argued, applying new technologies in the classroom does not automatically generate best learning outcomes. It is critical for language teachers to analyse how learners' interaction and collaboration can be stimulated through task design and implementation in multimodal online learning environments.

Despite emerging research interest in the fields of CALL and TBLT, few studies have examined how tasks can be designed to stimulate learners' collaboration and engagement in multimodal environments. This study was conducted to bridge the gap and to answer the following research questions:

- How do communicative tasks such as jigsaw, information gap, and decision-making tasks stimulate learners' collaboration and interaction in an online Chinese language class?
- What are learners' perceptions of the task design in the web-conferencing-based multimodal learning environment?
- What were the impacts of technology-mediated TBLT on their learning experience?

# 3 Context and Methodology

## *3.1 The Context of the Study*

The intention of this study was to explore the implementation of task-based language teaching in a web conferencing-based online setting, specifically in a beginners' online Chinese class. The web conferencing tool used in this study was Blackboard Collaborate, which includes features such as real-time audio, video, chat, interactive whiteboard, share screen, polling, emoticons, main room, and breakout rooms (Guo, 2013).

The study was conducted as a two-stage project. In the preliminary stage, participants attended two online sessions in which an information gap task and a jigsaw task were applied. Learners' feedback regarding their learning experiences were collected to improve the task design in this current study (Guo & Möllering, 2016). To further explore task design in multimodal learning environments, in this study, five online tasks—including two jigsaw tasks, one information-gap task, and two decision-making tasks—were designed and conducted with the participants.

Sixteen undergraduate students who enrolled in the second semester of an introductory Chinese language class participated in this study. They all attended the online sessions remotely using their PCs or laptops and a headset. Five of them (Students 1 to 4 and 9) had participated in the preliminary study. Before this study commenced, they had learned Chinese for one semester. In this study, the participants attended five one-hour online sessions throughout the semester. An online training session was conducted in a computer lab before the treatment started. In the training session, sample interactive tasks were assigned to participants in breakout rooms to familiarise them with the multimodal learning environment.

## *3.2 Task Design*

In this study, five fortnightly one-hour online sessions were conducted (see Table 1). The current study follows Pica et al.'s (1993) task typology. The five tasks included two jigsaw tasks, two decision-making tasks, and one information gap task. Since the information gap and jigsaw tasks were well received by students in the preliminary study, these were also implemented in this study. As learners' language abilities improved, decision-making tasks were also designed and introduced.

In the current study, the task design followed Willis's (1996a, 1996b, 1998) and Ellis's (2003) TBL framework, and it also took into account Hampel's (2006) task design framework in the audio-conferencing environment. In order to reinforce vocabulary and grammar learning and facilitate learners' collaborative learning, there were three stages in the one-hour online sessions (see Table 2).

**Table 1** Summary of tasks implemented in the current study

| | | Task type | Topic |
|---|---|---|---|
| 1 | Week 2 | Information gap | Applying for a Chinese visa |
| 2 | Week 4 | Decision-making | Buying clothes and sending them to China |
| 3 | Week 6 | Jigsaw task | Maps and showing directions |
| 4 | Week 10 | Decision-making | Planning for a trip |
| 5 | Week 12 | Jigsaw | Describing an accident |

**Table 2** Summary of task sequence and activities

| Stage | Room type | Activities | Duration |
|---|---|---|---|
| Pre-task | Main room | Warm-up | 20 minutes |
| Task | Breakout rooms | Tasks | 20 minutes |
| Reporting | Main room | Task presentation and teacher feedback | 20 minutes |

### *3.3 Data Collection*

This study employed mixed methods, which encompasses both qualitative and quantitative methods, to answer the three research questions. Three types of data resources were employed, including web conferencing archive collections, post-session interviews, and post-session questionnaires. All five online sessions were recorded using Blackboard Collaborate and Screenflow in order to capture learners' and their teacher's interactions in a range of small-group tasks. Learner/learner interaction in the process of task completion were transcribed for the purpose of discourse analysis to identify instances of collaborative learning. Post-session surveys and in-depth interviews were carried out after the fifth online session. In the interviews, relevant materials, such as the screenshot of the online sessions and task descriptions, were provided to the interviewees to help them recall the previous tasks conducted throughout the semester. Open-ended questions were designed in post-session surveys to elicit learners' perceptions on task design and the collaborative learning experience in the multimodal environment.

## 4 Results

### *4.1 Learners' Collaboration in the Online Sessions*

There were a great number of examples in the data which showed that collaborative learning took place both in pairs and in groups in the current study. In example 1, student 14 explicitly requested student 3′s help when she did not know how to say "study" and "apply" in the target language. When she could not finish her sentence, student 3 attempted to guess what she was about to say based on the information he had. Moreover, student 3 pointed out that student 14 should add "我要" in that sentence to express what she wanted to do. Although student 3 rendered his assistance in English, he successfully helped student 14 complete her sentence "我要去中国, 我要办签证。[I'm going to China. I want to apply for a visa.]".

Example 1. (the first online task).

| | |
|---|---|
| St 3: 你好,有什么可以帮您? | [Hello, how can I help you?] |
| St 14: How do you say study, apply?<br>St 3: 学习 and 办 | [Study, apply for] |
| St 14:我去中国学习, 办… | [I'm going to study in China. Apply…] |
| St 3: 签证? | [Visa] |
| St 14: Right 签证, 我办签证 | [Right, visa. I apply for a visa.] |
| St 3: I think you have to say I want to, like, 我要 | [I want to.] |
| St 14: Oh 我要去中国, 我要办签证。 | [I'm going to China. I want to apply for a visa.] |

In example 2, student 4 noticed that student 1 confused "天(day)" with "日(date)", so he asked her to clarify it. After that, he also corrected a grammar mistake in her sentence. On reviewing the current data, abundant examples of similar peer corrections were found. Part of the reason for this was because the participants were instructed by the teacher to help and provide correction to peers in the completion of the tasks. While observing the learners' group work, I consistently encouraged them to actively help their partners, which in turn fostered their own language and communication development.

Example 2. (the first online task).

| | |
|---|---|
| St 3: 出生年月日? | [Date of birth] |
| St 14: 1992年6月10天 | [10th June, 1992 (incorrect words for date)] |
| St 3: you mean 日? | [Day] |
| St 14: Yeah, How do I say this "I need to change money"? | |
| St 3: 换钱 | [Change money] |
| St 14: 我换钱 | [I change money.] |
| St 14: put 得, say 我得换钱 | [need to. I need to change money.] |
| St 3: Yeah, 我得换钱。 | [Yeah, I need to change money.] |

In example 3, students 5 and 6 were working on a whiteboard, showing directions on a campus map. Student 5 kept saying "往下边走 [go down]", which can be understood when using a map but is not appropriate in a face-to-face conversation. Although student 6 understood the instructions, he still elaborated the correct form of the expression to student 5. This type of negotiation was not triggered by a non-understanding or unknown lexical or syntactic item. However, throughout the collaboration, both the students' attention was drawn to language form, which was conducive to their SLA.

Example 3. (the third online session).

| | |
|---|---|
| St 5: 你往下边走。 | [You go down.] |
| St 6: 下边? | [Down?] |
| St 5: 下 | [Down.] |
| St 6: You mean down? | |
| St 5: Yeah, like back, down | |
| St 6: It's like 左, 前 | [Left, front] |
| St 5: like 你往下边走 | [You go down.] |
| St 6: You can say like 你往南走 | [You go south.] |
| St 5: Oh, ok, 你往南走, 往左拐, 教室在公园对面。 | [Ok, you go south, and then turn left. The classroom is opposite to the park.] |
| St 6: 教室. (Typing "教室"on the whiteboard) | [Classroom] |

In the third task, student 10's sentences "你往北走, 你往西走 [You go north, and go west]" were grammatically correct, but lacked conjunction words. Student 1 articulated the problem and suggested that student 10 use "再 [and then]" to link the two clauses. According to Smith (2003), "metalinguistic talk may prove helpful in uncovering the root of the problem" (p. 47), although it may divert time away from task completion. In this example, "先…再… [first…, and then…]" was the key grammar structure that the students learned in class. The negotiation process indicated that the acquisition of new learning did take place.

Example 4. (The third online session).

| | |
|---|---|
| St 10: 好。你参观图书馆 | [Ok. You're going to visit the library.] |
| St 1: 图书馆在哪儿? | [Where is the library?] |
| St 10: OK, 图书馆, 你往北走, 你往西走。 | [The library. You go north, go west.] |
| St 1: When you give the second direction, you need to say 再, like go again再往。 | [Then.] |
| St 10: Oh, so go first | |
| St 1: Yeah, so we start with 先往, and you want to give another command, 再往 direction go。 | [First, and then.] |
| St 10: Ok, 先往北走, 再往西走 | [Ok. Go north first, and then go west.] |

In addition, results from the interviews and post-session surveys showed that the implementation of tasks in the web conferencing-based online environment has great potential in stimulating collaborative learning.

In the post-session survey, participants were asked to choose which one they preferred between one-to-one and group tasks. Of the 16 participants, 14 preferred group work. Their reasons are shown in the following excerpts from the survey:

- Learning a language seems to be much more effective for me when doing it with a group. That way you can feed off the other students. It works very well in group situations.

- A group discussion is more preferable due to the possibility of creating a conversation in Chinese rather than a one-on-one, which may provide improvement in writing, reading and listening ability. Furthermore, a group discussion is less confronting, as you know the other students are at a similar level to you.

According to the participants' answers in the post-session survey, they believed that peer collaboration provided them with more opportunities for feedback and explanations. Peer interactions allowed them to support each other in a similar way. Working with other students helped to create a less pressured and more engaging environment, in which they felt less distracted and more willing to contribute to group discussion. In addition, group work seemed to contribute to creating a sense of community, in which the participants felt safe to share and help each other. As student 9 mentioned in the survey, having a partner that he could work with was conducive to his task completion:

> It's really great with [student 3], because I feel safe in a sense. I didn't feel a fool or anything. so maybe having someone that you do the tasks with all the time, it works. Like a buddy assisted [sic].

However, one student mentioned the disadvantage of group work; that is, having an unconfident, shy, or lower proficiency partner may not work as effectively as one-to-one (learner-teacher) interaction.

In this study as an instructor, I observed the learners' discussions in groups and occasionally intervened in their interaction when they needed technical or linguistic assistance. The degree of my intervention varied depending on the extent of the learners' participation and their achievement in the tasks. As Salmon (2003) states, the tutor's main role is to ensure "meaning making" rather than "content transmission" (p. 52). However, most of the time, I observed the learners only without intervention. Assistance was provided when it was requested by the learners or in situations when they were not able to resolve the problem by themselves.

## 4.2 Learners' Engagement in the Online Sessions

In terms of engagement, the majority of the participants believed that they felt engaged in the online sessions. Their feedback suggested that learners' participation increased when they were used to the online learning environment and when other participants' interactions made participation more appealing.

In the follow-up surveys, 11 out of the 16 students indicated "strongly agree" when asked if the tasks were engaging, while nine students also strongly agreed that they enjoyed doing the tasks with their peers. Results from the interviews showed that the participants enjoyed the last three sessions more as their language proficiency had gradually improved and they were more familiar with the multimodal environment. One student, for instance commented in the interview: "*I was particularly engaged talking in the last few sessions because I was more used to it*". Student 9 mentioned that he felt engaged doing the jigsaw since "*naturally you do one step, then the other person does the other step, then you swap. That was very good.*"

Moreover, in the post-task stage, all the groups were required to present their work in the breakout rooms, which made them pay more attention to the tasks. Student 12 commented in the interview: "*even [when] my task is finished and my presentation is finished in [the] online session, you still get to listen to other people's presentation[s] and you learn from them. So, it's useful in every way*". Student 2 mentioned that the teacher's questions after the presentation encouraged her and her partners to concentrate on the tasks, "*because we don't know when you're going to ask us questions. I always have to be listening*".

### *4.3 Learners' Perceptions on Task Design*

In the in-depth interviews, the participants were asked their preference regarding task types. The three types of tasks that were designed in this study all received positive feedback. Most of the participants believed that tasks requiring two ways of information exchange (jigsaw and information-gap tasks) were more straightforward. The decision-making task, on the other hand, were perceived to require more discussion on task planning.

- Decision-making is good, if you have a good partner and you are a team-working person, otherwise you'll just make the decision by yourself. Doesn't foster the communication well.
- The decision-making involves more work planning. I think the combination of all of them was good, but I particularly like the [task on giving] directions which is challenging.
- Information gap and jigsaw tasks: If you're both good learners, you know what you're doing and then it's [a] good way to foster communication.
- The jigsaw task is quite straightforward; we know what we have to do.

Moreover, in the interviews, learners' perceptions on task design confirmed that participation in the study was conducive to their target language learning and confidence building, which are summarised as follows:

- It created a less pressured environment to learn and practice the target language;
- Group work motivated the students to do more practice than individual study; and,
- The positive feedback and encouragement received from peers and the teacher made them feel more confident.

Student 10 commented in the interview: "*there is less pressure when you [are] in your room in front of [the] computer than in class*". Student 9 stressed that the positive feedback he received from the teacher and his partner was a key influence and played an important role in building his confidence in Chinese language learning:

*"Encouragement, exactly; you get that encouragement because obviously that feedback, that's sort of the presentation aspect. That's really good."* Student 3 confirmed that being able to complete a task without the teacher's assistance gave him a sense of achievement: *"It was positive feeling, because you just feel good about being able to complete the task, so if you complete with satisfaction, I guess, I can do it. Since you did it, you know you can do it, so you feel better because you can do the task"*. Student 8 stated: *"I learn better in groups. It's always good to have other people whom you can talk things through. You can ask questions just when everyone else is learning, like silly questions like "How do you say nine?' It's better just to ask someone next to you."*

# 5 Discussion and Conclusion

## *5.1 Collaborative Learning in Technology-Mediated TBLT*

Computer-supported collaborative learning (CSCL), derived from Vygotskyan cultural psychology, concerns learners' "collaborative learning" in a multimodal environment (Kirschner, 2002). Sociocultural theories stress the pivotal roles played by language and other tools, such as computers. Previous studies report that, compared to face-to-face interaction, collaboration supported by CMC is considered weak in social presence.

The findings in the current study prove that collaborative learning did occur in learner/learner interaction in group work. It takes place between a more competent learner and a lower proficiency learner. The findings in this study showed that when working together as a group, a more competent learner tended to help their partners by providing corrective feedback or even grammatical explanations, which echoes previous studies (e.g., Smith, 2003). It contributed to creating a less stressful learning environment and the participants felt more engaged when working with other students.

Secondly, when the students encountered breakdowns or problems that they could not resolve by themselves, it was important to have at least one teacher monitoring their interaction and providing timely assistance. The participants felt safe and comfortable having the teacher move around and liked to let the teacher know when they had language or technical issues.

## *5.2 Implications for Online Language Learning and Teaching*

To create an online collaborative learning environment, according to the findings of this study, language instructors need to consider the following aspects. Firstly, give students freedom to pair with others since some of them may have someone with whom they feel comfortable working. Secondly, teachers should act as moderators and need to keep an eye on students' interactions and provide assistance when necessary.

### 5.3 *Learners' Engagement and Task Design in Technology-Mediated TBLT*

According to the results, learners' participation in the tasks and task type were interrelated. The participants felt more engaged doing the tasks with their partners when they were able to use Chinese to communicate with others and were fully comfortable with the online environment and the interaction pattern.

### 5.4 *Implications for Online Language Learning and Teaching*

To design tasks that engage learners in technology-mediated TBLT requires language instructors to consider the following aspects. First of all, choose appropriate tasks that suit learners' language proficiency. Oxford (2006) states that "task-based teaching and learning potentially offer great riches if explored by teachers in their dual roles as instructor and action researcher" (p. 114). In the current study, as both teacher and researcher, I explored three different task types in the beginners' online Chinese task design. The results showed that the first-year students were more familiar with information gap and jigsaw tasks, which require producing only a certain outcome. However, tasks such as decision-making tasks, in which students can reach different outcomes, require relatively higher language proficiency. It may be challenging for beginners. Secondly, in the post-task stage, language teachers can ask students to present their work in groups. Similar to low-risk competition, students may feel less pressure in this scenario and hence are more likely to contribute more in their group work.

Moreover, the findings in the current research confirmed that the online tasks were conducive to learners' target language learning. To achieve such a goal, it may be useful for language teachers to consider implementing the following strategies. Firstly, creating a less pressured learning environment is vital for online language learning. Students, especially beginners, may feel more nervous at first when they attend online sessions. Working with students they already know or helping them become familiar with the tool can help alleviate their nervousness and boost their confidence. Secondly, providing more positive feedback and encouragement can make students, especially less competent learners, more willing to communicate and express themselves. Thirdly, teachers need to help familiarise students with the online learning environment as early as possible.

## 5.5 Contributions of this Study

This study aimed to explore the implementation of TBLT in a web conferencing-based beginners' online Chinese unit and to determine its influence on learners' collaboration and interaction. Moreover, it also shed light on how task design can stimulate peer-to-peer interaction to facilitate target language learning.

The findings confirmed the potential of technology-mediated TBLT for facilitating peer-to-peer collaboration and interaction. In other words, tasks designed in the web conferencing environment may provide learners with opportunities to modify their interaction when language breakdown takes place in conversation and in turn facilitates learners' SLA. Further, the study has notable implications on task design for online learning environments. However, since all the participants in the current study were on-campus students, a further study comparing the learning experiences of on-campus students and distance students may produce different results in terms of negotiation routines.

## References


Abrams, Z. I. (2016). Possibilities and challenges of learning German in a multimodal environment: A case study. *ReCALL, 28*(3), 343–363.

Berglund, T. Ö. (2009). Multimodal student interaction online: An ecological perspective. *ReCALL, 21*(2), 186–205.

Blake, R. (2000). Computer mediated communication: A window on L2 Spanish interlanguage. *Language Learning & Technology, 4*(1), 120–136. Retrieved from https://llt.msu.edu/vol4num1/blake/default.html.

Chapelle, C. (2001). *Computer applications in second language acquisition: Foundations for teaching, testing and research.* Cambridge, UK: Cambridge University Press.

Chapelle, C. (2003). *English language learning and technology: Lectures on teaching and research in the age of information and communication.* Amsterdam, The Netherlands: John Benjamins Publishing.

Ciekanski, M., & Chanier, T. (2008). Developing online multimodal verbal communication to enhance the writing process in an audio-graphic conferencing environment. *ReCALL, 20*(02), 162–182.

Doughty, C., & Pica, T. (1986). "Information-gap" task: Do they facilitate second language acquisition? *TESOL Quarterly, 20*(2), 305–325.

Doughty, C. J., & Long, M. H. (2003). Optimal psycholinguistic environments for distance foreign language learning. *Language Learning & Technology, 7*(3), 50–80.

Duran, G., & Ramaut, G. (2006). Tasks for absolute beginners and beyond: Developing and sequencing tasks at basic proficiency levels. In K. V. d. Branden (Ed.), *Task-based language education: From theory to practice* (pp. 47–75). Amsterdam, The Netherlands: John Benjamins Publishing.

Ellis, R. (2003). *Task-based language learning and teaching.* Oxford, UK: Oxford University Press.

Ellis, R. (2009). Task-based language teaching: Sorting out the misunderstandings. *International Journal of Applied Linguistics, 19,* 221–246.

Ellis, R., Tanaka, Y., & Yamazaki, A. (1994). Classroom interaction, comprehension, and the acquisition of L2 word meanings. *Language Learning, 44*(3), 449–491.

Gass, S. M., & Varonis, E. M. (1985). Task variation and nonnative/nonnative negotiation of meaning. In S. M. Gass & C. G. Madden (Eds.), *Input in second language acquisition* (pp. 149–161). Rowley, MA: Newbury House.

Guichon, N., & Cohen, C. (2014). The impact of the webcam on an online L2 interaction. *Canadian Modern Language Review/La Revue Canadienne Des Langues Vivantes, 70*(3), 331–354.

Guichon, N., & McLornan, S. (2008). The effects of multimodality on L2 learners: Implications for CALL resource design. *System, 36*(1), 85–93. https://doi.org/10.1016/j.system.2007.11.005

Guo, S. (2013). Applying web-conferencing in a beginners' Chinese class. In *Proceedings of the annual conference of ASCILITE, "Electric Dreams"* (pp. 345–349). Sydney, Australia: Australasian Society for Computers in Learning in Tertiary Education.

Guo, S., & Möllering, M. (2016). The implementation of task-based teaching in an online Chinese class through web conferencing. *System, 62,* 26–38. https://doi.org/10.1016/j.system.2016.07.003

Hampel, R. (2006). Rethinking task design for the digital age: A framework for language teaching and learning in a synchronous online environment. *ReCALL, 18,* 105–121.

Hampel, R. (2010). Task design for a virtual learning environment in a distance language course. In M. Thomas & H. Reinders (Eds.), *Task-basked language learning and teaching with technology* (pp. 131–153). Bloomsbury, UK: Bloomsbury Academic.

Hampel, R., & Hauck, M. (2004). Towards an effective use of audio conferencing in distance language courses. *Language Learning & Technology, 8*(1), 66–82.

Hampel, R., & Stickler, U. (2012). The use of videoconferencing to support multimodal interaction in an online language classroom. *ReCALL, 24*(02), 116–137.

Keller-Lally, A. (2006). *Effect of task-type and group size on foreign language learner output in synchronous computer-mediated communication (Unpublished doctoral dissertation).* Austin, TX: University of Texas.

Kirschner, P. (2002). Can we support CSCL? Educational, social and technological affordances for learning. In P. Kirschner (Ed.), *Three worlds of CSCL: Can we support CSCL* (pp. 7–34). Heerlen, The Netherlands: Open Universiteit Nederland.

Lai, C., & Li, G. (2011). Technology and task-based language teaching: A critical review. *CALICO Journal, 28*(2), 1–24.

Lai, C., Zhao, Y., & Wang, J. (2011). Task-based language teaching in online ab initio foreign language classrooms. *The Modern Language Journal, 95,* 81–103.

Levy, M., & Stockwell, G. (2013). *Call dimensions: Options and issues in computer-assisted language learning.* Oxon, UK: Routledge.

Lin, H. (2015). A meta-synthesis of empirical research on the effectiveness of computer-mediated communication (CMC) in SLA. *Language Learning & Technology, 19*(2), 85–117. Retrieved from https://llt.msu.edu/issues/june2015/lin.pdf.

Long, M. (1983). Native speaker/non-native speaker conversation and the negotiation of comprehensible input. *Applied Linguistics, 4*(2), 126–141.

Long, M. (1985). A role for instruction in second language acquisition: Task-based language training. In K. Hyltenstam & M. Pienemann (Eds.), *Modelling and assessing second language acquisition* (pp. 77–79). Clevedon, UK: Multilingual Matters.

Nunan, D. (2004). *Task-based language teaching.* Cambridge, UK: Cambridge University Press.

Ortega, L., & González-Lloret, M. (2014). *Technology-mediated TBLT: Researching technology and tasks.* Amsterdam, The Netherlands: John Benjamins Publishing.

Pica, T. (1994). Research on negotiation: What does it reveal about second language learning conditions, processes, and outcomes? *Language Learning, 44*(3), 491–527.

Pica, T., & Doughty, C. (1985). Non-native speaker interaction in the ESL classroom. In S. Gass & C. Madden (Eds.), *Input in second language acquisition* (pp. 115–132). Rowley, MA: Newbury House.

Pica, T., Holliday, L., Lewis, N., Berducci, D., & Newman, J. (1991). Language learning through interaction. *Studies in Second Language Acquisition, 13*(03), 343–376.

Pica, T., Kanagy, R., & Faloudon, J. (1993). Choosing and using communication tasks for second language instruction and research. In G. Crookes & S. M. Gass (Eds.), *Tasks and language learning: Integrating theory and practice* (pp. 9–34). Clevedon, UK: Multilingual Matters.

Pica, T., Lincoln-Porter, F., Paninos, D., & Linnell, J. (1996). Language learners' interaction: How does it address the input, output, and feedback needs of L2 learners? *TESOL Quarterly, 30*(1), 59–84. https://doi.org/10.2307/3587607

Rosell-Aguilar, F. (2005). Task design for audiographic conferencing: Promoting beginner oral interaction in distance language learning. *Computer Assisted Language Learning, 18*(5), 417–442.

Rouhshad, A., Wigglesworth, G., & Storch, N. (2016). The nature of negotiations in face-to-face versus computer-mediated communication in pair interactions. *Language Teaching Research, 20*(4), 514–534. https://doi.org/10.1177/1362168815584455

Salaberry, M. R. (2000). L2 morphosyntactic development in text-based computer-mediated communication. *Computer Assisted Language Learning, 13*(1), 5–27.

Salmon, G. (2003). *E-moderating: The key to teaching and learning online*. London, UK: Kogan Page.

Samuda, V., & Bygate, M. (2008). *Tasks in second language learning*. New York, NY: Palgrave Macmillan.

Smith, B. (2003). The use of communication strategies in computer-mediated communication. *System, 31*(1), 29–53.

Stickler, U., & Shi, L. (2013). Supporting Chinese speaking skills online. *System, 41*(1), 50–69.

Stockwell, G. (2010). Effect of multimodality in CMC tasks. In M. Thomas & H. Reinders (Eds.), *Task-based language learning and teaching with technology* (pp. 83–104). Bloomsbury, UK: Bloomsbury Academic.

Thomas, M. (2013). Task-based language teaching and CALL. In M. Thomas, H. Reinders, & M. Warschauer (Eds.), *Contemporary computer-assisted language learning* (pp. 341–358). London, UK: Continuum.

Van den Branden, K. (2006). *Task-based language education: From theory to practice*. Cambridge, UK: Cambridge University Press.

Vygotsky, L. S. (1978). *Mind in society: The development of higher psychological processes*. Cambridge, MA: Harvard University Press.

Wang, Y. (2004). Supporting synchronous distance language learning with desktop videoconferencing. *Language Learning & Technology, 8*(3), 90–121.

Wang, Y. (2006). Negotiation of meaning in desktop videoconferencing-supported distance language learning. *ReCALL, 18*(1), 122–145.

Wang, Y. (2008). *Distance language learning and desktop videoconferencing: A Chinese language case study*. Saarbrücken, Germany: VDM Verlag Dr. Müller.

Willis, J. (1996a). A flexible framework for task-based learning. In J. Willis & D. Willis (Eds.), *Challenge and change in language teaching* (pp. 52–62). Oxford, UK: Heinemann.

Willis, J. (1996b). *A framework for task-based learning*. Harlow, UK: Addison Wesley Longman.

Willis, J. (1998). Task-based learning? What kind of adventure? Retrieved from https://www.education.auckland.ac.nz/webdav/site/education/shared/about/centres/lipis/docs/Task-Based%20Learning-What%20Kind%20of%20Adventure.html.

Ziegler, N. (2016). Taking technology to task: Technology-mediated TBLT, performance, and production. *Annual Review of Applied Linguistics, 36*, 136–163.

**Sijia Guo** is a Lecturer in Chinese Studies, Department of International Studies at Macquarie University. Her PhD study was on online Chinese teaching in web-conferencing environments. Her research focuses on technology-enhanced language learning (TELL), technology-mediated task-based language teaching, and second language acquisition. She also published on captioned videos for L2 comprehension and design principles to enhance students' L2 acquisition in flipped classes.

# New Media in Education

# Learning in an Age of Participatory Culture: A Review of Informal Learning Through Social Media

**Jiahui Du and Yoo Kyung Chang**

**Abstract** Online participation through social media has been considered as having a great potential for promoting learning. However, the effectiveness depends on multiple factors. This paper identifies (1) the learner's motivation toward using social media to engage in informal learning and (2) the opportunities and challenges in implementing social media for informal learning through the review of 15 empirical articles. The review indicates that people are motivated to participate in learning based on their personal interests and experiences, the desire to share knowledge and exchange information. Their motivation is fundamental to the emergence of informal learning. Moreover, social media could facilitate informal learning as either personal learning or learning in a community. The considerations of this study can contribute to the current understanding of online informal learning and also provide implications to the future design of instructional technologies.

**Keywords** Participatory culture · Social media · Online participation · Informal learning

## 1 Introduction

### *1.1 Background of the Project*

Online participation is an important part of everyday life for much of the global population. Platforms such as social media allow users to make connections and interact through different functions such as sharing stories, communicating, and

J. Du (✉)
The University of Hong Kong, Hong Kong, China
e-mail: dujiah@connect.hku.hk

Y. K. Chang
Teachers College, Columbia University, New York, US
e-mail: ykc2110@tc.columbia.edu



W. W. K. Ma et al. (eds.), *Learning Environment and Design*, Educational Communications and Technology Yearbook,
https://doi.org/10.1007/978-981-15-8167-0_11

collaborating (Greenhow & Lewin, 2016). According to the Global Digital Report 2018, the number of social media users globally is 3.196 billion, which shows a 13% increase compared to last year (Chaffey, 2018). These online technologies simplify the procedure of communication and make it easier for people to create and share media content (Bers & Chau, 2010). Moreover, technology links the socially fragmented world together by gathering people in an imagined and virtual community (Ondrejka, 2008). Owing to its communicative and interactive design features, people can interact with one another by commenting on others' posts and simultaneously creating their own works to express themselves, carving paths to participatory culture. Such social nature of interaction encourages educators and researchers to consider the impact of social media on human communication learning.

### *1.2 Current and Potential of the Project*

Researchers have explored the learning potential of social media, mostly around the particular benefits that can be obtained from the use of social media, such as social skills in a participatory culture (Jenkins, 2006). However, there is a lack of a comprehensive understanding on how social media can support or challenge learning, such as the factors that may influence such effectiveness of informal learning. This paper identified the factors that can stimulate an individual's participation and learning in the online context, and the environment that social media provides for facilitating informal learning. The study uses social media as a representation of the participatory culture, and demonstrates the conditions, opportunities, and challenges of using social media for informal learning. Social constructionism is used as the theoretical framework to interpret the factors and their implications on learning.

This research is significant in three ways. First, with the rapid development of the technology, it is essential to understand the effects, opportunities, and limitations of social media on informal learning. Second, knowing the ways that social media could be supportive of informal learning could guide designers, educators, and researchers in supporting effective online participation. Finally, based on the characteristics of informal learning in social media, educators can think about the potential of implementing such traits into various educational activities. By reviewing previous literature, more questions will be raised in this area, and pathways will be provided for future research.

## 2 Definition of Key Terms

### *2.1 Participatory Culture*

Scholars argue that interactive media has gradually replaced traditional mass-media format (Gillespie, 2012). Massanari (2015) claims that this well-documented shift can be explained as the transfer of role played by the audience in the media production chain—from passive receptors to the active participants, who are able to create and produce their own media content. As what Jenkins (2006) indicated, the age of mass media in which some produce the content and many consume is moving away, and the age of participatory culture has come. In this new era, each contributor matters, and the chances to interact and express provide participants a sense of "social connection" (Bers & Chau, 2010).

Jenkins (2006) defined participatory culture as a low barrier engagement with informal mentorship which allows individuals to create and share their projects. In this paper, participatory culture appears as the context to study the opportunities and challenges that social media displays on promoting informal learning. Within this context, the conflicts between formal institutional learning and new participatory culture appear, which lead to the discussion of informal learning settings and the potential of learning within a social media context.

According to Bers and Chau (2010), the spectrum of participation ranges from socializing to creating. Some people are motivated to participate by socializing, such as communicating and interacting with others, while others participate actively based on their interests and desire to create and share their individual works or personal stories (Bers & Chau, 2010). The ranges of the spectrum indicate people's active participation in various activities, which is considered as one of the most important characteristics of participatory culture by Jenkins (2008).

### 2.1.1 Social Media and Participatory Culture

Social media, defined as a communication technology, offers powerful features that enable users to connect, share, or update their life and stories with others in the same online community (Greenhow & Lewin, 2016). A simple explanation of the relationship between social media and participatory culture is that social media is a representation and expression of participatory culture (Jenkins, 2008). Social media appears as a platform where people create and share the collective meaning by participating actively in online social activities (Jenkins, 2006). The unique and powerful features of social media effectively facilitate people's engagement by enriching the forms of participation, creating the connections with others in the community, and more (Greenhow & Lewin, 2016). Based on these arguments, this paper considers social media as a representation of participatory culture, which is used to study its potential of promoting informal learning in the later sections.

## 2.2 *Informal Learning*

### 2.2.1 Definition

Informal learning is generally described as an unstructured, experiential, and flexible learning activity (Tannenbaum et al., 2010). Some commonly used elements to define informal learning include location of learning, and purpose of learning (Green, 2004). In terms of location, learning experience happens outside formal institutions such as schools, while the purpose of learning refers to the fact that learners are intrinsically driven to learn (Green, 2004). However, due to the complex nature of learning and the diversity of conditions such as learning needs, interest, and preference that stimulate learning, it is difficult to find a shared rule that can best define informal learning (Green, 2004). In this study, informal learning is defined as the learning format that is directed and controlled by learners rather than educational institutions (Greenhow and Lewin, 2016). In particular, this paper focuses on informal learning facilitated by online participation through social media outside the school context.

### 2.2.2 Characteristics of Informal Learning

There are some defined characteristics of informal learning. First, to some extent, informal learning should reflect learners' active participation, which refers to the participation without any mandatory requirements or external requirement. Second, informal learning occurs not purely in an educational context, but also through action and doing (Tannenbaum et al., 2010). In other words, instead of being a luker, individual should demonstrate the participation through real action. Finally, informal learning is more likely to occur in a diverse, flexible, and learner-centered environment. Diverse and flexible settings are integrated with great opportunities provided for learners to explore whatever knowledge or information they are interested in, while learner-centered environment indicates the possibilities for learners to control their own learning activities (Park et al., 2011).

### 2.2.3 Significance

According to Chen and Bryer (2012), formal learning takes up only a small fraction throughout humans' lifelong learning. Moreover, this percentage will reduce gradually from twelfth grades to the graduate years. Therefore, as individuals grow up, informal learning becomes increasingly vital. As Park et al. (2011) explain, two-thirds of adults engage in intentional informal learning in their everyday life, outside traditional educational institutions. Research also finds that an incidental and spontaneous environment promotes people to learn significantly more as they get older compared to traditional settings (Park et al., 2011).

## 2.3 *Informal Learning, Social Media, and Participatory Culture*

The significance of studying informal learning through social media through the participatory culture context can be explained from two aspects. First, in the age of participatory culture, a more diverse and personalized learning environment is required by learners. However, traditional educational institutions have difficulty fulfilling this demand. This promotes learners and researchers to seek for other tools and learning environments that can help accomplish the learning goals. Second, based on the previous literature, social media demonstrates its potential in promoting informal learning. Additionally, the defined characteristic of informal learning provides availability of its emergence in a social and cultural context. The opportunities and challenges that are displayed make the study of particular ways that social media can be supportive for informal learning more worthwhile to be conducted.

### 2.3.1 Limitations of Formal Learning Settings

With worldwide connections increasing, diversity plays an important role in both life and workplace. In this case, the homogenous schooling experience fails to take advantage of what the heterogeneous society offers and cannot fulfill the demand of various learners (Ondrejka, 2008). Since the emergence of a new participatory culture, Internet-enabled media has become a part of people's everyday life (Deuze, 2009). The barriers, such as geographical locations, no longer exist. People from all over the world with different experiences and cultural backgrounds can share, play, and collaborate in an online environment. Many researchers consider this new participatory culture as ideal learning environments (Jenkins, 2006). With this in mind, schools have tried to apply this new participatory culture into their teaching in order to update their traditional teaching and learning systems. However, conflicts cannot be avoided when it comes to traditional education with new popular culture. Reports show that schools are slow to react and adopt the new participatory culture (Jenkins, 2006). In this case, the potential of taking advantage of participatory culture in afterschool programs or informal learning communities becomes worthy of researching.

### 2.3.2 Potential and Challenges of Informal Learning Through Social Media

The implementation of social media in education has long been discussed in various studies. Dabbagh and Kitsantas (2011) examined whether the use of social media, such as microblogs, could facilitate the learning process in informal settings. Study shows that informal learning occurs as people actively participating in social activities or meeting with new people and coming up with new ideas (Marsick & Volpe, 1999). However, not all kinds of social media or participation can promote informal learning. Chen and Bryer (2012) suggest that informal learning could occur only when social media and network services are used appropriately. The implementation of social media in formal learning shows some problems such as some instructors being reluctant to the use social media in their classrooms, while others adopt social media slowly in their teaching.

The defined characteristics of informal learning are conductive to social media which provides an open, diverse, and flexible environment that encourages people to interact and make connections with one another. However, as less restrictions and instructions are involved in the process of informal learning, learners might be easily distracted by various irrelevant contents on the social media. Challenges like this can also be seen readily. In this case, research on specific opportunities and challenges that social media have for promoting informal learning becomes extremely worthy of further study.

## 3 Inquiry Questions

By viewing previous studies, this review suggests the factors shaping people's online participation and learning process. The three questions that are used to formulate the study are as follows:

(1) What motivates people to engage in social media and the online informal learning?
(2) In what ways could social media facilitate informal learning through online participation?
(3) What are the challenges of using social media as an instructional tool for informal learning?

A comprehensive review of empirical articles was conducted to address the above questions.

## 4 Theoretical Framework

The theoretical framework that is used in this paper is social constructivism. The fundamental claim of social constructivism is that individuals' understanding, and knowledge of the world are mutually "co-constructed" (Schrader, 2015). According to social constructivism, knowledge is derived from social processes, such as daily interactions with others. Therefore, any kind of social interactions are vital for a social constructivist to study (Burr, 1995). Vygotsky (1978) first presented the sociocultural approach to learning. According to Vygotsky (1978), learning is a process of making meaning by integrating and sharing various perspectives or experiences within the community. In the course of interacting with others, the shared ideas, values, or even distinctions could promote learners to engage and think, and as a result it facilitates the growth of individuals' knowledge and understanding (Pritchard & Woollard, 2010).

Social constructivist theory can be useful for this paper in many ways. First, social constructivists consider learning as an active process that encourages learners to continuously explore and build their knowledge (Whitty & Anane, 2014). This argument corresponds with the defined characteristics of informal learning in this paper, which can be an effective support on interpreting the emergence of informal learning. Moreover, social constructivist theory demonstrates the great potential of stimulating informal learning within a participatory context, in that the process of co-constructing knowledge might occur during their active participation.

Finally, according to social constructivism, learning is a social and interactive process (Vygotsky, 1978), and virtual community is in fact a collection of these characteristics (Taylor, 2015). In this case, virtual community can be considered as "an implementation of social constructivism approach to knowledge" (Whitty & Anane, 2014, p.1). Social constructivist theory is important to support the categorization and arguments that were made in later sections, which helps demonstrate the potential of social media for supporting informal learning.

## 5 Literature Search Strategy

This review examines 15 empirical articles since 2014, and these articles are selected based on their relevance to informal learning in social media context. The key words that are used for selection include informal learning, social media, and participatory culture. According to Fink's (2014) two screen selection criteria, 30 articles were initially selected from Google Scholar database based on the first practical screening criteria of data, content, languages, and so on. Among the initially selected articles, each one is examined based on its methodological quality, which indicates the quality of the research design and sample (Fink, 2014). Finally, 15 articles were chosen to be used to analyze different perspectives of social media and informal learning (see Appendix A for more information on the titles, authors, purposes, and findings of each article). The literature review is developed based on three questions, and the arguments as well as findings in this literature will be used to answer these questions. Following a synthesis, there is a discussion section of all the selected literature works, which focuses on the contributions of these works to the understanding in the field, arguments, and considerations that can be concluded after reviewing the articles, and the answers to the three inquiry questions will be discussed.

# 6 Considerations for Participatory Learning Through Social Media

Social media has been defined as personal communication technology which emphasizes people's ability to share and cooperate (Fuchs, 2014). In this case, social media breaks through the limitations of time and space in communication and provides users a place to engage in online social activities. Social constructivism identifies learning as a "social product", which occurs in a social context via different forms of online participation such as communicating, sharing, and other forms of social interactions (Woo & Reeves, 2007). The characteristics of both social media and learning demonstrate the potential of using social media for informal learning.

## *6.1 What Motivates People to Engage in Social and the Online Informal Learning?*

Motivations are important to study in that most of the online informal learning occurs through people's self-motivated participation. Active participation is important for both participatory culture and informal learning, and is also conceptualized by social constructivist theory, which demonstrates that learning is an "active process of making meaning" (Luo & Clifton, 2017). The literature suggests that the motivations can be explained from various aspects using models.

### 6.1.1 Personal Interest and Experience

Lundgren and Crippen (2009) identify the personal context as a component of Contextual Model of Learning (CMoL). It suggests that one of the most acknowledged motivations for people to participate and learn in the online settings is their personal interests (Lundgren & Crippen, 2009). These self-motivated users visit or use social media purposefully, and their engagement in the informal online settings depends on personal interest or experience. Sackey et al. (2015) analyze from the nature of informal learning, which indicates that learners' interests and excitement are the primary factors driving informal learning. In other words, a user's initial interest or previous experience might lead them to engage in the online content or to use a particular online platform. This statement can be used to explain Lundgren and Crippen's (2009) argument that people's willingness to actively engage in the online informal learning settings is driven by their initial motivations.

### 6.1.2 Knowledge Sharing and Information

Gilbert (2016) focuses on a Twitter-based community and studies members' motivation on participating in an online community. Findings indicate that knowledge sharing is a fundamental motivation for people to participate in the online community via social media (Gilbert, 2016). People are motivated to join the community and voluntarily share knowledge with others or learn from what other people share. Zhang et al. (2016) also claim that exchanging information is a primary motivation for users to engage in social media. From another perspective, according to Brewer (2007), dating back to the ancient times, humans used to cooperate with one another for survival. As time passed, they created dependence on sharing information and collecting knowledge. In this case, except for gaining information from others, humans tend to share what they know with their peers as well. This motivation also demonstrates great potential of promoting informal learning supports by social constructivism, which indicates that the process of sharing and exchanging information is in fact a process of knowledge co-construction (Schrader, 2015).

### 6.1.3 Frequently Used Model in Research

Three of the selected articles mention the Technology Acceptance Model (TAM) while explaining people's motivations for participating in online informal learning. This model could be used to explain the growing trend of using social media for online participation and informal learning from a user's perspective. TAM is an information system theory which demonstrates the "learner's motivation for technology acceptance" (Sarwar et al., 2019). Al-Rahmi and Zeki (2017) claim that the usefulness and ease of use are the two main aspects to answer the question about people's new technology adoption choices. TAM is widely used in research for explaining the people's motivation toward using social media and other technologies, and it is applicable to analyze the problem of people's acceptance or rejection on choosing or using a particular technology (Al-Rahmi & Zeki, 2017).

## *6.2 In What Ways Could Social Media Facilitate Informal Learning Through Online Participation?*

The next step is to study how these online activities could support informal learning. While people are participating in the social media context, what learning environments or advantages that social media offer display the potential of facilitating informal learning?

### 6.2.1 Personal Learning Environment

A Personal Learning Environment (PLE) demonstrates the availability for learners to direct their learning via using various social media tools (Whitty & Anane, 2014). Within the PLE environment, learners can choose their preferred learning tools or features in order to meet their demands. The personalization of PLE makes it inherently related to informal learning (Whitty & Anane, 2014). Sarwar et al. (2019) indicate that compared with the traditional institutional learning modes, one of the biggest differences in informal learning is that it enables learners to control learning themselves. Learner's control is one of the important aspects of informal learning according to Song and Lee (2014), which includes the control on learning goals, management of learning plans, the environment, and so on (Song & Lee, 2014). Social media makes it possible for individuals to create or acquire information and knowledge, and it also provides the opportunities to learn through various activities and communications. In other words, the PLE created by social media enables learners to control their informal learning activities (Song & Lee, 2014). Therefore, social media shows its great potential of facilitating informal learning by providing this flexible and diverse online environment for users.

### 6.2.2 Learning Within a Community

A virtual community in social media has been considered as having a great potential for promoting informal learning. According to Chunngam et al. (2014), a virtual community enables individuals to build and share their knowledge informally in terms of its virtual environment. Informal learning within a community is also supported by social constructivist theory, which states that the society creates the informal learning environment in the way that it enables people within the community to communicate, share, and interact with whoever they want (Woo, 2007).

There are two ways to learn within a community. First, learning can be gained through participation. Gilbert (2016) demonstrates that social media such as Twitter provides users with a comfortable place and egalitarian environment to make contributions. Participants are able to let their voice be heard through participation, which greatly encourages their activeness to contribute. Twitter and other social media platforms afford various degrees of participation and also enable the "contribution" that everyone makes to be seen and shared (Gilbert, 2016). From the social constructivism perspective, participating in the activity and getting connected with others indicate great potential of co-constructing the knowledge so that learning can occur during the experience (Hammond, 2001).

Second, people can learn through interaction within a community. Social constructivism believes that knowledge is derived from daily interactions and other social processes (Burr, 1995), such that any form of social interaction is worthy of being better explored (Pritchard, 2010). As Song and Lee (2014) indicate, social media supports creation and dissemination of knowledge among various online communities. As the design features of social media and the forms of participation become more diverse, using such technologies for social interaction is considered to have an increasing potential for supporting informal learning (Song & Lee, 2014).

Lange (2019) studies how YouTube supports informal learning. In his study, informal learning occurs through interaction. According to Lange (2019), video makers and viewers can learn through interacting with each other by exchanging feedback. For video makers, they can learn about viewers' thoughts on their videos and how to improve the quality of videos from different perspectives in order to fulfill the demands of different viewers. This process of reproduction also indicates the emergence of social constructivist learning, in that personal understanding and the knowledge they gained from others are mutually co-constructed (Schrader, 2015). The process for viewers to review, comment, and critique could also facilitate informal learning within YouTube. These interactions between makers and viewers are observed as a way of constituting productive online participation (Lange, 2019). Social constructivism demonstrates that individuals' knowledge and understanding is improved through interaction such as sharing ideas and interpreting distinctions, which encourages learners to engage and think from different perspectives (Pritchard, 2010).

### 6.2.3 Frequently Used Model in Research

Social media platforms enable people to contribute knowledge and exchange information quickly and easily by sharing and interacting with one another, and it also supports traditional engagement models such as Communities of Practice (CoPs) (Gilbert, 2016). This model was used in three of the selected articles. CoPs describe a community of people who share an interest in common and gather to learn more about that interest (Gilbert, 2016). Within this community, people are motivated by shared interests and willingness to exchange best practices and experiences (Chunngam et al., 2014). Norman et al. (2015) also points out how CoPs could support informal learning in a social media context. The previous members develop a "share knowledge bank", which includes the content such as learning experience and methods. In this case, the "knowledge flow" is supported between the members within the community, and meanwhile members are able to engage with other learners in the community by interacting or working collaboratively such as completing tasks in the community (Chunngam et al., 2014).

## *6.3 What Are the Potential Challenges of Using Social Media for Informal Learning?*

Although social media has demonstrated its great potential to support informal learning, the challenges for both learners and educators who are interested in learning or researching in the informal settings still exist. According to the 15 articles, the challenges for informal learning through social media can be categorized into the following four aspects.

i. **Cybersecurity**

While the potential of social media for facilitating online informal learning settings attracts more users to participate, one of the biggest challenges is how users can participate and learn safely within this environment. Sarwar et al. (2019) claim that with the increasing use of social media nowadays, the problem of cybercrime is on the rise. Issues such as cyberbullying and hacking servers might have a great influence on users while participating in online activities. Shen et al. (2017) also demonstrate the concerns for an individual's safety and privacy while using social media to participate in the online learning environment. Whitty and Anane (2014) point out that the obstacles for users seeking to adopt social media as their learning tool include the lack of privacy as well as confidentiality. These weaknesses of online participation would lower learners' motivation to participate, and would affect their learning experiences as well.

### 6.3.1 Content and Design

The unprofessional characteristic of learning through social media can be explained from two perspectives. Sackey et al. (2015) discuss that the particular challenge for an informal learning designer is to figure out how to provide the learning experiences which could fulfill the different demands of various learners. As Lundgren and Crippen (2009) argue, some social media platforms are designed for marketing purposes, and their strategies are marketing related instead of learning related. In this case, the activities and communities that are provided via this platform might have little impact on contributing to the users' knowledge construction. From the user's perspective, their challenge is how to deal with all kinds of information they encounter online. According to Lange (2019), there are lots of videos that lack quality in YouTube, and learners need to be aware of not being affected by these contents. Similar to Lange (2019), Kind and Evans (2015) demonstrate the challenges for learners to differentiate the online content—what are the professional and credible content versus the nonprofessional ones.

### 6.3.2 Information and Communication

The rapid development of technologies brings convenience but also challenges for people's lives and learning experiences. According to Zhang et al. (2016), some people indicate that they have difficulty with following information in real time which makes them feel confused while using social media. Luo and Franklin (2015) argue that based on their study, some learners find it difficult to capture the "central convergence" of the information flow in social media. The conversations and information flow easily anywhere, which improves the difficulty for learners to catch up. Additionally, due to the lack of face-to-face communication, misunderstandings might easily occur in either communicating with others or acquiring the specific content (Zhang et al., 2016).

### 6.3.3 Entertainment

As more features and functions are being added to social media, the challenges for using social media to learn become more serious than ever before. As what Sarwar et al. (2019) mention, study implies many students still consider social media as mainly a source of entertainment, which makes the boundary between play and learning unclear. Besides the purpose of using social media, some students take advantage of the convenience and rich resources of social media to find answers without devoting time to think or discuss with their peers (Sarwar et al., 2019). Moreover, the various kinds of information and designs can easily distract learners' attention no matter what activity they are participating in. As a challenge for both educators or developers and learners, educators and social media developers need to think about how to ensure the users are motivated by particular contents instead of purely leisure online activities, and learners have to remind themselves of their learning purpose and to concentrate during their learning experiences (Whitty & Anane, 2014).

## 7 Conclusion

Traditional institutional learning is facing challenges to keep up with the new culture, tools, and learners' various needs. In this case, informal learning shows its great potential of adopting the new culture because of its characteristics. As a medium of participatory culture, social media is conductive to online participation and informal learning. The paper explored the potential of social media for informal learning by addressing three key questions through a review of 15 articles. Considerations in the previous section reveal that people are motivated to participate and engage in the informal learning settings because of their personal interest and experience, the desire to share knowledge and exchange information. Meanwhile, social media promotes informal learning in the way that it creates personal learning environments and the virtual community, which include various features and activities that can potentially facilitate the emergence of informal learning demonstrated by the social constructivist theory. Finally, the challenges of using social media for learning also suggest that critical and comprehensive understanding is needed for both educators and learners.

## 7.1 Limitation and Future of the Project

This study explores the ways that social media can support or challenge informal learning. The relationship between characteristics of social media and learning can guide future design of informal learning using social media. However, there are limitations in this study. The number of literature articles that were examined in this paper is limited. Thus, the considerations and arguments may not be sufficiently generalizable. Future studies could be more precise on researching the features of social media and the variables that affect informal learning.

# Appendix

Table 1

Social Media and Informal Learning Studies for Literature Review

| Title | Authors | Purpose | Findings |
|---|---|---|---|
| Acceptance of social media fellow groups for learning: extension of technology assessment model (TAM) | Nawaz, Shakeel, Nawaz, Hamza (2017) | To explore the acceptance of fellow groups among students with the help of technology assessment model (TAM) (p.1). | This research shows that student's perception about usefulness of fellow groups has positive impact on attitude towards the use (p.5). |
| A model of using social media for collaborative learning to enhance learners' performance on learning | Al-Rahmi & Zeki (2017) | To explore the impact of several factors on collaborative learning and students' satisfaction which lead to a better learners' performance (p.9). | The results highlighted that both collaborative learning and students' satisfaction have a positive influence on learners' performance. It also revealed the high satisfaction by students using social media enhances collaborative learning which leads to a better performance (p.9). |
| Analysis of Social Media Influencers and Trends on Online and Mobile Learning | Shen, Kuo & Ly (2017) | To investigate the social media messages related to online or mobile learning, find insights in the public perception about online and mobile learning and response to enhance participants' e-learning experience (p.2). | • The discussion about online learning was more popular than mobile learning on Twitter (p.13).<br>• Twitter users who concerned about mobile learning were more enthusiastic on reposting related information than the ones who concerned about online learning (p.14). |
| Constructing learning spaces: what we can learn from studies of informal learning online | Sackeya, Nguyenb & Grabill (2015) | To study the conditions that are necessary for learning to occur in online spaces, and the best practices associated with effective learning these environments (p.2). | Technologies can act as facilitators. How people made use of the digital infrastructure to promote participants engaged in sharing experiences and objects, making their ideas public, and were attentive and accountable to each other matters. Facilitation is what made these conditions possible (p.11). |
| Exploring the Roles of Social Participation in Mobile Social Media Learning: A Social Network Analysis | Norman, Nordin, Din, Ally, & Dogan (2015) | To investigate roles of social participation in mobile social media learning. Focused on investigating participation of learning in mobile social media where learners are mobile and use social media on mobile devices (p.1). | Findings have revealed that role shifts of social participation can occur where student can become more central and active in learning discussion or become more less central and passive the discussion (p.16). |
| Has Web2.0 revitalized informal learning? The relationship between Web2.0 and informal learning | Song & Lee (2014) | To investigate the relationship between Web 2.0 levels and the evaluation of informal learning (p.1). | • Current informal learning websites have moderately adopted the most heavily promoted features of Web 2.0 (p.15).<br>• Correlation analyses showed a positive relationship between Web 2.0 features and informal learning website ratings (p.15). |

(Table 1 continued)

| | | | |
|---|---|---|---|
| Informal learning on YouTube | Lange (2019) | To study what forms of learning could occur on YouTube and how people learning through their participation (p.1). | YouTube's open-ended structure and the vast array of its content matter encourage self-directed exploration and a sense of sometimes unexpected discovery, which facilitates learning on the basis of shared interests and passions (p.2). |
| Learning and the practice of social media in informal science education centers | Lundgren & Crippen (2019) | Analysis the social media practices across informal science learning centers through the lens of the Contextual Model of Learning (p.31). | Social media sites are essential for informal learning, but some science learning centers use marketing-related strategies for social media as opposed to learning related strategies (p.49). |
| Learning in a Twitter-based community of practice: an exploration of knowledge exchange as a motivation for participation in #hcsmca | Gilbert (2016) | Exploring the role of learning in #hcsmca. Specifically, to examine learning as a driver of participation in the community, how the platform supports learning, and what aspects of learning are most important to community members (p.3). | • Twitter was found to be a platform that enables learning for #hcsmca members, albeit not universally. Four thematic elements of learning in #hcsmca were identified (p.2).<br>• For interviewees, learning was a motivation for participating in the community (p.2). |
| Membership, participation and knowledge building in virtual communities for informal learning | Chunngam, Chanchalor & Murphy (2014) | To identify how members' interest in the domain or subject combined with access to expert knowledge play a role in promoting membership, participation and knowledge building in a virtual community for informal learning (p.2). | Members' access to expert knowledge positively influences participation and knowledge building in a virtual community for informal learning; interest in the subject of the community positively influences membership, participation and knowledge building in a virtual community for informal learning (p.1). |
| Social media for lifelong learning | Kind & Evans (2015) | To provide a review of the use of social media for lifelong learning (p.1). | Engagement with social media can parallel engagement in the learning process over time, to the extent that online social networking fosters feedback and collaboration (p.1). |
| Social network enhancement for non-formal learning | Whitty & Anane (2014) | To study the informal interactions of social networks, the structure of learning programme and the multiple facets of collaborative approaches, in order to enhance student engagement (p.1). | The potential of social network can be explained based on the integrated framework, which could support informal learning and facilitate knowledge creation and sharing through socialization, externalization and combination (p.1). |
| The role of social media in collaborative learning: a coordination perspective | Zhang, Chen, Sun & Wang (2016) | This study aims at investigating the effects of social media activities on teamwork outcomes through the process of coordination (p.8). | • The information processing activities use does not significantly affect communication quality (p.8).<br>• Social media communication is simpler, so misunderstandings often occur (p.8). |

(Table 1 continued)

| | | | |
|---|---|---|---|
| Tweeting and blogging – moving towards education 2.0 | Luo & Franklin (2015) | To bridge the gap between the conceptual research and the empirical studies to help scholars better understand this emergent agenda while shedding light on this area (p.3). | Almost all of the students (11 out of 12) in the survey and interviews believed in the considerable potential of incorporating social media, either Twitter or blog, into educational settings (p.27). |
| Usage of social media tools for collaborative learning: the effect on learning success with the moderating role of cyberbullying | Sarwar, Zulfiqar, Aziz & Chandia (2019) | Using Technology Acceptance Model (TAM) and Constructivism Theory to study the usage of social media sites for students' collaborative learning. Explore the impact of cyberbullying for student's learning (p.4) | • Positive relationship between Perceived usefulness and collaborative learning; Perceived ease of use doesn't affect learning attitude directly; Negative relationship between Perceived enjoyment and collaborative learning. (p.21).<br>• Social media site encourages collaborative learning to promote the development of learning environment (p.23). |

*Note:* this form provides the title, authors, purpose and the findings of the study. All 15 literature articles are included in this form. And they are arranged in an alphabetic order. All the purposes and findings are cited by paper numbers.

## References


Al-Rahmi, ssW. M., & Zeki, A. M. (2017). A model of using social media for collaborative learning to enhance learners' performance on learning. *Journal of King Saud University—Computer and Information Sciences, 29,* 526–535.

Bers, M. U., & Chau, C. (2010). New media and technology: Youth as content creators. *New Directions for Youth Development,* 78–87.

Brewer, M. B. (2007). The importance of being we: Human nature and intergroup relations. *American Psychologist,* 728–738.

Burr, V. (1995). *An introduction to social constructionism.* Routledge.

Cacioppo, J. T., & Patrick, W. (2008). Loneliness: Human Nature and the Need for Social Connection. *Am J Psychiatry*.

Chaffey, D. (2018). Global social media research summary 2018. *Smart Insights*. Retrieved March 28, 2018.

Chen, B., & Bryer, T. (2012). Investigating instructional strategies for using social media in formal and informal learning. *The International Review of Research in Open and Distributed Learning, 13*(1).

Chunngam, B., Chanchalor, S., & Murphy, E. (2014). Membership, participation and knowledge building in virtual communities for informal learning. *British Journal of Educational Technology, 45*(5), 863–879. https://doi.org/10.1111/bjet.12114

Deuze, M. (2009). *Corporate appropriation of participatory culture* (pp. 27–40). Participation and Media Production: Critical Reflections on Content Creation.

Fink, A. (2014). *Conducting research literature reviews: From the internet to paper* (4th ed.). Sage.

Fuchs, C. (2014). What is social media. *Social media: A critical introduction what is social media?* (pp. 31–51). London: SAGE Publications.

Gilbert, S. (2016). Learning in a Twitter-based community of practice: An exploration of knowledge exchange as a motivation for participation in #hcsmca. *Information, Communication and Society, 19,* 1214–1232.

Gillespie, R. (2012). The art of criticism in the age of interactive technology: Critics, participatory culture, and the avant-garde. *International Journal of Communication, 6,* 56–75.

Green, J. S. (2004). *Literature review in informal learning with technology outside school*. Futurelab series (Report 7). Futurelab, London, UK.

Greenhow, C., & Lewin, C. (2016). Social media and education: Reconceptualizing the boundaries of formal and informal learning. *Learning, Media and Technology, 41*(1), 6–30. https://doi.org/10.1080/17439884.2015.1064954

Hammond, L. D., Austin, K., Orcutt, S., & Rosso, J. (2001). How people learn: Introduction to learning theories. *The Learning Classroom: Theory into Practice A Telecourse for Teacher Education and Professional Development*.https://web.stanford.edu/class/ed269/hplintrochapter.pdf

Jenkins, H. (2008). *Convergence culture*. New York: New York University Press.

Jenkins, H. (2006). *Confronting the challenges of participatory culture: Media education for the 21st Century*. The MIT Press.

Kind, T., & Evans, Y. (2015). Social media for lifelong learning. *International Review of Psychiatry, 27,* 124–132. https://doi.org/10.3109/09540261.2014.990421

Lange, P. G. (2019). Informal learning on YouTube. *The International Encyclopedia of Media Literacy*. https://doi.org/10.1002/9781118978238.ieml0090

Lundgren, L., & Crippen, K. (2009). Learning and the practice of social media in informal science education centers. *International Jl. On E-Learning, 1,* 31–52.

Luo, T., & Franklin, T. (2015). Tweeting and blogging: Moving towards education 2.0. *STEMPS Faculty Publications, 18*.

Luo, T., & Clifton, L. (2017). Examining collaborative knowledge construction in microblogging-based learning environment. *Journal of Information Technology Education: Research, 16,* 365–390.

Marsick, V. J., & Volpe, M. (1999). Chapter 1 the nature and need for informal learning. *Informal Learning on the Job, 1*(3), 1–9.

Massanari, A. L. (2015). *Participatory Culture, Community, and Play: Learning from Raddit*. Peter Lang.

Norman, H., Nordin, N., Din, R., Ally, M., & Dogan, H. (2015). Exploring the roles of social participation in mobile social media learning: A social network analysis. *International Review of Research in Open and Distributed Learning, 16*.

Ondrejka, C. (2008). Education unleashed: Participatory culture, education, and innovation in second life. *The Ecology of Games: Connecting Youth, Games, and Learning, 229–252*. https://doi.org/10.1162/dmal.9780262693646.229

Park, Y., Heo, G. M., & Lee, R. (2011). Blogging for informal learning: Analyzing bloggers' perceptions using learning perspective. *Educational Technology & Society, 14*(2), 149–160.
Pritchard, A., & Woollard, J. (2010). The precepts of social constructivism and social learning theory. In *Psychology for the classroom: Constructivism and social learning* (pp. 34–43). Routledge.
Sackey, D. J., Nguyen, M. T., & Grabill, J. T. (2015). Constructing learning spaces: What we can learn from studies of informal learning online. *Computers and Composition, 35*, 112–124.
Sarwar, B., Zulfiqar, S., Aziz, S., & Chandia, K. E. (2019). Usage of social media tools for collaborative learning: The effect on learning success with the moderating role of cyberbullying. *Journal of Educational Computing, 57*(1), 246–279. https://doi.org/10.1177/0735633117748415
Schrader, D. E. (2015). Constructivism and learning in the age of social media: Changing minds and learning communities. In *Constructivism Reconsidered in the Age of Social Media: New Directions for Teaching and Learning, Number 144* (1st ed.).
Shen, C. W., Kuo, C. J., & Ly, P. T. (2017). Analysis of social media influencers and trends on online and mobile learning. *International Review of Research in Open and Distributed Learning, 18*.
Song, D., & Lee, J. (2014). Has Web 2.0 revitalized informal learning? The relationship between Web 2.0 and informal learning. *Journal of Computer Assisted Learning, 30*, 511–533.
Tannenbaum, S. I., Beard, R. L., McNall, L. A., & Salas, E. (2010). Informal learning and development in organizations. In S. W. J. Kozlowski & E. Salas (Eds.), *SIOP organizational frontiers series. Learning, training, and development in organizations* (p. 303–331). Routledge/Taylor & Francis Group.
Taylor, M. (2015). Leveraging social media for instructional goals: Status, possibilities, and concerns. In C. Stabile & J. Ershler (Eds.), *Constructivism reconsidered in the age of social media: New directions for teaching and learning*. Wiley.
Vygotsky, L. (1978). *Mind in society*. London: Harvard University Press.
Whitty, C., & Anane, R. (2014). *Social network enhancement for non-formal learning*. IEEE Computer Society.
Woo, Y., & Reeves, T. C. (2007). Meaningful interaction in web-based learning: A social constructivist interpretation. *Internet and Higher Education, 10*, 15–25.
Zhang, X., Chen, H., Sun, Y., & Wang, N. (2016). The Role of Social Media in Collaborative Learning: A Coordination Perspective. *Wuhan International Conference on E-Business 2016 Porceedings*. https://core.ac.uk/download/pdf/301368636.pdf

**Jiahui Du** is a PhD student from the Faculty of Education, in the Academic Unit of Human Communication, Development, and Information Sciences (CDIS) of HKU. Jiahui received her bachelor's degree in journalism from Indiana University, Bloomington. Then Jiahui went to Teachers College, Columbia University, for master study, specializing in instructional technology and media. Jiahui's current research interest is on using technologies to promote students' learning performance, with a goal to help educators and tech developers design better educational technologies, and assist students to choose and use technologies scientifically for their learning experience.

**Yoo Kyung Chang** Ph.D. is a lecturer at the program of Communication, Media, and Learning Technologies Design, Teachers College. Her research and professional expertise lies in the design and development of technology to support and understand learning and human development. Her current research focuses on the cognitive, affective, and metacognitive implications of data-driven design in a diverse context ranging from online learning to media consumption behavior. Previously, Yoo Kyung Chang engaged in the design and research of playful science learning at New York Hall of Science, interactive simulation-based science learning, and formal and informal medical education.

# Advancing Digital and Social Inclusion: Developing Intergenerational Learning Through Digital Creativity Program—A Pilot Case Study

**Yun Yi Tan and Nabilah Abd Raman**

**Abstract** The aging society's population around the world is continuing to increase due to the advance in technology and healthcare. This article outlines an initiative to address the digital inclusion for the Malaysian aging society through an intergenerational learning approach and describe the methodology used to develop the collaborative program. The initiative is a unique collaboration between a public and private university and also experts from the industry in Malaysia, in which seven modules related to digital creativity and literacy, are presented through a face-to-face mode and mentoring partnerships between the young and old. The modules are developed with considerations of learning strategies preferred by older individuals and the local cultural context. This study conducted a qualitative approach in exploring to what extent can the program enhance digital literacy of the aging society through digital creativity learning. Suggestions for improvement and implications for promoting this initiative are also discussed.

**Keywords** Intergenerational interaction · Digital literacy · Digital creativity · Lifelong learning · Educational gerontology

## 1 Introduction

The aging society consists of people who are at the age of 60 and above. The people within the aging society are usually referred to as senior citizens and most of them are the retirees. The United Nations' World Population Prospects (2017) reported that the total number of elderly is 962 million and compromises 13% of the world's

Y. Y. Tan (✉)
Universiti Sains Malaysia, Centre for Instructional Technology and Multimedia, 11800 Gelugor, Penang, Malaysia
e-mail: yunyi.tan@usm.my

N. A. Raman
Sekolah Menengah Kebangsaan Taman Selesa Jaya, 81300 Skudai, Johor, Malaysia
e-mail: nabilah.abdraman94@gmail.com



W. W. K. Ma et al. (eds.), *Learning Environment and Design*, Educational Communications and Technology Yearbook,
https://doi.org/10.1007/978-981-15-8167-0_12

population. As life expectancy of human increases along with the advance of the technology, the number of senior citizens population is to be more than double by 2050. Since the population of the aging society will keep growing within the upcoming years, their rights and needs in many aspects such as healthcare, technology, income security and social service should be thoroughly considered.

In Malaysia, senior citizens are classified as those aged 60 and older. The current population of older Malaysians aged 65 and above is expected to double from 2.4 million to 5.5 million between 2020 and 2030. According to the 2018 Internet User Survey conducted by the Malaysian Communication and Multimedia Commission (MCMC), it is estimated that 6.5% of the Malaysian population who uses internet are older adults. The survey also indicated that the Malaysian older adults are gradually adjusting to the use of internet and digital platforms as the number of older adults using the internet has increased, compared with only 2.6% in 2016. In 2019, the ICT Use and Access by Individuals and Household Survey conducted by the Department of Statistics Malaysia reported that the percentage of internet access has increased to 90.1% compared with 87% in 2018, with participating in social network as the highest online activity.

The advent of the internet has provided access to resources and services at our fingertips. However, is everyone being given equal opportunities to take up such technology? As more service providers shifting their businesses from offline to online, the number of older adults in need of assistance to learn technology is also rising. In fact, the majority of older adults needed support to help them learn how to adapt technology to their needs (Anderson & Perrin, 2017). The gap faced by older adults is more toward the second-level digital divide where not only access to technology should be considered but also on the skills in utilizing technology (Springate, Atkinson, & Martin, 2008). Insufficient digital skills is the most significant factor that affects technology usage, which refers to the second-level digital divide (McDonough, 2016). As a result of the lack of necessary technological knowledge and skills, elderly users are not able to use digital applications persistently (Yang & Huang, 2015). Although the middle-aged adults are likely to be more digitally aware as time goes by, there would still be the need for continuous training to maintain digital literacy and keep pace with the ever-evolving technological advances (Rusdi, Sahari, & Noor, 2017). How do we help to maintain their confidence in accessing and utilizing digital services?

Digital divide such as the lack of technology access still exists in Malaysia because of the demographic factors. The Malaysia Ministry of Women, Family and Community Development has initiated support systems for older adults on various aspects such as recreation, transportation, facilities, healthcare, media, housing, social security, education and training. There are also community centers in Malaysia that offer ICT training to elevate older adults' technology skills, keeping them mentally active (Wong, 2011). Nevertheless, study shows that older adults require special needs to grasp the use of different types of control and functions when using technology. More research effort should be put into increasing the transparency and usability of technology to encourage their participation in the information society. Indeed, the learning capabilities of older adults are different from the younger learners due to the declination in their physical ability. The health problems encountered by older adults such as physical disabilities may inhibit various activities in their daily life including learning to use current technologies (Raistrick, 2016).

Another factor that could also interfere with the learning process of older adults is their cognitive ability. Older adults encounter cognitive limitations in terms of their working memory and spatial ability (Heaggans, 2012). The declination of working memory within older adults would affect their learning performance (Pliatsikas et al., 2019). According to Wolfinbarger et al. (2005), older people are the most motivated in using technology when they had help. Nonetheless, they might have experienced embarrassment when they attempted to use technological devices as they tend to act slower due to their limitations. Being in their golden age, older adults tend to be more conscious about their pride as they do not want to depend on others too much especially those who are younger than them and prefer to learn it from their peers. It might also be difficult for them to find learning peers due to different interest or physical restrictions (Chou et al., 2013). Therefore, it is important to highlight the need of including appropriate support from younger family members, friends and also youth from their local community in creating better learning opportunities to bridge the digital divide.

Intergenerational learning is an interactive learning process between people of different generation where they can both learn from one another or learn from a more knowledgeable person. It is regarded as one of the earliest processes of an informal learning, emphasizing on the "family learning" context. It is also viewed from the perspective of systematic knowledge and skill transmission, including sharing of expertise, traditions and beliefs between generations (Newman & Hatton-Yeo, 2008; Hoff, 2007). Usually, the older family members such as parents or grandparents would share their life experiences and wisdom with their younger generation, such as children and grandchildren to preserve family traditions and culture. Intergenerational learning then evolved into formal intergenerational programs, which focuses on the participation of older people in enhancing the performance of young people through professional education and training courses. The early models of intergenerational learning focus on extrafamilial relations by emphasizing on the role of older people as mentors or teachers. Such intergenerational programs have helped the young people to pave their path to professional occupations, especially occupations that would require them to work with older adults. Young people who have opportunities to have early contacts and learn with the elderly also demonstrated optimistic attitudes toward the older generation (McGuire & Mefford, 2007).

The generational pattern changed significantly toward the arrival of twentieth century. The shift of society from industrial to knowledge has resulted in the difference of life experiences between younger generation and the elderly. The exponential pace of technological change has also created distance between the generations where they lead different lifestyles. With the societal transition, it is no longer common for the younger generation to live with their parents or grandparents, leaving them vulnerable to social isolation and loneliness when they spend most of their time alone at home. This calls for more awareness on the value of intergenerational program to help reconnect generations and create better understanding and tolerance. Findings from prior studies suggested that there are growing interests among older adults to participate in intergenerational learning opportunities where knowledge sharing could be done both ways. Intergenerational learning is no longer just about providing opportunities for different generations to learn from each other, but also to nurture social cohesion in a community (Fonseca, Lukosch, & Brazier, 2019). The studies reported that the motivation of older adults to engage in lifelong learning programs has increased because they enjoyed the meaningful social interaction with other learners (Brady, Cardale, & Neidy, 2013) and learning on college or campuses where the majority is youth learners (McWilliam & Barrett, 2018). Besides that, there are also improvements in attitudes toward the young adults as they are able to understand them better especially on their current challenges and approaches (Andreoletti & Howard, 2018).

There are several key principles underpinning intergenerational programs. Reciprocity is one of the important components that should guide intergenerational efforts (Formosa, 2014). For instance, the activity should be designed to benefit both young and older participants, where they could exchange information to support each other. Hatton-Yeo (2006) highlighted that there are eight principles that could ensure successful initiatives of intergenerational programs—namely (1) sound project planning and management, (2) partnership working, (3) developing the capacity of communities, (4) promoting social inclusion, (5) capturing the commitment of champions, (6) demonstrating the benefits to participants, (7) the importance of preparing the generations and (8) challenging stereotypes and ageism. The combination of these key principles forms the basis of our intergenerational program. To fill the gap in research and practice, we explored the opportunities for intergenerational learning particularly on how the age differences between generations could potentially enrich content and interaction in learning processes. Our interests are in promoting intergenerational learning through digital creativity to facilitate active engagement and knowledge sharing among public and private universities, expertise of industry players and also the aging community. In early 2019, we embarked on a program to engage more young adults and older adults in technology-oriented communication, content producing and art designing. We proposed a program that encourages the use of existing latest mobile applications to create creative content. At the same time, the instructional strategies used in the program are designed to cater the perceptual, cognitive and motor functions of older people and older adults.

## 2 This Program Is Aimed to Address the Following Issues

1. Older adults should be encouraged to gain access to technologies, it allows them to communicate, share and obtain information. Based on the study by Chou et al. (2013), senior users hope that they could participate in their children's lives through internet. Social isolation can cause depression among the older adults and it would affect their health condition. Hence, utilization of technology could be one of the solutions for older adults to overcome loneliness (Chopik, 2016).

Therefore, the older adults' accessibility for ICT should be prioritized including the educational program that are designed for them and special content based on their needs (Mordini et al., 2009).

2. As the elderly population is increasing, there is also a growing interest in seeking lifelong learning opportunities. The lifelong learning society is formed when individuals regardless of their age are motivated to engage in continuous learning throughout their life in formal and informal settings. Lifelong learning became closely connected to intergenerational learning when its focus shifted to social capital and social cohesion. Effective intergenerational programs resulted in positive outcomes for both young and older generations. The interaction resulted in improving elderly's health and well-being and elevates self-esteem of the younger adults (Pliatsikas et al., 2019). There are still limited intergenerational programs in Malaysia despite the community interest. Although intergenerational learning is seen to have the potential to foster social inclusion and reciprocal sharing of expertise between learners of all ages, it still has its concerns and challenges.
3. Learning programs for older adults should consider appropriate design and instructional strategies that align with their physical and cognitive abilities (Ahmad, Razak, Zainal, Kahar, & Adnan, 2013). Although studies that address the digital gap among older adults in Malaysia do exist, the studies mostly discuss about older adults' perspective toward the technology and website interface for senior users (Nor, Hashim, Husin, & Aziz, 2015; Rusdi etal., 2017). There are still limited studies that focus on designing learning modules and instructional strategies that teach older adults technology according to their needs and interests.

## 3 Pilot Case Study

The program entitled Senior Academy Workshop or "Bengkel Teknologi Senior" is a 3-year project with a series of learning programs that are specifically developed for the target group of over 55 years old. Initiated by The Centre of Instructional Technology & Multimedia, Universiti Sains Malaysia (CITM-USM), this program is a unique community engagement partnership that was formally developed with a private university—Multimedia University (MMU) and professionals from the online media industry—Amanz Media Sdn. Bhd. The structure of the program includes two full-day workshop sessions with the local language, Bahasa Melayu as the main medium of instruction.

The main goal of this program is to increase the technological capacity, skills and competency older adults through creative art learning experiences. It is also aimed to support and enhance their participation in lifelong learning and sustain independence through the acquired skills in digital technology.

The objectives pursued are:

1. Promote potential use and benefits of mobile applications and internet to enhance older adults' life quality.

2. Enhance confidence, independence and digital skills of seniors in using and applying technologies in their daily life.
3. Encourage older adults to curate and produce digital creative content.
4. Encourage social inclusion and active online citizenship without feeling fear of new technology.
5. Provide opportunities for intergenerational dialogue and knowledge exchange.
6. Identify the learning strategies of older adult learners and explore the influencing factors.

Instead of just conducting virtual learning sessions and providing learning content online, this program proposed a face-to-face mode, where older adults are also able to learn at their own pace. This is because we would like to address the importance of social interaction for older adults in enhancing their mental health.

a. ***Instructional Strategies for Older Adults***

This program also acknowledges that older adults learn differently, and they have special learning needs. Despite the number of online tutorials and learning resources available online, there are still limited materials that could cater the need of older adults. Older adults tend to experience more difficulty in learning how to use technologies compared with the younger users as they tend to take a longer time in adjusting to the features of systems. Their cognitive processes are also affected by the deteriorating vision, hearing and cognition, which could lead to memory lapses (Rusdi et al., 2017). Younger users would not face much problems with their daily use of technology. However, older adults tend to find it difficult to adapt to the use of such modern technologies due to cognitive declination (Wolfson et al. 2014). During the natural aging process, older adults would experience loss of vision and cognitive abilities (Pliatsikas et al., 2019). The cognitive declination would also affect their working memory performance, especially when information presented does not fit their learning needs (Heaggans, 2012).

Older adults are able to learn effectively with adequate assistance and support such as providing training with appropriate instruction designs that are catered to their cognitive abilities, working memory and sensory deficits. Therefore, instructional strategies and teaching method play a significant role in encouraging older adults to engage in learning activities and learn independently with others or on their own. Based on the recommendations from the existing literature (Jones & Bayen, 1998; Uechi, 2010; Ahmad et al., 2013; Wolfson et al., 2014), a set of instructional strategy that could potentially help older adults learn is developed to guide the design of the program modules and instructions (Table 1).

Upon analyzing the most popular criteria for the acquisition of knowledge and skill of older adults, instructional strategies with three key criteria are proposed. They are: (1) Instructions, (2) Content to be Learned and (3) Application.

b. ***Program Delivery***

The delivery of the program is focused on maximizing engagement opportunities for the older adult participants and the younger generation who have volunteered as

**Table 1** Strategies for module development and instructions for older adults

| Criteria | Strategies |
|---|---|
| Instructions | (a) The instruction should be simple and meaningful where the unnecessary information will not be included and to minimize the amount of reading needed.<br>(b) Include clear and explicit learning objectives.<br>(c) Provide a clear flow for the learning objectives prior the instruction.<br>(d) Use language that can be easily understood by older adults and avoid jargon terms. |
| Content to be Learned | (a) Break down the instructional content into small meaningful units with specific goals.<br>(b) Incorporate guiding images or videos to assist the learning process.<br>(c) Choose appropriate font styles and size with appropriate background contrast that are easy to read. |
| Application | Provide practice for each unit of the instructions |

instructors and facilitators. They consist of university students, working adults and also lecturers from both private and public universities.

2 days of intensive face-to-face instruction and learning activities are provided to the participants. The official language of Malaysia, Bahasa Melayu, is used as the main medium for learning materials and instructions to address the digital language divide and facilitate social inclusion of local older adults. The lesson notes and content are also designed according to the proposed strategies mentioned in Table 1. Each participant is given a complete package of printed learning materials that consist of five selected learning modules, a pen and a notebook for private journals.

i. **Modules Developed**

The program was structured into two days. Seven modules were developed for this program, namely (1) Digital Graphics, (2) Digital Photography, (3) Video Production, (4) Fundamental Techniques of Video Editing, (5) Cyberthreats & Cybersecurity Issues, (6) Creative Writing and (7) Mobile Applications. Five out of seven modules were selected based on the learning needs of each session's participants. The first day was dedicated to the exploration of design and creativity and the second day focused on IT literacy such as information security, technological understanding and mobile application utilization. Each module is structured to be 2.5 hours long, with 1 hour of introduction and instruction and 1.5 hours of practice and application exercises.

c. ***Program Participants***

The rationale behind this program was to create engagement opportunities among older adults and the younger generation. It is aimed primarily to empower older adults aged 60 and above and also preretirees aged between 55–59 years old through intergenerational learning.

Using the snowball technique, the recruitment process for the participants is conducted as follows: first, in identifying the location of where we planned to organize the program, we reach out to the senior citizen organizations, community housing and centers and within the targeted neighborhood among our acquaintances. As this is a voluntary-based program, a "call for participation" is also disseminated to the residents through various means. We distributed printed publications such as program brochure in popular hangout areas for older adults. We also identified the key local informant and leaders within the community and send out e-invitations through messenger apps.

The criteria included: (a) older adults aged 55 years and above, (b) have moderate experience in using any technological devices and (c) able to read and write in Bahasa Melayu.

## 4 Methodology

The aim of this paper is to explore and gain insights into the older adult participants' experiences during the program. Specifically, this paper intends to shed light on to what extent can the program help older adults in advancing their technological skills through digital creativity learning. Hence, a qualitative approach is selected because it could provide in-depth data within the studied structure (Mills & Gay, 2019) and allowed researchers to explore more on participants' real-world experience (Creswell, 2014). The data used in the analysis comprised of conversational interviews, observation and artifacts from participants such as artworks and outcomes produced upon completion of modules.

Our data are drawn from four-module sessions that focus on the creative skills. As a pilot study, we included experiential data of participants who have volunteered for the study over the 2-day program. In addition, we also acquired information on their basic demographics and brief overall technology experience before attending the program. The questions focused on their experiences and level of confidence when using technological devices such as mobile phones and computers. For artifacts, we downloaded and documented the artworks produced by the participants as evidences of improvement and skills demonstrated. Finally, the final part of the data collected was from interviews conducted at the end of the 2-day program. Open-ended questions are designed for the in-depth interview to delve into participants' perspectives and experiences they had during the learning process. The interviews are digitally recorded upon permission by the participants. They are conducted in both English and Malay languages.

## 5 Participants' Learning Experiences

A total of 35 preretirees and older adults have participated in our program since 2019. According to the demographic survey filled out by the participants, their ages ranged from 56 to 73. 56% of the participants owns at least a Diploma and possessed various work experiences.

The program team has developed four modules under the digital creativity category to teach older adults on how to make their own graphic artwork, take better photographs and create their own video. The modules are: (a) Digital Graphics, (b) Digital Photography, (c) Video Production and (d) Fundamental Techniques of Video Editing.

Prior to our program session, the participants were introduced to the instructors and youth volunteers. All participants were given a printed module book where they can easily refer to when they could not keep up with the instructor's guide. To encourage socializing and increase intergenerational contact, participants were given an hour to mingle around with other participants and have conversations with the youth volunteers. Through this ice-breaking activity, the participants became familiar with the team.

d. ***Learning digital graphic***

Digital graphic is a module that offers an introduction to 2-D computer graphics and encourages unbounded exploration. In this module, participants were introduced to a computer graphic software—Adobe Photoshop. Participants were also taught on ways to express their creativity freely and create their own digital graphics with the tools and features available. During this session, participants were required to draw a scenery of nature that consisted of six components, namely, (a) hills, (b) sun, (c) sky, (d) cloud, (e) grasses and (f) flowers. The majority of participants was interested in learning beyond what has been taught in the module.

56-year-old Shaif (pseudonym) tended to add additional components to his artwork. For example, he placed two images of cats on the hill and an airplane in the sky (Fig. 1). He was very inquisitive and sought assistance from the youth volunteers in incorporating his design ideas as the instructions were not included in the module book. He asked if he could use existing images and integrate them in his artwork. He said:

> I could not draw well but I would like to have a dinosaur, tiger and an airplane in my artwork. Are there any possibilities of adding existing images from the Internet?

The youth volunteer in charge, Hazwan (pseudonym) had responded to his inquiries in a calm manner. With years of experience as a graphic designer in a media company, he was able to help Shaif efficiently by demonstrating simple yet clear steps to adding the requested elements. He said:

> No worries, sir. It is actually very easy to do so, and I believe you could do this at home on your own too. Firstly, let's visit Google Images. Next, type the name of the object you're searching for, followed by the word "PNG". PNG means Portable Network Graphics format. Images with transparent background are usually stored in this format. Next, choose your preferred image and save it into your computer. Finally, drag the images into your artwork.

Figure 1 below shows the final outcome Shaif's artwork.

**Fig. 1** Shaif's Artwork

Upon seeing Shaif's artwork, other participants were motivated to improve their artwork as well. Hazwan's instructional guide was being distributed among other youth volunteers and helped enhancement of the seniors' artwork. Toward the end of the session, almost all participants managed to add images to their existing artwork. One female participant in her early 60s, Sally (pseudonym), wished to create a sunset scenery instead. She told the youth volunteer:

> I think I prefer to use warm colors for my artwork. Can I paint a sunset scenery with orange and red glow? (Figure 2).

e. ***Learning digital photography***

In this session, participants are taught on how to master their camera (smartphones, digital camera) and enhance their photos with free photo editing applications on their smartphones. They are exposed to tips and tricks on taking better photos, the basic settings and functions available on devices, and learn components of photography such as composition, lighting and exposures. One male participant in his mid-70s, Zarif, was a reserved participant at first. However, he was very driven to capture a photograph that would differ from other participants during this session. When everyone else was taking photographs on the same floor, he requested permission from the instructor to conduct his photoshoot session at another floor. He said:

> I love the wall on the Ground Floor. I think my photograph would turn out nice with such pattern as my background.

He also attempted the application of Rule of Third in the photograph taken, combined with the brick pattern on its background. He added that although he has owned a smartphone for more than a year, there are still many camera features that he had yet to explore. Several other participants demonstrated their enthusiasm in

**Fig. 2** Sally's Sunset Artwork

photography by searching for their own subjects out of the classroom, instead of staying in. Noraini (pseudonym) is a 57-year-old medical officer who will be retiring from her current profession soon. She mentioned that she enjoyed the session very much. She said:

> After listening to the instructor's sharing on photography components, I now understand why my younger children like to take photographs in front of random walls. It actually turned out nice! No wonder they often told me that my photographs are quite old-fashioned. I usually snap the photographs without considering the latest trend (Fig. 3).

f. ***Learning video production***

**Fig. 3** Photograph taken by Zarif

**Fig. 4** Participants using their mobile devices on a tripod

This module introduces participants to the basic theories behind creative video production processes. They were also exposed to techniques on how to record professional videos via their personal smartphones and digital camera. Prior to this session, all participants were given a tripod to help them produce their own video. The program committee had also provided video cameras to participants who were not in possession of any camera that could do video recordings (Fig. 4).

Hana (pseudonym), a female participant aged 56 expressed on how her video shooting techniques could be improved so that she could film smoothly while walking. She selected a location where she could move around freely before she started filming and sought advice from the facilitators on the walking techniques.

She said:

> I need to know the right way to walk while filming. Please teach me how should I walk steadily without shaking my camera! I could get more interesting shots by doing so.

From the above, we may infer that older adults were quite critical in ensuring the quality of the digital content although it was their first attempt.

Some of the participants were quite cautious in handling their video cameras. They expressed their concerns on the durability of the devices if they accidentally fall off the tripod. Nevertheless, that did not stop them from exploring the functions available on their camera. Ismael (pseudonym) is a retired policeman. He is 75 years old. He asked Heng (pseudonym), one of the youth volunteers, on ways to increase the audio quality. As a student majoring in Cinematography, Heng provided a vast explanation on the portable microphones that could be purchased easily from retail shops.

Ismael responded:

> Thank you for sharing that with me. No wonder we have so many young people who have become Youtubers with great content quality now. All accessories are easily available.

**Fig. 5** Youth volunteers guiding participants in using video editing applications

g. ***Learning video editing techniques***

In this module, participants were introduced to the fundamental techniques used in professional video editing. The instructors demonstrated how the techniques could be simplified and achievable by using free video editing applications downloadable in their devices (Fig. 5).

Malia (pseudonym) is a housewife. She is now 64 years old. Although she often required extra assistance in using her mobile phone, she expressed strong preferences in her own design choice. She asked if she could add transitions to her videos when editing them. She said:

> Transitions like zooming and sliding would make my video more interesting. A long video without transitions would be boring.

At the end of the session, she was able to produce a 1-minute long video with multiple transition effects.

This module is also designed to inform senior citizens about image and video fraud. By explaining how video footages could be manipulated easily with the current existing video editing applications, participants were taught to be more cautious of misinformation and fake news. For instance, the instructor demonstrated how the use of green screen easily deceives audiences. Participants were also given awareness of the extent to which video editing techniques could change the digital content they see in the media. Idris (pseudonym) mentioned that he was not aware of the technology advancement and fake videos could be made so real and authentic. He said:

> I have always thought that all movies must be filmed in real sets. I have never thought of how easy video manipulation could be done with free downloadable applications! With this, I could fake a vacation to Paris with my wife too. All I need is a green screen now. Hahaha.

Upon completion of this module, all participants were able to publish their own video and share it into the internet for public viewing.

## 6 Reflections: Engaging Older Adults in Digital Creativity Programs

This pilot study shows older adults actively participating in digital creativity programs although they were not exposed to such learning activities before. The majority of participants found that the program has given them the opportunity to unleash their creative side. Preliminary results of this study indicate that learning modules that focus on digital creativity skills mentioned earlier in this paper could be seen as potential lifelong learning programs that address social inclusion for older adults. Although there are still limited studies on the impact of creative involvement on older adults' active aging, prior research on enhancing active aging through creativity and art-related activities found that art making is able to promote physical and cognitive health effectively especially among older adults (Abraham, 2005; Noice, Noice, & Kramer, 2014; Ravid-Horesh, 2004; Stewart, 2004). Similarly, Cohen et al. (2006) in his study indicated that older adults' overall health could be improved by participating in creative arts programs.

A number of important insights emerged from the interviews: the motivation and engagement of the older adults' learning experience. The participants report being motivated by their own desire to learning about digital creative content and enjoyed their learning process with the youth volunteers. This is in line with the previous literature on online learning (Brady et al., 2013) and socially focused activities (Brady et al., 2013; LorekDattilo et al., 2012). Throughout the sessions, we have learned that the participants were very curious to know how contents that are shared on social media platforms such as Facebook, YouTube and Instagram were made. They were also self-motivated in exploring how they could produce similar outcomes to be shared on their personal platforms too. Even though they are only required to complete the given task according to the instructor's guide, it was clear that many participants had taken the extra mile to achieve better outcomes.

Furthermore, they mentioned that they wished to share their self-produced content with the younger generation in their family. This is consistent with findings from earlier research (Carstensen et al., 2003; Kim & Merriam, 2004), which indicated that social and emotional goals could serve as a strong motivation factor for older adults to engage in activities. The participants also portray positive emotions upon completion of the learning activities, which could further enhance their motivation for learning (Boulton-Lewis & Buys, 2015; Shuck, Albornoz, & Winberg, 2007). Nevertheless, there were also participants who have expressed their insecurities toward technology usage especially when it involves online money transactions and privacy.

Findings of this study also provided better perspectives to understand the technological learning process of older adults, specifically through intergenerational approach. Participants indicated that generation gap is one of the main barriers that limits their communication with younger family members, especially when it comes to technology use. According to them, the younger generation would have a different outlook and needs than the elder generation, which often cause hardship in finding common ground. As late adopters of technology, older adults are usually afraid of exploring new technology and the fear could contribute to disconnections with the younger generations. However, the individual support offered by the youth volunteers and instructional strategies that are developed for elder learning is likely to have enhanced participants' persistence in accomplishing the programs. This study revealed that there were positive interactions among the participants and youth volunteers. Narrative excerpts indicated that some of the participants have developed a close relationship and bonding with the youth volunteers. The benefits of intergenerational programs are similar to findings from previous studies including enabling older adults to understand the younger people better (Andreoletti & Howard, 2018) and sharing of perspectives between different generations (Dauenhauer et al., 2016).

## 7 Conclusion

This study is not without its limitations. The sample is relatively low in number and could not be generalized to other older adults. It is important to note that this study is not meant for generalization purposes but rather to explore on how older adults engage in digital creativity programs and their learning experiences when taking each creative module. This study has provided some evidences that older adults should be given opportunities to learn and demonstrate their talent through creative art programs that are age inclusive. The unique collaboration between CITM-USM, MMU and Amanz Media has provided a way to enhance social inclusion and, more importantly, improve lifelong learning opportunities for older adults in Malaysia. The three organizations have different roles and responsibilities in this program. CITM-USM is responsible for developing instructional strategies, determining course content and developing units that are appropriate for older adults, marketing the program and select participants into the program. On the other hand, the major responsibility of MMU is to develop course content based on their expertise in nurturing talents for the local creative industry while Amanz Media is responsible in ensuring course structure and content being delivered with updated information and in parallel with real-world needs.

While this program has provided the older adults opportunities to continue their learning in campus environment, program organizers should also be mindful of their physical and cognitive capabilities as some older adults may require assistance in getting to the program venue. Related issues such as traveling time, transportation or health challenges should be put into consideration when deciding the program sessions. It is also suggested that a session dedicated to the introduction of technological devices be conducted before delivering the modules, as older adults need more time to get themselves used to the technology devices and digital applications especially if it was their first time using them.

In terms of practical implication, findings from this study may help instructional designers and course coordinators to develop learning content for older adults especially in digital art-making courses. Based on the lessons we learned from this study, we postulate that the digital knowledge gap for the older generation could be substantially narrowed if we embrace the idea of essentialism—each generation adapts and uses technology differently according to their own goals. We intend to expand the program to other states of Malaysia and include new learning activities that support the principle of control and choice of older adults. Our future research may examine whether older adults of different background prefer different course content and instructional strategies that we have suggested.

**Acknowledgements** This project is made possible with funding from Amanz Media Sdn Bhd under the AmanzNXT initiative. We would like to thank all volunteers for their dedicated participation and efforts. We would also like to show our gratitude to Research Creativity and Management Office (RCMO), Universiti Sains Malaysia for the research support provided under the Bridging-Incentive Grant (BI0026) (304 / PMEDIA / 6316437).

## References

Abraham, R. (2005). *When words have lost their meaning: Alzheimer's patients communicate through art.*Praeger Publishers/Greenwood Publishing Group.

Ahmad, N. A., Razak, F. H. A., Zainal, A., Kahar, S., & Adnan, W. A. W. (2013). Teaching older people using web technology: A case study. *International Conference on Advanced Computer Science Applications and Technologies, 2013,* 396–400.

Anderson, M., & Perrin, A. (2017). Tech adoption climbs among older adults. *Pew research center, 2017,* 1–22.

Andreoletti, C., & Howard, J. L. (2018).Bridging the generation gap: Intergenerational service-learning benefits young and old.*Gerontology & Geriatrics Education, 39*(1), 46–60.

Boulton-Lewis, G. M., & Buys, L. (2015).Learning choices, older Australians and active ageing.*Educational Gerontology, 41*(11), 757–766.

Brady, E. M., Cardale, A., &Neidy, J. C. (2013).The quest for community in Osher lifelong learning institutes.*Educational Gerontology, 39*(9), 627–639.

Carstensen, L. L., Fung, H. H., & Charles, S. T. (2003). Socioemotional selectivity theory and the regulation of emotion in the second half of life. *Motivation and Emotion, 27*(2), 103–123.

Chopik, W. J. (2016). The benefits of social technology use among older adults are mediated by reduced loneliness. *Cyberpsychology, Behavior, and Social Networking, 19*(9), 551–556.

Chou, W. H., Lai, Y. T., & Liu, K. H. (2013). User requirements of social media for the elderly: A case study in Taiwan. *Behaviour & Information Technology, 32*(9), 920–937.

Cohen, G. (2006). Research on creativity and aging: The positive impact of the arts on health and illness. *Generations, 30*(1), 7–15.

Creswell, J. W. (2014). *A concise introduction to mixed methods research.*SAGE publications.

Dauenhauer, J., Steitz, D. W., & Cochran, L. J. (2016).Fostering a new model of multigenerational learning: older adult perspectives, community partners, and higher education.*Educational Gerontology, 42*(7), 483–496.

Fonseca, X., Lukosch, S., & Brazier, F. (2019). Social cohesion revisited: a new definition and how to characterize it. *Innovation: The European Journal of Social Science Research, 32*(2), 231–253.

Formosa, M. (2014). Four decades of Universities of the Third Age: past, present, future. *Ageing & Society, 34*(1), 42–66.

Hatton-Yeo, A. (2006). *Intergenerational Practice: Active Participation Across the Generations.* Stoke-on-Trent: Beth Johnson Foundation.

Heaggans, R. C. (2012). The 60's are the new 20's: Teaching older adults technology. *SRATE Journal, 21*(2), 1–8.

Hoff, A. (2007). Intergenerational learning as an adaptation strategy in aging knowledge societies. *Education, Employment, Europe*, 126–129.

Jones, B. D., &Bayen, U. J. (1998).Teaching older adults to use computers: Recommendations based on cognitive aging research.*Educational Gerontology: An International Quarterly, 24*(7), 675–689.

Kim, A., & Merriam, S. B. (2004). Motivations for learning among older adults in a learning in retire-ment institute. *Educational Gerontology, 30,* 441–455.

LorekDattilo, A. E., Ewert, A., &Dattilo, J. (Spring 2012). Learning as leisure: Motivation and outcome in adult free time learning. *Journal of Park and Recreation Administration, 30*(1), 1–18.

McDonough, C. C. (2016). The effect of ageism on the digital divide among older adults. *Journal of Gerontology and Geriatric Medicine, 2,* 1–7.

McGuire, S., & Mefford, L. (2007). Growing up and growing older. *The Journal of School Nursing, 23*(2), 80–85.

McWilliams, S. C., & Barrett, A. E. (2018). "I hope I go out of this world still wanting to learn more": Identity work in a lifelong learning institute. *The Journals of Gerontology: Series B, 73*(2), 292–301.

Mills, G. E., & Gay, L. R. (2019).*Educational research: Competencies for analysis and applications.* Pearson. One Lake Street, Upper Saddle River, New Jersey 07458.

Mordini, E., Wright, D., Wadhwa, K., De Hert, P., Mantovani, E., Thestrup, J., &Vater, I. (2009). Senior citizens and the ethics of e-inclusion. *Ethics and Information Technology, 11*(3), 203–220.

Nations, U. (2017). World population prospects: The 2017 revision, key findings and advance tables. *Department of Economics and Social Affairs PD, editor*. New York: United Nations.

Newman, S., & Hatton-Yeo, A. (2008). Intergenerational learning and the contributions of older people. *Ageing horizons, 8*(10), 31–39.

Noice, T., Noice, H., & Kramer, A. F. (2014). Participatory arts for older adults: A review of benefits and challenges. *The Gerontologist, 54*(5), 741–753.

Nor, N. F. M., Hashim, F., Husin, S., & Aziz, J. (2015). Connecting The Senior Citizens with ICT: The Development Of Portal Komuniti Kita. *e-Bangi, 12*(3).

Pliatsikas, C., Veríssimo, J., Babcock, L., Pullman, M. Y., Glei, D. A., Weinstein, M., et al. (2019). Working memory in older adults declines with age, but is modulated by sex and education. *Quarterly Journal of Experimental Psychology, 72*(6), 1308–1327.

Raistrick, C. (2016). Social media for senior citizens: An introductory course. *Instructional Design Capstones Collection. 16*

Ravid-Horesh, R. H. (2004). "A temporary guest": the use of art therapy in life review with an elderly woman. *The Arts in Psychotherapy, 31*(5), 303–319.

Rusdi, R., Sahari, N., & Noor, S. F. M. (2017). Usability guidelines for elderly website interface. *Asia-Pacific Journal of Information Technology and Multimedia, 6*(2).

Shuck, B., Albornoz, C. and Winberg, M. 2007. Emotions and their effect on adult learning: A constructivist perspective. *Proceedings of the Sixth Annual College of Education Research Conference: Urban and International Education Section,* 108–113.

Springate, I., Atkinson, M., & Martin, K. (2008). *Intergenerational Practice: A Review of the Literature. LGA Research Report F/SR262.* National Foundation for Educational Research. The Mere, Upton Park, Slough, Berkshire, SL1 2DQ, UK.

Stewart, E. G. (2004). Art therapy and neuroscience blend: Working with patients who have dementia. *Art therapy, 21*(3), 148–155.

Uechi, N. (2010). An instructional video module on spreadsheets accommodating the needs of senior citizens.*Presented at the Technology, Colleges, and Community Worldwide Online Conference.*

Wolfinbarger, M., Gilly, M. C., & Schau, H. J. (2005). *Keeping up with the Times: Innovation and Usage of the Internet Among Later Adopters.* Berkeley: Center for Research on Information Technology and Organizations.

Wolfson, N. E., Cavanagh, T. M., & Kraiger, K. (2014). Older adults and technology-based instruction: Optimizing learning outcomes and transfer. *Academy of Management Learning & Education, 13*(1), 26–44.

Wong, C. Y. (2011). Exploring the relationship between mobile phone and senior citizens: A Malaysian perspective. *International Journal of Human-Computer Interaction*, 65–77.

Yang, M., & Huang, H. (2015). Research on interaction design of intelligent mobile phone for the elderly based on the user experience. *International Conference on Human Aspects of IT for the Aged Population,* 528–536.

**Tan Yun Yi** is currently a Senior Lecturer at Universiti Sains Malaysia. She graduated from Multimedia University with a Bachelor degree in Digital Media (Hons) in 2009. She has worked in various positions in agencies related to ICT and creative industries. She was later awarded a fellowship from Universiti Sains Malaysia to further her postgraduate studies. In 2018, she received her PhD in Educational Technology from The University of Hong Kong, Hong Kong. Her research includes educational technology, multimedia design and gerontechnology. Her research interest is also driven by her curiosity in creative media and technology and also how they could be used to enhance learners' learning experiences.

**Nabilah Abd Raman** is currently working as a Teacher at Sekolah Menengah Kebangsaan Taman Selesa Jaya. She graduated from Universiti Sains Malaysia in 2018 with a degree in Bachelor of Education (TESOL) (Hons). After graduating, she continued her postgraduate study in Master of Instructional Multimedia that also took place in Universiti Sains Malaysia. She finished her master's degree within 1 year and graduated in 2019. She is highly interested in instructional design and technology fields which she believes will greatly elevate learning and teaching experience.

# Motivational Effects of Immersive Media on Adolescents' Engagement in Cantonese Opera

**Lee Cheng and Bo-Wah Leung**

**Abstract** Recent advancements in communication and computer technology have made available and affordable the immersive technology necessary for broader applications beyond industrial use. Facing growing competition from the emergence of new media as an attractive option for youngsters' leisure time and cultural enjoyment, traditional arts such as Cantonese opera have undergone important transformations in a bid to survive the rapidly changing sociocultural landscape. With a dearth of research on the stimulating effects of new media, this chapter presents the findings of a project that examined the motivational effects of immersive media on adolescents' engagement in Cantonese opera. Thirty-two undergraduate students from Hong Kong were invited to participate in a study that required them to watch a Cantonese opera theatre performance using both conventional and immersive media; to that end, a 360-degree Cantonese opera theatre performance was produced and viewed using a virtual reality (VR) headset, after which the participants' viewing experiences and perceived motivational effects were compared with a conventional 2-D screen display via a questionnaire survey. Findings revealed that the immersive media was found to be more interactive and motivational than the conventional approach, revealing the possibility of better engaging the younger generation in the appreciation and learning of this particular genre.

**Keywords** Virtual reality · Cantonese opera · Immersive video · Motivation · Intangible cultural heritage

---

L. Cheng (✉) · B.-W. Leung
Department of Cultural and Creative Arts, The Education University of Hong Kong, Hong Kong, China
e-mail: lcheng@eduhk.hk

B.-W. Leung
e-mail: bwleung@eduhk.hk



W. W. K. Ma et al. (eds.), *Learning Environment and Design*, Educational Communications and Technology Yearbook,
https://doi.org/10.1007/978-981-15-8167-0_13

# 1 Introduction

The transmission of traditional arts has never been more difficult. On the one hand, artistic practices, especially those of Asian origin, frequently have to contend with the dominance exerted by Westernization and modernization (Matsunobu, 2018); on the other, the emergence of new media is an attractive option for youngsters' leisure time and cultural enjoyment. Many traditional arts have undergone various kinds of transformations in order to survive in the modern world, including their integration in community settings and formal education, the shift from an oral to a conservatory tradition, and the use of online mediums for self-directed learning (Leung, 2018; Matsunobu, 2018; Waldron, 2016). The rise of new media has also played a transformative role in the adaptation of the traditional arts as part of the assimilation process within a rapidly changing sociocultural landscape.

Cantonese opera is one such traditional art that can potentially benefit from these kinds of transformations. Originating in Guangdong Province, where the Cantonese dialect serves as the lingua franca among the region's population, Cantonese opera has existed as one of China's regional operatic genres since the Qing Dynasty (1644-1911). It reached its golden age in the 1920s-1950s, during which many state singing styles were developed by a number of eminent and innovative artists (Leung, 2019). After the Cultural Revolution, the popularity of Cantonese opera began to decline, especially with the introduction of Western pop music and movies in the 1970s (Chan, 1991), resulting in a gradual diminution of interest in the genre among young people. A recent study on the motivation to learn Cantonese opera found that primary students are the demographic most interested in this genre, while junior secondary students are reluctant to assume the identity of a 'Cantonese opera admirer', regarding the genre as outdated and oldfashioned (Leung & Leung, 2010).

The United Nations Educational, Scientific and Cultural Organization (UNESCO) listed Cantonese opera among its Intangible Cultural Heritage (ICH) items in 2009, thereby recognizing the art form's cultural value and representativeness within southern China. In response to the necessity of preserving and transmitting Cantonese opera locally, the Hong Kong SAR Government has advocated for and supported the development of Cantonese opera through a variety of different channels. These include the establishment of the Cantonese Opera Advisory Committee (COAC) and the Cantonese Opera Development Fund (CODF), the venue construction of the West Kowloon Cultural District's (WKCD) Xiqu Centre, and an emphasis on Cantonese opera and local traditional cultures within school education (Chan, 2012; Curriculum Development Council, 2017).

Despite this support, the transmission of Cantonese opera still faces many challenges. In addition to the findings by Leung and Leung (2010), Hong Kong music teachers lack both confidence and competence when it comes to teaching the art form at school due to a lack of prior training in teacher education (Leung, 2014). Clearly, in order for the art form to remain both sustainable and viable, more effort is needed to help the younger generation engage with Cantonese opera.

Recent advancements in communication and computer technology have increased both the availability and the affordability of the immersive technology necessary for broader applications beyond industrial use. Such technology enables greater interaction between the user and the immersive environment, offering novel and appealing ways of presenting and accessing information in ways that were either inefficient and/or cost-prohibitive in the past. Commercialized platforms such as virtual reality (VR) headsets, gaming consoles, and 360-degree cameras are all widely available, leading to an increase in opportunities for the general public to view and generate immersive media content.

Previous studies have reported positive motivational effects of immersive technology for teaching and learning purposes, covering both formal and informal learning approaches across different disciplines (*c.f.* Howard & Gutworth, 2020). Applications of immersive technology for the promotion and preservation of cultural heritage are also growing rapidly (e.g. Jung & Tom Dieck, 2017; Lercari, Shiferaw, Forte, & Kopper, 2017; Roberto et al., 2017), yet attempts that involve Cantonese opera have been few and far between. Fung Cheung Hung Troupe's (2019) production of *The Legend of Mulan* (木蘭傳説) is a recent attempt to incorporate immersive technology into Cantonese opera, making use of three-dimensional projections onto a transparent screen in order to demonstrate how the actor interacts with the scene. The computerized assessment and learning system for Cantonese opera movement developed by Leung et al. (2018) is another recent example.

With a dearth of research on the potential engagement of new media, combined with the need to cultivate interest among the younger generation, a project was implemented to produce an immersive Cantonese opera video and examine its motivational impact on adolescents. This paper reports the findings from a research study that compared the perceived effects on the participants of immersive media on the one hand and the conventional two-dimensional display method on the other of the same Cantonese opera content. It is hoped that the findings from this study might better help align the use of immersive media and the development of Cantonese opera in order to achieve the ultimate goals of transmission and preservation of this traditional art form as an ICH.

## 2 Literature Review

Recent scholarly work with a research interest in the use of immersive media for engagement can be divided into four themes: (1) immersion; (2) interaction; (3) imagination; and (4) motivation, all of which have provided experimental evidence informing and shaping the current study's theoretical framework.

## 2.1 Immersion

Immersiveness determines how realistic the virtual environment feels when the individual is situated inside it. In common with many other developers dedicated to ensuring a more effective application of the media, Choi et al. (2018) developed a novel authoring framework and play-based learning system for marine biology education that combined a multilayer representation of VR content with image-based interactions. Survey results revealed that more fully realized elements within the VR content could improve the immersive experience, thereby allowing users to become more actively involved. Han's (2020) study was concerned with comparing individuals' feelings of 'presence' between an immersive virtual field trip using VR headsets and a conventional one using a two-dimensional display method; the findings revealed a greater sense of overall enhancement with the former.

## 2.2 Interaction

Previous research has also investigated the distinctive characteristics of immersive media when it comes to the levels of user interactivity. Zhang et al. (2019) created and compared an immersive storytelling experience for immunology learning based on three different levels of interactivity and found a positive influence between interactivity and students' attention, focus, and engagement in learning. Sah and Peng (2019) experimented with the immersive experience in a hunting simulation game with two groups of players participating in interactive and non-interactive settings. The study revealed that interactive media generated a greater influence on the construction of concrete identities than non-interactive media.

## 2.3 Imagination

Higher affordance for imagination enables one to better experience and understand the subject content. Paletta et al. (2019) created a serious game for elderly people in which affective and sensory experiences were induced within real urban environments. Alatta and Freewan's (2017) study echoed the same qualities, in the process revealing the enhancement of architectural students' imaginations and the ability of immersive media to better understand and realize their ideas.

### 2.4 Motivation

Liu et al. (2019) conducted quantitative research in order to determine the differences between immersive and non-immersive media on young adults' motivation to undertake physical activity. Higher levels of intrinsic motivation were found among those individuals who participated in immersive cycling, suggesting the positive effect of such media for the promotion of greater participation and engagement. Another study by Makransky et al. (2019) also revealed greater increases in intrinsic motivation when it came to delivering laboratory safety training through immersive simulation compared with conventional settings.

## 3 Theoretical Framework

Two theoretical frameworks support this study. The first, that of the potentially stimulating effect of immersive media, is grounded in self-determination theory. This posits that the engagement of an activity stems from intrinsic motivation, which in turn depends on whether or not the activity is inherently interesting or enjoyable (Ryan & Deci, 2000). While empirical studies have revealed the enjoyment and imagination of an immersive environment in diverse research and knowledge areas (Rubio-Tamayo, Barrio, & García, 2017), its potential benefits, especially with regards to stimulating young people's interest in traditional arts, additionally make it a worthwhile area of study for the purposes of cultural preservation.

Second, the use of immersive and interactive media to achieve the educational aim of cultural transmission is guided by the constructivist paradigm, which holds that the acquisition of knowledge is based on active and interactive experiences with the learning environment (Dewey, 1916; Piaget, 1967). The immersive environment, such as the 360-degree Cantonese opera theatre performance described in this study, has the potential to contribute to the active participation of learning in ways that are rarely encountered in traditional media. Finally, Orbach's (1979) theoretical framework—that initial motivation can be prompted by simulation for active participation—supplements the theoretical foundations of this study by bridging the aforementioned intrinsic motivation framework and constructivist approach to learning.

## 4 Aim

Aimed at examining the motivational effects of immersive media on adolescents' engagement in Cantonese opera, the following research questions guided this study:

1. To what extent does the use of immersive media motivate adolescents' engagement in Cantonese opera?
2. What are the motivational differences between adolescents' experience of different viewing methods?

## 4.1 Participants

Thirty-two Cantonese-speaking students aged 18–22 from The Education University of Hong Kong were invited to participate in the study; 19 were female and 13 were male. As students of a virtual reality (VR) course, which was offered as an elective for all undergraduate students, most of them had little or no experience with VR headsets. They were shown a 5-minute excerpt of a Cantonese opera theatre performance using two viewing methods, after which they were invited to complete a questionnaire survey.

## 4.2 Questionnaire Survey

The questionnaire survey consisted of two parts. The first part collected demographic information including age, gender, and previous experience with immersive technology; the second part assessed participants' interest in Cantonese opera after viewing the performance through both an immersive VR headset and a conventional two-dimensional display. The items in the second part were adapted from the questionnaire developed and validated by Huang et al. (2010), which was used to measure attributes of students' attitudes towards virtual reality learning environments. There were 12 items grouped into four dimensions of measurement, evaluated using a 7-point Likert scale.

## 4.3 Production

A 360-degree Cantonese opera theatre performance was produced in collaboration with the School of Chinese Opera at the Hong Kong Academy for Performing Arts. Two 10-minute excerpts from a famous Cantonese Opera, *Dream of the West Chamber* (西樓錯夢), performed by students from the School using a specially constructed and decorated stage, were filmed twice with a high-end 360-degree camera (see Fig. 1). An instrumental group was placed at one side of the stage, serving as both a practical reason for the actors to interact with the music and an educational one by deepening the audience's understanding of Cantonese opera performance practices. The first take was filmed placing the camera at the centre of the stage, and

**Fig. 1** Setting of the stage for filming the 360-degree Cantonese opera performance

the second take (in order to achieve the first-person 360-degree view) was filmed by placing the camera on the main character. Unfortunately, the second take was aborted in the latter part of the research study due to overwhelming camera shake during the recording process.

## 4.4 Procedure

Most of the data collection was implemented before and after the VR course in the computer classroom, while the rest took place within the University environment according to the availability of the participants and the researcher. All research procedures followed the guidelines prepared by the University ethics committee, with a formal ethical review application being submitted and approved. After an explanation of the purpose of the study and the data collection process, participants were assured of their right to confidentiality and anonymity. Upon receipt of their formal agreement, participation commenced with a brief introduction regarding the various experimental procedures.

Participants were asked to view the same Cantonese opera theatre performance twice, the only difference being the viewing method. The first was to view the videotaped performance using a conventional 2D display on an LCD screen (hereafter, "2D method"); the second was to view the aforementioned 360-degree video using a VR headset in an immersive way, which allowed them to control the viewing direction via head movements (hereafter, 'VR method'). Half of the participants started with the conventional method and the other half with the immersive method in order to avoid any preconceived bias regarding the different media forms. They were then asked to spend 5 min completing the questionnaire survey.

# 5 Findings and Discussion

Findings of the questionnaire survey were divided into four themes, based on the aforesaid theoretical framework: (1) immersion; (2) interaction; (3) imagination; and (4) motivation. A paired sample *t*-test was used to determine the differences between participants' perceived motivations towards Cantonese opera according to the two different viewing methods.

## 5.1 *Immersion*

Participants were asked to rate and compare the immersive aspect of the two different viewing methods (Table 1). Of the three listed elements making up this category, the VR method rated significantly higher than the 2D method in terms of the realism and the immersive viewing experience ($p < 0.05$). However, the VR method rated higher when it came to the amount of attention paid, a finding that was not supported by the significance of the *t*-test ($p > 0.05$).

Participants' positive opinions about the degree to which the VR method could enhance realistic responses and immersive feelings within the virtual environment echo those of previous studies (Choi et al., 2018; Han, 2020). They also supplement

**Table 1** *t*-Test results for participants' perceived immersion

| | | Mean | Standard deviation | *P*-value |
|---|---|---|---|---|
| The viewing method creates a realistic look of the theatre environment | VR | 4.88 | 1.00 | 0.00 |
| | 2D | 3.97 | 1.00 | |
| I pay more attention when using the viewing method | VR | 3.84 | 0.68 | 0.06 |
| | 2D | 3.47 | 0.76 | |
| I feel immersed with the viewing method | VR | 5.41 | 0.91 | 0.00 |
| | 2D | 3.75 | 0.88 | |

**Table 2** *t*-Test results for participants' perceived interaction

| | | Mean | Standard deviation | *P*-Value |
|---|---|---|---|---|
| The viewing method can enhance learner's interaction | VR | 4.91 | 0.78 | 0.00 |
| | 2D | 4.13 | 0.94 | |
| The viewing method can enhance viewer's interaction | VR | 5.06 | 0.91 | 0.02 |
| | 2D | 4.41 | 0.91 | |
| I would like to share my experience with other people | VR | 3.84 | 0.68 | 0.60 |
| | 2D | 3.75 | 0.76 | |

the existing body of the literature that hypothesizes that the degree of immersion and the realism of the immersive media can apply to the domains of theatre performance and traditional arts. The increasingly narrow distinctions between realistic and virtual environments form the conditional basis for the developmental applications of immersive media in ways that were not previously possible, while simultaneously exceeding presently limited applications for leisure and entertainment among existing practices.

## 5.2 Interaction

Participants rated whether they thought the two viewing methods could enhance learners' interaction, either as a learner or as a viewer (Table 2). The VR method rated significantly higher than the 2D method for both items ($p < 0.05$), thereby suggesting the advantages of interactivity afforded by immersive technology in terms of learning and appreciation (Christou, 2010; Sah & Peng, 2019). The similar rating scores for participants' willingness to share their experiences for both viewing methods are perhaps due to the fact that there already exists widespread and prior knowledge of VR among adolescents. When compared with commercially available immersive products, such as VR games and design applications, the interactivity provided by the 360-degree Cantonese opera theatre performance used in this study lags behind in its ability to control nothing more than the viewing direction.

## 5.3 Imagination

Few differences between the VR and 2D methods were recorded among the participants for all three items within the category of imagination (Table 3). Similarly, no significant effects were found in any of the *t*-tests ($p > 0.05$). Despite the absence of statistical evidence, similar rating scores between the two viewing methods suggest that immersive media may struggle to facilitate participants' imagining the theatre

**Table 3** *t*-Test results for participants' perceived imagination

| | | Mean | Standard deviation | *p*-Value |
|---|---|---|---|---|
| The viewing method can help me to better engage in the understanding of the Cantonese opera theatre | VR | 5.38 | 1.16 | 0.81 |
| | 2D | 5.31 | 0.74 | |
| I feel the viewing method can help me to better imagine and understand the content of the Cantonese opera | VR | 4.63 | 0.91 | 0.89 |
| | 2D | 4.59 | 1.04 | |
| I feel the viewing method can help me to better understand by imagination the Cantonese opera theatre environment | VR | 4.44 | 1.05 | 0.91 |
| | 2D | 4.41 | 1.19 | |

environment within the context of this research study. The level of imagination afforded by the immersive theatre environment may not be as competent as those provided in other contexts, such as those in the aforesaid previous studies (Alatta & Freewan, 2017; Paletta et al., 2019).

## 5.4 *Motivation*

Comparatively high rating scores and significant effects were revealed when it came to the impressiveness and enhancement of interest for viewing Cantonese opera theatre performance using the VR method ($p < 0.05$). This result is consistent with previous studies that compared the effect on stimulation between immersive and conventional media (Liu et al., 2019; Makransky et al., 2019). The use of immersive media positively leveraged the degree of intrinsic motivation, whereby the participants felt more impressed by, and showed a greater interest in, the subject content (Table 4).

**Table 4** *t*-Test results for participants' perceived motivation

| | | Mean | Standard deviation | *p*-Value |
|---|---|---|---|---|
| It is impressed to view Cantonese opera theatre performance through the viewing method | VR | 5.63 | 0.94 | .00 |
| | 2D | 4.03 | 1.20 | |
| The viewing method can enhance my interest in Cantonese opera theatre performance | VR | 4.44 | 1.11 | .01 |
| | 2D | 3.66 | 0.94 | |
| The viewing method can enhance my motivation in knowing and learning more about Cantonese opera | VR | 4.66 | 0.94 | .25 |
| | 2D | 4.34 | 1.04 | |

While a previous study has shown that the younger generation considers Cantonese opera to be outdated and old-fashioned (Leung & Leung, 2010), the use of immersive technology to reveal rarely seen innovative practices could prove to be a positive mode of presentation, especially when it comes to adolescents. The use of stimulating effects with immersive media might similarly assist in enabling the younger generation to know and learn more about Cantonese opera, especially those who lack any kind of prior knowledge and experience of the art form.

The difference between the rating scores of the two groups of participants with regards to the enhancement of motivation in knowing and learning more about Cantonese opera was not as great as for the other two items. This might be explained by the lack of extrinsic motivation in the traditional art afforded by the viewing method. In order to sustain and synergize the stimulating effects of immersive media and in order to achieve the ultimate goal of transmitting and preserving Cantonese opera, more research remains to be undertaken that allows future generations to understand the cultural values and representativeness of traditional arts to society as a whole.

## 6 Concluding Remarks

Findings from this study have revealed the potential of immersive media to better engage adolescents in Cantonese opera theatre performance. The use of immersive media was found to be more realistic, interactive, impressive, and motivational than conventional media, supporting the hypothesis that it could be used to promote the appreciation and learning of Cantonese opera or other traditional arts in the future.

Based on the success of this pilot project, further dissemination has been initiated in order to share the experience of immersive media with the wider general public through community engagement (see Fig. 2). With continuing efforts to integrate modern technological practices with Cantonese opera, the current, negative impression of this genre as being outdated and old-fashioned might soon be removed. If younger and future generations are more willing to engage in the appreciation and learning of Cantonese opera, the objective of sustaining the traditional arts through transformation will have been achieved. Further developments leading to the 'immersification' of Cantonese opera could include interactive and educational elements between the viewer and the media and a dedicated Cantonese opera repertoire designed specifically for the 360-degree and VR screening approach.

With the ever-increasing and pervasive reach of entertainment and communication technology proving a significant barrier to encouraging the younger generation to engage with the traditional arts, the challenges facing transmission have never been more daunting. In order to both survive and remain sustainable in modern society, conventional art forms have to transform themselves to fit the cultural needs and social changes brought about by the Digital Era. Early examples date back to the operative performances of the 1950s, both from the Western and Chinese arts tradition, when the shift from stage to screen first began (Cheng, 2018; Citron, 2010). Further adaptions

**Fig. 2** Public engagement with the Cantonese opera VR experience

should not simply be encouraged, but must be embraced if the traditional arts are to remain competitive in the rapidly changing sociocultural landscape of the twenty-first century and beyond.

## References

Alatta, R. A., & Freewan, A. (2017). Investigating the effect of employing immersive virtual environment on enhancing spatial perception within design process. *ArchNet-IJAR: International Journal of Architectural Research, 11*(2), 219–238.

Chan, C. Y. (2012). Safeguarding intangible cultural heritage in Hong Kong: A lesson to learn from Cantonese opera. *Master's dissertation. The University of Hong Kong*.. https://doi.org/10.5353/th_b4834428.

Chan, S. Y. (1991). *Improvisation in a ritual context: The music of Cantonese opera.* The Chinese University Press.

Cheng, L. Y. (2018). 香港摩登:五十年代都市發展與香港粵劇發展脈絡[A modern city: The urban development of Hong Kong and the local Cantonese opera productions in the 1950s]. *Journal of Chinese Ritual, Theatre and Folklore, 199,* 213–262.

Choi, K., Yoon, Y. J., Song, O. Y., & Choi, S. M. (2018). Interactive and immersive learning using 360° virtual reality contents on mobile platforms. *Mobile Information Systems, 2018*.https://doi.org/10.1155/2018/2306031.

Christou, C. (2010). Virtual reality in education. In A.Tzanavari, & T. Tsapatsoulis (Eds.), *Affective, Interactive and Cognitive Methods for E-Learning Design: Creating an Optimal Education Experience* (pp. 228–243). IGI Global. https://doi.org/10.4018/978-1-60566-940-3.ch012.

Citron, M. J. (2010). *When opera meets film. Cambridge University Press.* https://doi.org/10.1017/CBO9780511781834.

Curriculum Development Council. (2017). *Arts Education Key Learning Area Curriculum Guide (Primary 1 – Secondary 6).* Hong Kong Government Printer.

Dewey, J. (1916). *Democracy and education.* The Free Press.

Fung Cheung Hung Troupe. (2019, August 9–25). *The legend of mulan.* Live performance in Xiqu Centre, Hong Kong.

Han, I. (2020). Immersive virtual field trips in education: A mixed-methods study on elementary students' presence and perceived learning. *British Journal of Educational Technology, 51*(2), 420–435. https://doi.org/10.1111/bjet.12842.

Howard, M. C., &Gutworth, M. B. (2020). A meta-analysis of virtual reality training programs for social skill development. *Computer and Education, 144.*https://doi.org/10.1016/j.compedu.2019.103707.

Huang, H. M., Rauch, U., & Liaw, S. S. (2010). Investigating learners' attitudes toward virtual reality learning environments: Based on a constructivist approach. *Computer and Education, 55*(3), 1171–1182. https://doi.org/10.1016/j.compedu.2010.05.014.

Jung, T. H., & Tom Dieck, M. C. (2017). Augmented reality, virtual reality and 3D printing for the co-creation of value for the visitor experience at cultural heritage places. *Journal of Place Management and Development, 10*(2), 140–151. https://doi.org/10.1108/JPMD-07-2016-0045.

Lercari, N., Shiferaw, E., Forte, M., & Kopper, R. (2017). Immersive visualization and curation of archaeological heritage data: Çatalhöyük and the Dig@IT App. *Journal of Archaeological Method and Theory, 25*(2), 368–392. https://doi.org/10.1007/s10816-017-9340-4.

Leung, B. W. (2014). Teachers' transformation as learning: Teaching Cantonese opera in Hong Kong schools with a teacher-artist partnership. *International Journal of Music Education, 32*(1), 119–131. https://doi.org/10.1177/0255761413491174.

Leung, B. W. (2018). A proposed model of transmission of Cantonese opera for Hong Kong higher education: From oral tradition. *Arts and Humanities in Higher Education,* In press, 1–23. https://doi.org/10.1177/1474022218791465.

Leung, B. W. (2019). Conception of creativity as personal style of Cantonese opera artists in Hong Kong: A socio-cultural perspective. *Journal of Artistic and Creative Education, 13*(1), 1–13. https://jace.online/index.php/jace/article/view/183.

Leung, B. W., & Leung, E. C. K. (2010). Teacher-artist partnership in teaching Cantonese opera in Hong Kong schools: Student transformation. *International Journal of Education & The Arts, 11*(5). http://www.ijea.org/v11n5/.

Leung, B. W., Mok, M. M. C., Kuo, B.-C., Liu, Z.-Y., Lam, S. M., Ng, G. W. T., Cheng, L., Ma, K. H. H., & Choi, T. W. (2018). An assessment of learning Cantonese opera movement in Hong Kong: Application of the computerised kinetic chain assessment and learning system. In K. J. Kennedy, & J. C. K. Lee (Eds.), *Routledge international handbook on schools and schooling in Asia* (pp. 220–233). Routledge. https://www.routledgehandbooks.com/doi/10.4324/9781315694382-20.

Liu, W., Zeng, N., Pope, Z. C., McDonough, D. J., & Gao, Z. (2019). Acute effects of immersive virtual reality exercise on young adults' situational motivation. *Journal of Clinical Medicine, 8*(11), 1947–1956. https://doi.org/10.3390/jcm8111947.

Makransky, G., Borre-Gude, S., & Mayer, R. E. (2019). Motivational and cognitive benefits of training in immersive virtual reality based on multiple assessments. *Journal of Computer Assisted Learning, 35*(6), 691–707. https://doi.org/10.1111/jcal.12375.

Matsunobu, K. (2018). Cultural policy and the transmission of traditional music in Japan. In B. W. Leung (Ed.), *Traditional musics in the modern world: Transmission, evolution, and challenges* (pp. 41–56). Springer. https://doi.org/10.1007/978-3-319-91599-9_4.

Orbach, E. (1979). Simulation games and motivation for learning: A theoretical framework. *Simulation and Games, 10*(1), 3–40.https://doi.org/10.1177/0037550079101001.

Paletta, L., Grabher, A., Dini, A., Mayr, M., Taberhofer, A., Pokvic, E., Gunzer, W. & Pilz, R. (2019). Immersive imagination in urban oases of mindfulness: The VR-SenseCity toolbox for

sensible, emotional and measurable experiences in future smart cities. In *Proceedings of the 24th international conference on urban planning and regional development in the information society geomultimedia* (pp. 391–400). Competence Center of Urban and Regional Planning.
Piaget, J. (1967). *Biologie et connaissance* [Biology and knowledge]. Gallimard.
Roberto, A., Angela, C., Armando, F., Delfina, M., Donato, P., Mirta, R., et al. (2017). A framework to design, develop, and evaluate immersive and collaborative serious games in cultural heritage. *Journal on Computing and Cultural Heritage, 11*(1), 1–22. https://doi.org/10.1145/3064644.
Rubio-Tamayo, J. L., Barrio, M. G., &García, F. G. (2017). Immersive environment and virtual reality: Systematic review and advances in communication, interaction and simulation. *Multimodal Technologies Interact, 1*(4). https://doi.org/10.1037/0003-066X.55.1.68.
Ryan, R. M., & Deci, E. L. (2000). Self-determination theory and the facilitation of intrinsic motivation, social development, and well-being. *American Psychologist, 55,* 68–78. https://doi.org/10.1037/0003-066X.55.1.68.
Sah, Y. J., & Peng, W. (2019). How does interactivity in a video game render people susceptible? A mediating role of concrete-level action identity in immersive media. *Cyberpsychology: Journal of Psychosocial Research on Cyberspace, 13*(4).http://dx.doi.org/10.5817/CP2019-4-4.
Waldron, J. (2016). An alternative model of music learning and 'Last Night's Fun1': Participatory music making in/as participatory culture in Irish traditional music. *Action, Criticism and Theory in Music Education, 15*(3), 86–112.
Zhang, L., Bowman, D. A., & Jones, C. N. (2019). Exploring effects of interactivity on learning with interactive storytelling in immersive virtual reality. In *Proceedings of the 11th international conference on virtual worlds and games for serious applications* (VS-Games) (pp. 33–40). IEEE. https://doi.org/10.1109/VS-Games.2019.8864531.

**Lee Cheng** is currently an Assistant Professor of the Department of Cultural and Creative Arts at The Education University of Hong Kong (EdUHK), who also serving as the director of EdUHKiLOrk, a laptop orchestra and mobile device ensemble. He is a digital artist-teacher and researcher, with his works featured as part of, amongst others, International Symposium on Electronic Art (ISEA), International Computer Music Conference (ICMC), World Stage Design, and Mana Contemporary Chicago; his research articles have been published by top-ranked journals *including Technology, Pedagogy and Education, Journal of Knowledge Management*, and *Journal of Computer Assisted Learning.*

**Bo-Wah Leung** is currently a Professor and Head of the Department of Cultural and Creative Arts and Director of Research Centre for Transmission of Cantonese Opera at The Education University of Hong Kong. He received the prestigious Musical Rights Award from the International Music Council in 2011 for his project entitled 'Collaborative Project on Teaching Cantonese Opera in Primary and Secondary Schools'. At present, he is Chair of Asia-Pacific Symposium for Music Education Research, Chief Editor of *Asia-Pacific Journal for Arts Education.*

# The Efficacy of the Mobile Application-Based Mindfulness Mediation on Adjustment of College Students in Hong Kong: A Randomized Controlled Trial

**Raymond Chi Fai Chui, Nicolson Yat-Fan Siu, Wai-Man Tsui, and Chloe Yuet-Tung Wong**

**Abstract** Given the rapid development of information technologies, there is an increasing number of mobile application-based mindfulness interventions. Overseas studies have confirmed the positive effect of these interventions on well-being of both clinical and healthy samples. This study aims to explore whether the same effect can be produced on the adjustment of Hong Kong college students. A mobile-based randomized-controlled trial was conducted in this study for 21 days. College students were invited and randomly assigned to engage with the mindfulness intervention ($n = 27$) or a control group listening to classical music ($n = 28$). Results of repeated measures ANOVA indicated a small positive effect on college adjustment but no statistically significant effect on stress and mindfulness. Following the instruction of the application is positively associated with mindfulness and college adjustment. The results suggest that the usage of mobile application-based mindfulness intervention can only generate limited benefits to participants without support from instructors or therapists. The lack of motivation in self-guided well-being application may discourage the regular involvement in the intervention and fail to guarantee substantial positive outcomes.

---

R. C. F. Chui (✉) · W.-M. Tsui
Department of Social Work, Hong Kong Shue Yan University, 10 Wai Tsui Crescent, Braemar Hill, North Point 999077, Hong Kong
e-mail: cfchui@hksyu.edu

W.-M. Tsui
e-mail: wmtsui@hksyu.edu

N. Y.-F. Siu · C. Y.-T. Wong
Department of Counselling and Psychology, Hong Kong Shue Yan University, 10 Wai Tsui Crescent, Braemar Hill, North Point 999077, Hong Kong
e-mail: yfsiu@hksyu.edu

C. Y.-T. Wong
e-mail: yuettung0923@gmail.com



W. W. K. Ma et al. (eds.), *Learning Environment and Design*, Educational Communications and Technology Yearbook,
https://doi.org/10.1007/978-981-15-8167-0_14

**Keywords** Mindfulness · Adjustment · Stress · College students · Mobile application

## 1 Introduction

Hong Kong is a fast-paced city with her culture greatly shaped by Chinese philosophy and value. Because academic studies are regarded as the noblest pursuit and a path to success, expectation of academic achievement is prominent in society. Under this cultural context, the education system in Hong Kong is competitive and selective (Kwan, 2010); the achievement-oriented school experience possibly causes stress salient in students (Long, Huebner, Wedell, & Hills, 2012). Adolescents who earn their entry to college face additional challenges on top of academic stress in a new environment.

An adjustment phase is almost inevitable for freshmen upon college entry for a number of reasons (Dyson & Renk, 2006). Contrary to the previously restrained role trajectories as secondary school students, new experiences and opportunities in college create greatly diverse life paths while most students also gain flexibility and self-direction (Schulenberg, Sameroff, & Cicchetti, 2004). Students have to negotiate and establish autonomy or become independent in living that these demands and pressure often cause anxiety and stress(Gerdes & Mallinckrodt, 1994). Such change in role is one of the developmental challenges in the transition and forms pressure that can easily lead to mental disorders if not relieved in time (Liu, Ping, & Gao, 2019). New expectations also emerge with the transition from society, parents, friends and also themselves. In Hong Kong, traditional Chinese culture values the development of career and the provision of care for one's parents after college. To meet these expectations in Hong Kong of limited social mobility and opportunities, students might be stressed to equip themselves early by developing a plan for career and locate resources and opportunities throughout the college years. These expectations may be another cause of psychological distress (Wang & Heppner, 2002).

College students are emerging adults who are transitioning from adolescence to adulthood (Ramasubramanian, 2017). Emerging adulthood is a transition brief in time that has the potential to extensively change nearly all aspects of an individual's life (Schulenberg et al., 2004). The developmental challenges exist across emotional, personal, social and academic contexts. Students' personal responsibilities, social environments and institutional academic support all undergone rapid and considerable changes (Dvořáková et al., 2017). All these call for major developmental changes in multiple aspects of life; students vary in their pace of development in the adjustment phase. For some, these multiple demands can contribute to alteration in mental health and psychopathology (Schulenberg et al., 2004). For instance, maladaptive coping to these stress often give consequences that are not healthy (DeBerard, Spielmans, & Julka, 2004). Stress can lead to poor academic achievement among adolescent (Schraml, Grossi, & Makower, 2012). The sufferings from heavy academic work make college students at risk for developing anxiety disorders and in turn affect their psychological well-being (Cho, Ryu, Noh, & Lee, 2016). In Hong Kong, there is an alarmingly high prevalence of depression, anxiety and stress symptoms among first-year student in college (Wong, Cheung, Chan, Ma, & Tang, 2006). Measures facilitating college adjustments are necessary to help students to cope with these challenges in a new stage of life.

The mindfulness mediation may help release the stress and enhance the well-being of college students. Mindfulness refers to a state of "awareness that emerges through paying attention in a particular way: on purpose, in the present moment, non-judgementally" (Kabat-Zinn, 1994, p. 4). The mindfulness mediation has been incorporated into interventions for depression treatment, stress reduction and relapse prevention for addictive behaviour, such as Mindfulness-Based Cognitive Therapy, Mindfulness-Based Stress Reduction programme and Mindfulness-Based Relapse Prevention programme. There is a wealth of recent empirical studies that confirm the positive effect of mindfulness mediation on individual well-being (Eberth & Sedlmeier, 2012; Lomas, Medina, Ivtzan, Rupprecht, & Eiroa-Orosa, 2019; Spijkerman, Pots, & Bohlmeijer, 2016). The research question of this study is to examine the effectiveness of the mobile application-based mindfulness intervention in enhancing the adjustment of college students in Hong Kong.

The goal of mindfulness practice is to enhance self-awareness. Paying attention to the present moment purposefully and non-judgmentally can enhance the awareness of physical sensations, cognitions and emotions (Epstein, 1999). Mindfulness practice develops the capability of participants to respond to mental processes skillfully by accepting their thoughts and emotion non-judgmentally as well as observing them objectively (Cho et al., 2016). It cultivates internal awareness by sustainably attending to present moment with an attitude of openness, curiosity and acceptance (Hayes-Skelton & Graham, 2013). This perspective of decentering enables individuals to adopt a broader awareness toward their experience and distance themselves from anxious thoughts and negative emotions. Accepting the emotion can reduce negative emotions and sadness (Wolgast, Lundh, & Viborg, 2011). Mindfulness-based intervention has been confirmed as a means to reduce stress for both clinical and healthy populations (Gotink et al., 2015; Khoury, Sharma, Rush, & Fournier, 2015).

The mindfulness-based intervention focuses on reducing maladaptive and automatic avoidance of negative affect and enhancing distress tolerance (de Dios et al., 2012). It is designed to facilitate the mindfulness mediation practice of individuals to enhance their awareness of triggers, habitual patterns and automatic or reactive behaviour in the everyday life (Bowen, Chawla, & Marlatt, 2011). The intervention is an attentional control training which expects to promote the individual awareness of the sense of choice, compassion and freedom in dealing with negative reactions to experience, decentering oneself from the automatic patterns of thought and redirecting to the present moment (Bowen et al., 2011; Segal, Williams, & Teasdale, 2018). Concepts and practices of body scan, mediation, decentering, loving kindness, acceptance and self-care are usually introduced in mindfulness-based interventions (Segal et al., 2018).

Given the rapid development of information and communication technologies, there is an increasing number of mindfulness intervention delivered through internet or mobile application. Numerous studies have confirmed the positive effect of web-based mindfulness interventions of both clinical and healthy samples. A meta-analysis of 15randomized controlled triad studies revealed that web-based mindfulness intervention has been confirmed with positive and small impact on depression, anxiety, well-being and mindfulness and moderate effect on stress (Spijkerman et al., 2016). Another systematic review supported the positive effects of web-based mindfulness intervention on general health (physical functioning, disability or quality of life) and psychological well-being (depression, anxiety or self-efficacy) among patients with chronic illness (Mikolasek, Berg, Witt, & Barth, 2018). Apart from individual well-being, the web-based mindfulness intervention can also promote social connectedness (Aspy & Proeve, 2017), enhance self-compassion, reduce caregiver burden (Stjernswärd & Hansson, 2017) and pain intensity (Henriksson, Wasara, & Rönnlund, 2016).

Recently, mobile technology has been adopted to deliver mindfulness intervention, which makes the intervention more assessable and conveniently for participants to follow. A brief application-based mindfulness intervention has been proved with effectiveness to protect participants against perceived stress for healthy British adults (Champion, Economides, & Chandler, 2018). In the study, participants were randomized to the self-guided mindfulness mediation group and wait-list control group. Participants were encouraged to follow the application to practice 10–20 minutes of mindfulness exercise for a total of 30 days. The mindfulness group has shown with positive impact on life satisfaction, stress and resilience at day 10 and day 30 of the intervention. Another study revealed that the mindfulness mediation application is effective to deliver the practice to healthy individuals to improve their mental well-being, reduce distress and job strain (Bostock, Crosswell, Prather, & Steptoe, 2019). 288 healthy employees at two UK companies were randomized to a mindfulness mediation practice application group and a wait-list control group. Participants were asked to practice one 10–20-minutes mediation per day for 8 weeks. Results showed that the mediation improved employees' global well-being, positive affect, anxiety, depressive symptoms and job strain. The brief self-guided training of these two studies encouraged the participants to follow and practice mediation at any time convenient to them in each day.

Other studies with shorter intervention period have also concluded with positive outcomes. A cross-countries study revealed that a 10-day brief 10-min mindfulness exercise has medium positive influence on positive affect and small effect on depressive symptoms among participants (Howells, Ivtzan, & Eiroa-Orosa, 2016). In this study, participants in the experimental group were instructed through a smartphone application to follow the daily mindfulness exercises for 10 minutes a day over 10 days while those in the control group were asked to write out an outline of what they did on each day. Results showed that participation in mindfulness application led to gains in positive affect and reduction in depressive symptoms. Even in one brief session of web-based mindfulness has been proved with effect on enhancing mindfulness. In an empirical study, participants were asked to carry out a 5-min mindfulness body scan under the guideline of a mindfulness audio file while participants in the control were asked to take breaths. This brief single session web-based mindfulness intervention was effective to increase the levels of mindfulness of US residents (Mahmood, Hopthrow, & de Moura, 2016).

Results of the above studies indicated that the effect of these web- or mobile-based self-guided applications is comparable to other traditional forms of brief mindfulness training (Cavanagh et al., 2013; Howells et al., 2016). Different from the conventional mindfulness intervention, mobile application-based mindfulness intervention is mainly self-help, less intensive, lack of interaction with an instructor and sharing with group mates. However, they are more assessable, cost-effective, user-friendly and feasible tool to disseminate intervention instruction (Howells et al., 2016). The instruction from these applications is also convenient and easy to follow. Positive results of the mobile application-based mindfulness intervention have been supported by numerous studies (Stjernswärd & Hansson, 2017). Studies have confirmed the positive influence of the mobile application-based intervention on mindfulness(Rosen, Paniagua, Kazanis, Jones, & Potter, 2018), positive affect (Howells et al., 2016), depression (Bostock et al., 2019; Howells et al., 2016), sleep quality (Lengacher et al., 2018), craving-rated eating and overeating behaviour (Mason, Jhaveri, Cohn, & Brewer, 2018).

There are several advantages of the web-based or mobile application-based interventions as compared with traditional workshop-based interventions. They are easy to access and available at any time; require fewer personnel and economic resources for delivery; allow participants to work at their own pace and in any venue where is comfortable to them and permit participants to be anonymous in seeking the service (Henriksson et al., 2016; Mikolasek et al., 2018; Spijkerman et al., 2016). Even a brief single session of web-based mindfulness has been proved effect on enhancing mindfulness(Mahmood et al., 2016). However, there is a lack of Asian studies to support the effectiveness of the mobile application-based mindfulness intervention. Only two studies have been found to confirm the benefits of web-based mindfulness intervention. One study has revealed a small and temporary effect of the web-based mindfulness intervention in reducing depressive symptoms of clinical samples (Noguchi, Sekizawa, So, Yamaguchi, & Shimizu, 2017) and another found a medium positive influence on positive affect and small effect on depressive symptoms among healthy participants (Howells et al., 2016).

Mindfulness interventions are developed from Western Psychology for psychological distress and maladaptive behaviour (Shapiro, 2009). However, the Eastern and Western conceptualizations of mindfulness may have significant differences (Christopher, Charoensuk, Gilbert, Neary, & Pearce, 2009). The Western conceptualization of mindfulness may not generalize to Asian cultures. A study revealed that American college students have higher levels of mindfulness than Turkish students (Özyeşil, 2012). A strong influence from social value under collectivist culture prevents individuals to develop a clear sense of awareness and non-judgemental attention that promote mindfulness. Since the mindfulness experience differs between Eastern and Western cultures, the present study is of significance to confirm the efficacy of the mobile application-based mindfulness intervention on the psychological well-being of Chinese college students in an Asian cultural context.

The mindfulness mediation application is effective to deliver the practice to healthy individuals to improve their job strain (Bostock et al., 2019) and general stress(Carissoli, Villani, & Riva, 2015; Champion et al., 2018; Yang, Schamber, Meyer, & Gold, 2018). Empirical studies have also confirmed the effect of the mindfulness-based interventions for youth. These interventions can improve subjective and objective sleep quality (Blake et al., 2016) and positive thinking (Cho et al., 2016) as well as reducing general stress and anxiety (Blake et al., 2016; Call, Miron, & Orcutt, 2014) and examination-related anxiety (Cho et al., 2016). Therefore, it is anticipated that the mobile application-based mindfulness intervention can help college students to alleviate their general stress and improve psychological well-being.

Based on the results of previous studies, the following hypotheses are formulated:

1. The mobile application-based mindfulness intervention is effective to reduce the stress of college students
2. The mobile application-based mindfulness intervention is effective to enhance the mindfulness and adjustment of college students.

## 2 Methods

### *2.1 Participants*

Our sample of undergraduate students was recruited from Hong Kong Shue Yan University through posted notices, broadcast emails and direct invitations from faculty staff. Individuals with regular practice of mindfulness were excluded from the study. 110 participants were first consented to voluntarily participate in the present study. Then they were randomly assigned into the control group (n = 55) or the experimental group (n = 55). Participants in the control group were required to listen to the classical music from a mobile application, while for those who were in the experimental group were required to follow the instructions from a mobile application to practice mindfulness. Both groups were required to spend 5–8 minutes with the corresponding mobile application daily for 3 consecutive weeks. Only 30 participants from the control group and 31 participants form the experimental group have completed the intervention and the online survey for the pre–post-assessment. Two from the control group and four from the experimental group were further excluded from the present study due to either a failure to adhere to the instructions or a completion of the online survey in less than 5 minutes. Therefore, a final sample of 28 participants from the control group and 27 participants from the experimental group was included for data analysis. The demographic information is shown in Table 1. The control group consisted of 17 female and 11 male participants with a mean age of 19.29 (SD = 1.46), while the experiment group consisted of 20 female and 7 male participants with a mean age of 19.30 (SD = 3.12). There was no significant difference between the groups in terms of age and gender.

**Table 1** Demographic characteristics of control and experimental groups

| | Control (n = 28) | Experimental (n = 27) | $t/\chi 2$ | $p$ |
|---|---|---|---|---|
| Age | 19.29 (1.46) | 19.30 (3.12) | −0.016 | 0.987 |
| Gender | | | 1.114 | 0.291 |
| – Female | 17 | 20 | | |
| – Male | 11 | 7 | | |

## 2.2 *Procedures*

Participants were invited to participate in the present study through posted notices, broadcast emails and direct invitations from faculty staff. Online informed consent, demographic information and details of mindfulness experience were first obtained from all participants. Except those who have regular mindfulness practice, the rest were randomized into control and experimental groups. A briefing session was held for each group and participants must attend in order to be qualified to join the study. During the briefing section, the participants were required to fill the pre-assessment survey, download the mobile application and register an account. Experimenter also showed how to use the mobile application and practised the first session with the participants. For the control group, we have adopted the mobile application, Muzik Air—Classical Music, which was available on both Android (https://play.google.com/store/apps/details?id=com.muzik&hl=en) and iOS (https://apps.apple.com/us/app/muzik-air-classical-music/id683134818) platform. Participants were required to use this mobile application and choose a piece of classical music to listen for 8 minutes per day for 3 consecutive weeks. A systemic review of empirical studies has confirmed the positive effect of brief music listening interventions on individual well-being (Daykin et al., 2018).

While for the experimental group, the mobile application, Newlife330, which was also available on both Android and iOS platform, was adopted (for further information, access the website https://newlife330.hk/?section=3). This mobile application is developed by New Life Psychiatric Rehabilitation Association to promote mindful experiences in our everyday lives and achieve psychological and spiritual well-being. It guides the user to practice mindfulness activities, such as mindful breathing practice, body scan, mindful stretching, sitting mindful meditation and mindful walking. Similarly, participants from the experimental group were required to practice mindfulness under the guidance of the mobile application for 8 minutes per day for 3 consecutive weeks. Both groups were suggested to choose a tranquil and comfortable environment during the intervention sessions. A record form was also distributed, and participants were required to fill in their experiences and reflections. At last, a concluding session was held for each group and experimenter conducted the last intervention session with the participants. Participants were required to fill in the post-assessment survey and asked a few questions concerning their experiences on using the mobile application as a guided intervention. To compensate the time and effort spent by the participants, they were award HK$40 and a chance for lucky draw. The prizes included HK$10, food and drink coupons.

Previous studies have argued that no-intervention control group (passive control) will make the result difficult to interpret because placebo effects and related factors cannot be ruled out (Au, Gibson, Bunarjo, Buschkuehl, & Jaeggi, 2020). Therefore, an active control design has been adopted in the present study in which participants are engaged in an alternate intervention—listening to classical music. Listening to classical music can bring positive emotions like calm and peace and this is the reason for using it as an active control group. The research questions of this study are to investigate whether the particular ingredients of the mindfulness intervention and not the contact aspects of the intervention can change students' adjustment; and whether mindfulness intervention can bring a greater effect on positive emotions compared with listening to classical music (Brigham, Feaster, Wakim, & Dempsey, 2009). Indeed, it is common that listening to classical music is used as an active control group for mindfulness-based intervention (Gu, Cavanagh, & Strauss, 2018).

## 2.3 *Measures*

Data for the present study were collected using self-reported online survey for pre–post-assessment. A demographic questionnaire was done before the pre-assessment to collect the age, gender and mindfulness experience from participants. A record form was included for participants to record the date and time of using the mobile application as well as the brief description of their experience of listening to music or practising mindfulness. A few questions were also raised at the end of the post-assessment to survey their subjective experience and perception of the helpfulness and the feasibility of using mobile application as guidance for mindfulness practice. The pre–post-assessment included the following three inventories to explore the benefits of the interventions on perceived stress, attention and awareness of present moment experiences in daily functioning and college adjustment.

**Perceived Stress Scale**. The Perceived Stress Scale (PSS; Cohen, Kamarck, & Mermelstein, 1983, 1988) is a widely used, reliable and valid self-reported psychological instrument for measuring the perception of stress. It included 10 items to assess the extent to which individuals appraised life events as stressful in the previous month. Questions were rated on a 5-point Likert scale ranging from 0 (never) to 4 (very often), with total score ranged from 0 to 40. A higher score indicates a greater perceived stress. Sample items include, 'In the last month, how often have you felt nervous and stressed?', 'In the last month, how often have you found that you could not cope with all the things that you had to do?' and 'In the last month, how often have you felt difficulties were piling up so high that you could not overcome them?'. The PSS has shown a satisfactory internal consistency and good test–retest reliability in previous studies with college student samples (Al kalaldeh & Abu Shosha, 2012; Mitchell, Crane, & Kim, 2008). The Cronbach's alpha coefficient for the Chinese scale reported in the previous study was 0.86 and 0.84 for the pre-assessment and post-assessment, respectively (Chen, Yu, & Xiong, 2019).

**Mindfulness Attention Awareness Scale**. The Mindfulness Attention Awareness Scale (MAAS; Brown & Ryan, 2003) is a 15-item self-reported instrument that measures individuals' attention and awareness of present moment experiences in daily functioning. The MAAS focuses on two aspects of mindfulness: awareness and attention, as opposed to other attributes that have been associated with mindfulness, for example, empathy, gratitude and acceptance (Brown & Ryan, 2003). Items were rated upon a 7-point Likert scale, ranging from 1 (almost never) to 7 (almost always). The higher the score, the lower the levels of dispositional mindfulness. Sample items include, 'I could be experiencing some emotion and not be conscious of it until some time later'.'It seems I am "running on automatic," without much awareness of what I'm doing', and 'I drive places on 'automatic pilot' and then wonder why I went there'. The MAAS has been validated in college and working adults (Brown & Ryan, 2003, 2004) and the Cronbach's alpha coefficients were all above 0.80.

**College Adjustment Test**. The College Adjustment Test (CAT; Pennebaker, Colder, & Sharp, 1990) is a 19-item survey to measure coping in relation to college adjustment. It aimed to measure the extent of students' experience in various thoughts and feelings about being in college. Major life events that affected all aspects of individuals' functioning, including emotions, eating habits, physical health, motivation, social behaviours and even views about oneself were assessed on a 7-point Likert scale, ranging from 1 (not at all) to 7 (a great deal). Apart from the overall adjustment rating, three sub-scales emerged and they were negative affect, positive affect and home sickness. The higher the overall score and the sub-scale scores, the higher the level of the corresponding attributes. The negative affect scale consists of nine items (e.g.'to what degree have you worried about being in college in general?'); the positive affect consists of six items (e.g. 'to what degree have you liked college in general?')and homesick scale consists of six items (e.g. 'To what degree have you missed your parents and other family members?'). The CAT has exhibited a satisfactory internal consistency in previous studies with college student samples (Datu, 2012). The Cronbach's alpha coefficients for negative affect sub-scale are 0.86–0.87, positive affect sub-scale is 0.74–0.76 and homesick scale is 0.79 (Palai & Kiumar, 2016; Ranney & Troop-Gordon, 2012).

In the present study, Cronbach's alpha coefficients for the pre- and post-assessments of all scales were all above 0.07 and the internal consistency is acceptable. For the PSS, the Cronbach's alpha coefficient is 0.86 for pre-assessment and 0.89 for post-assessment. Regarding the MAAS, the Cronbach's alpha coefficient is 0.80 for pre-assessment and 0.87 for post-assessment. While for CAT, the Cronbach's alpha coefficient for the total score is 0.76 for pre-assessment and 0.77 for post-assessment. The Cronbach's alpha coefficient for sub-scales of positive affect is 0.81 and 0.73; negative affect is 0.87 and 0.84 and homesick is 0.71 and 0.79, respectively.

## 3 Results

All statistical analyses were conducted with SPSS 25.0 (IBM Corp., Armonk, N.Y., USA). Both groups of participants have participated in the mobile application intervention for 3 weeks and completed the pre–post-assessment online survey. A repeated measure analysis of variance (ANOVA) was conducted on the three measures (i.e. PSS, MAAS and CAT) with time (i.e. Pre- and post-assessment). There was a significant interaction effect found for CAT, $F\ (1, 53) = 5.40, p = 0.024$, but not for PSS, $F\ (1, 53) = 2.66, p = 0.109$ and MAAS, $F\ (1, 53) = 0.174, p = 0.678$ (Fig. 1). There was also no significant main effect found for MAAS, $F\ (1, 53) = 0.149, p = 0.701$ and PSS, $F\ (1, 53) = 1.77, p = 0.190$. Further analyses have been conducted to explore the interaction effect of CAT with time. Paired sample *t*-test results showed a significant increase in the CAT overall adjustment score for the experimental group, $t\ (26) = -2.70, p = 0.012$, but not the control group, $t\ (27) = -2.70, p = 0.012$. For the experimental group, CAT sub-scale analyses showed homesick, $t\ (26) = 2.35, p = 0.027$, was significantly decreased while the changes in positive, $t\ (26) = -1.46, p = 0.157$, and negative affect, $t\ (26) = 1.96, p = 0.060$, were not significant (Table 2).

Analyses have been done on the experience of using the mobile app for both control and experimental groups (Table 3). Majority of the participants are using the mobile app every day and there were no differences between control and experimental groups, $\chi^2\ (3, 55) = 4.16, p = 0.245$. In addition, both groups have a similar level of interest, $t\ (53) = 0.352, p = 0.726$, perception of the usefulness of the intervention, $t\ (53) = 0.862, p = 0.392$, and the mobile App $t\ (53) = -1.28, p = 0.206$. However, there was a significant difference on the ease of mastering the intervention through the mobile application, $t\ (53) = 0.4.28, p = 0.000$. More participants from the experimental group agree that it was easy to acquire the required skills through the mobile

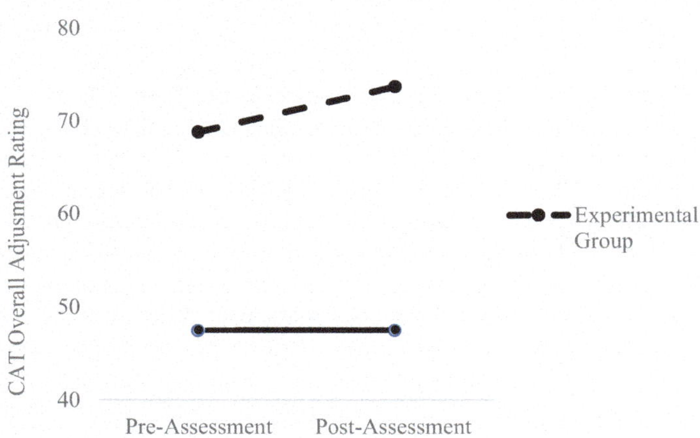


**Fig. 1** Interaction effect of CAT overall adjustment score with pre–post-assessment

**Table 2** Mean, standard deviations and within-group t-test scores for variable scores before and after the intervention

| | Control group (n = 28) | | | | | Experimental group (n = 27) | | | | |
|---|---|---|---|---|---|---|---|---|---|---|
| | Pre-Ax | Post-Ax | *T* | *p* | *d* | Pre-Ax | Post-Ax | *t* | *p* | *d* |
| Perceived Stress Scale[a] | 29.54 (6.21) | 29.71 (5.51) | −0.29 | 0.773 | 0.029 | 33.81 (5.42) | 32.07 (5.72) | 1.72 | 0.098 | −0.312 |
| Mindfulness Attention Awareness Scale[b] | 47.54 (9.96) | 47.57 (12.29) | −0.02 | 0.982 | 0.003 | 50.44 (9.06) | 49.52 (8.83) | 0.54 | 0.594 | −0.103 |
| College Adjustment Scale—Overall[c] | 82.36 (12.95) | 82.11 (12.89) | 0.19 | 0.848 | −0.019 | 68.78 (11.90) | 73.67 (11.50) | −2.70 | 0.012 | 0.418 |
| – Positive Affect[d] | 28.14 (5.78) | 28.29 (5.03) | −0.23 | 0.819 | 0.028 | 25.63 (5.85) | 26.56 (4.75) | −1.46 | 0.157 | 0.175 |
| – Negative Affect[e] | 32.57 (8.86) | 33.07 (8.31) | −0.55 | 0.586 | 0.058 | 42.11 (10.02) | 39.11 (9.09) | 1.96 | 0.060 | −0.314 |
| – Homesick[f] | 24.14 (5.32) | 24.14 (5.50) | 0.00 | 1.000 | 0.000 | 27.74 (4.37) | 25.93 (4.51) | 2.35 | 0.027 | −0.408 |

*Note*
[a]The score ranged from 0 to 40, the higher the score, the higher the perceived stress
[b]The score ranged from 15 to 105, the higher the score, the worse the performance of attention and awareness on present-moment experiences in daily functioning
[c]The score ranged from 19 to 133, the higher the score, the better the college adjustment
[d]The score ranged from 6 to 42, the higher the score, the more the positive affect one felt
[e]The score ranged from 9 to 63, the higher the score, the more the negative affect one felt
[f]The score ranged from 6 to 42, the higher the score, the more the homesick one felt

**Table 3** Experience of using the mobile application for both control and experimental groups

| | Control (n = 28) | Experimental (n = 27) | $t/\chi^2$ | $p$ |
|---|---|---|---|---|
| Practice Adherence | | | 4.16 | 0.245 |
| – Everyday | 18 | 10 | | |
| – 15–20 days | 3 | 6 | | |
| – 7–14 days | 2 | 3 | | |
| – Less than 7 days | 5 | 8 | | |
| Perceive the intervention as useful | | | 0.862 | 0.392 |
| – Strongly disagree | 0 | 0 | | |
| – Disagree | 0 | 1 | | |
| – Little disagree | 2 | 1 | | |
| – Neutral | 10 | 8 | | |
| – Little agree | 9 | 6 | | |
| – Agree | 7 | 9 | | |
| – Strongly agree | 0 | 2 | | |
| Perceive the application as useful | | | −1.28 | 0.206 |
| – Not at all | 0 | 0 | | |
| – Little | 3 | 5 | | |
| – Somewhat | 19 | 9 | | |
| – Much | 4 | 8 | | |
| – Very much | 2 | 5 | | |
| Ease of mastering the invention through mobile application | | | 4.28 | 0.000 |
| – Strongly disagree | 0 | 0 | | |
| – Disagree | 5 | 1 | | |
| – Neutral | 18 | 7 | | |
| – Agree | 5 | 18 | | |
| – Strongly agree | 0 | 1 | | |
| Interested in the program | | | 0.352 | 0.726 |
| – Not at all | 0 | 0 | | |
| – A little bit | 8 | 8 | | |
| – Somewhat | 18 | 15 | | |
| – Much | 1 | 3 | | |
| – Very much | 1 | 1 | | |

application compared with the participants from the control group. Correlation analyses have been done to further examine the relationship between the experience of using the mobile application with PSS, MAAS and CAT. There were no associations found between all variables for the control group. While for the experimental group, associations were found between practice adherence and CAT and MAAS but not PSS. Those who were more adhered to the use of the mobile application for mindfulness practice were more likely to have greater increase in the CAT overall adjustment score, $r_\tau = 0.416, p = 0.007$, and a greater decrease in the MAAS score, $r_\tau = -0.348, p = 0.025$.

## 4 Discussion and Conclusion

The present study examined the effects of mobile application-based mindfulness intervention of adjustment of college students. In contrast to the finding of previous studies (Blake et al., 2016; Call et al., 2014), the present results failed to support the hypotheses that the mindfulness mediation application is effective to reduce stress and enhance mindfulness. However, the data confirmed the small effect of the mobile application-based intervention in improving well-being of college students. There was no significant difference between the intervention and the control group in the stress and mindfulness measures. A lack of direct contact with instructors or therapists may hinder the effects of mobile-based intervention (Boggs et al., 2014). Though change in the outcomes is not significant for the two measures, statistical trends suggest that there is a small improvement from pre-test to post-test, implying a longer period of participation in the intervention may create more promising effects of psychological functioning (Spijkerman et al., 2016; Stjernswärd & Hansson, 2017). The increased levels of college adjustment in participants may indicate that the intervention is effective.

The authors only reminded participants to practice the intervention via email once a week. This may affect the level of engagement of participants. The lack of motivation of college students reduced their time and effort in practice and in turn reduced the effects on individual well-being. When conducting sub-group analyses for the experimental group, those who followed the required instruction of the mobile-based intervention to practice every day appeared to have higher levels of improvements. This implies that the mobile-based mindfulness intervention is effective when used regularly.

Results of the study suggest that the usage of mobile application-based mindfulness intervention may not be able to generate substantial benefits for participants without support from instructors or therapists. A Lack of motivation in self-guided well-being application may discourage the regular involvement of non-clinical individuals in the intervention and fail to guarantee positive outcomes. Mindfulness cannot be facilitated without regular practice (Glück & Maercker, 2011). The intervention also tends to have small effects on healthy population who have less room for improvement (Spijkerman et al., 2016). Further study can be conducted to explore why the intervention is unable to enhance mindfulness and reduce stress.

One possible improvement to the mobile-based intervention is to provide a support function. Both human support and automated support in web-based intervention have proven effective in improving the effectiveness (Spijkerman et al., 2016). Reminders and online support can be added in the mobile-based intervention to enhance the retention rates and satisfaction of participants (Boggs et al., 2014). Further studies should be conducted to examine the effect of automatic support or human support in improving the effectiveness of mobile-based interventions.

Although the present study has found the positive effects of the mobile application as guidance of mindfulness intervention, there are several research caveats that warrant attention. First, even though the participants were randomly assigned into the control or the experimental group, the two groups exhibited some degree of differences on the baseline performance on PSS, MAAS and CAT. The effect of listening to classical music appeared to be less effective when compared with mindfulness intervention. However, this may due to the fact that the participants from the control group were having a better well-being at baseline. Therefore, cautions have to be taken when considering the results from between group comparison. Second, it was difficult to monitor whether the participants have used the mobile application to either listen to classical music or practising mindfulness everyday. The present research relied on their subjective report to gather the information on their usage and involvement in the intervention. Last but not least, it is important to examine the motivation of the participants in joining the mobile application intervention program. The present research did not look into how motivation may affect the effectiveness of using the mobile app intervention program. From the existing literature about adopting evidence-based mobile application interventions, motivation is an important factor that facilitates the effectiveness of the invention (Pagoto, Schneider, Jojic, DeBiasse, & Mann, 2013). Thus, it is believed that strategies that help improving motivation and assist with problem-solving while using the application may reduce the participants' resistance and increase their willingness in using the mobile application.

## References

Al kalaldeh, M. T., & Abu Shosha, G. G. (2012). The application of perceived stress scale in health care studies. An analysis of literature. *International Journal of Academic Research, 4*, 45–50. https://doi.org/10.7813/2075-4124.2012/4-4/B.6.

Aspy, D. J., & Proeve, M. (2017). Mindfulness and loving-kindness meditation: Effects on connectedness to humanity and to the natural world. *Psychological Reports, 120*(1), 102–117. https://doi.org/10.1177/0033294116685867.

Au, J., Gibson, B. C., Bunarjo, K., Buschkuehl, M., & Jaeggi, S. M. (2020). Quantifying the difference between active and passive control groups in cognitive interventions using two meta-analytical approaches. *Journal of Cognitive Enhancement*. https://doi.org/10.1007/s41465-020-00164-6.

Blake, M., Waloszek, J. M., Schwartz, O., Raniti, M., Simmons, J. G., Blake, L., … Allen, N. B. (2016). The SENSE study: Post intervention effects of a randomized controlled trial of a cognitive–behavioral and mindfulness-based group sleep improvement intervention among at-risk adolescents. *Journal of Consulting and Clinical Psychology, 84*(12), 1039–1051. https://doi.org/10.1037/ccp0000142.

Boggs, J. M., Beck, A., Felder, J. N., Dimidjian, S., Metcalf, C. A., & Segal, Z. V. (2014). Web-based intervention in mindfulness meditation for reducing residual depressive symptoms and relapse prophylaxis: A qualitative study. *Journal of Medical Internet Research, 16*(3), 332–343. https://doi.org/10.2196/jmir.3129.

Bostock, S., Crosswell, A. D., Prather, A. A., & Steptoe, A. (2019). Mindfulness on-the-go: Effects of a mindfulness meditation app on work stress and well-being. *Journal of Occupational Health Psychology, 24*(1), 127–138. https://doi.org/10.1037/ocp0000118.

Bowen, S., Chawla, N., & Marlatt, G. A. (2011). *Mindfulness-based relapse prevention for addictive behaviors*. New York: The Guilford Press.

Brigham, G. S., Feaster, D. J., Wakim, P. G., & Dempsey, C. L. (2009). Choosing a control group in effectiveness trials of behavioral drug abuse treatments. *Journal of Substance Abuse Treatment, 37*(4), 388–397.

Brown, K. W., & Ryan, R. M. (2003). The benefits of being present: The role of mindfulness in psychological well-being. *Journal of Personality and Social Psychology, 84,* 822–848.

Brown, K. W., & Ryan, R. M. (2004). Perils and promise in defining and measuring mindfulness: Observations from experience. *Clinical Psychology: Science and Practice, 11,* 242–248.

Call, D., Miron, L., & Orcutt, H. (2014). Effectiveness of brief mindfulness techniques in reducing symptoms of anxiety and stress. *Mindfulness, 5*(6), 658–668.

Carissoli, C., Villani, D., & Riva, G. (2015). Does a meditation protocol supported by a mobile application help people reduce stress? Suggestions from a controlled pragmatic trial. *Cyberpsychology, Behavior, and Social Networking, 18*(1), 46–53. https://doi.org/10.1089/cyber.2014.0062.

Cavanagh, K., Strauss, C., Cicconi, F., Griffiths, N., Wyper, A., & Jones, F. (2013). A randomised controlled trial of a brief online mindfulness-based intervention. *Behaviour Research and Therapy, 5*(9), 573–578.

Champion, L., Economides, M., & Chandler, C. (2018). The efficacy of a brief app-based mindfulness intervention on psychosocial outcomes in healthy adults: A pilot randomised controlled trial. *PLoS ONE, 13*(12).

Chen, W., Yu, S., & Xiong, D. (2019). Effects of Tai Chi intervention on perceived stress, anxiety, and sleep in college students. *Advances in Physical Education, 10*(1), 54–67.

Cho, H., Ryu, S., Noh, J., & Lee, J. (2016). The effectiveness of daily mindful breathing practices on test anxiety of students. *PLoS ONE, 11*(10).

Christopher, M. S., Charoensuk, S., Gilbert, B. D., Neary, T. J., & Pearce, K. L. (2009). Mindfulness in Thailand and the United States: A case of apples versus oranges? *Journal of Clinical Psychology, 65*(6), 590–612.

Cohen, S., Kamarck, T., & Mermelstein, R. (1983). A global measure of perceived stress. *Journal of Health and Social Behavior, 24,* 386–396.

Cohen, S., Kamarck, T., & Mermelstein, R. (1988). Perceived stress in a probability sample of the United States. In S. Spacapan & S. Oskamp (Eds.), *The social psychology of health* (pp. 31–67). Newbury Park, CA: Sage.

Datu, J. A. D. (2012). Drawing predictive determinants of college adjustment: Perspectives from two private sectarian collegiate institutions. *Researchers World, 3*(2), 16.

Daykin, N., Mansfield, L., Meads, C., Julier, G., Tomlinson, A., Payne, A., … Burnett, A. (2018). What works for wellbeing? A systematic review of wellbeing outcomes for music and singing in adults. *Perspectives in public health, 138*(1), 39–46.

de Dios, M. A., Herman, D. S., Britton, W. B., Hagerty, C. E., Anderson, B. J., & Stein, M. D. (2012). Motivational and mindfulness intervention for young adult female marijuana users. *Journal of Substance Abuse Treatment, 42*(1), 56–64.

DeBerard, M. S., Spielmans, G., & Julka, D. L. (2004). Predictors of academic achievement and retention among college freshmen: A longitudinal study. *College Student Journal, 38*(1), 66–80.

Dvořáková, K., Kishida, M., Li, J., Elavsky, S., Broderick, P. C., Agrusti, M. R., & Greenberg, M. T. (2017). Promoting healthy transition to college through mindfulness training with first-year college students: Pilot randomized controlled trial. *Journal of American College Health, 65*(4), 259–267. https://doi.org/10.1080/07448481.2017.1278605.

Dyson, R., & Renk, K. (2006). Freshmen adaptation to university life: Depressive symptoms, stress, and coping. *Journal of Clinical Psychology, 62*(10), 1231–1244. https://doi.org/10.1002/jclp.20295.

Eberth, J., & Sedlmeier, P. (2012). The effects of mindfulness mediation: A meta-analysis. *Mindfulness, 3*(3), 174–189.

Epstein, R. M. (1999). Mindful practice. *JAMA, 282*(9), 833–839.

Gerdes, H., & Mallinckrodt, B. (1994). Emotional, social, and academic adjustment of college students: A longitudinal study of retention. *Journal of Counseling & Development, 72*(3), 281–288. https://doi.org/10.1002/j.1556-6676.1994.tb00935.x.

Glück, T. M., & Maercker, A. (2011). A randomized controlled pilot study of a brief web-based mindfulness training. *BMC Psychiatry, 11*. https://doi.org/10.1186/1471-244X-11-175.

Gotink, R. A., Chu, P., Busschbach, J. J. V., Benson, H., Fricchione, G. L., & Hunink, M. G. M. (2015). Standardised mindfulness-based interventions in healthcare: An overview of systematic reviews and meta-analyses of RCTs. *PLoS ONE, 10,* e0124344. https://doi.org/10.1371/journal.pone.0124344.

Gu, J., Cavanagh, K., & Strauss, C. (2018). Investigating the specific effects of an online mindfulness-based self-help intervention on stress and underlying mechanisms. *Mindfulness, 9*(4), 1245–1257.

Hayes-Skelton, S., & Graham, J. (2013). Decentering as a common link among mindfulness, cognitive reappraisal, and social anxiety. *Behavioural and Cognitive Psychotherapy, 41*(3), 317–328.

Henriksson, J., Wasara, E., & Rönnlund, M. (2016). Effects of eight-week-web-based mindfulness training on pain intensity, pain acceptance, and life satisfaction in individuals with chronic pain. *Psychological Reports, 119*(3), 586–607. https://doi.org/10.1177/0033294116675086.

Howells, A., Ivtzan, I., & Eiroa-Orosa, F. J. (2016). Putting the 'app' in happiness: A randomised controlled trial of a smartphone-based mindfulness intervention to enhance wellbeing. *Journal of Happiness Studies: An Interdisciplinary Forum on Subjective Well-Being, 17*(1), 163–185. https://doi.org/10.1007/s10902-014-9589-1.

Kabat-Zinn, J. (1994). *Wherever you go, there you are: Mindfulness mediation in everyday life.* New York: Hyperion.

Khoury, B., Sharma, M., Rush, S. E., & Fournier, C. (2015). Mindfulness-based stress reduction for healthy individuals: A meta-analysis. *Journal of Psychosomatic Research, 78*, 519–528. https://doi.org/10.1016/j.jpsychores.2015.03.009. PMID: 25818837.

Kwan, Y. K. (2010). Life satisfaction and self-assessed health among adolescents in Hong Kong. *Journal of Happiness Studies, 11*(3), 383–393. https://doi.org/10.1007/s10902-009-9146-5.

Lengacher, C. A., Reich, R. R., Ramesar, S., Alinat, C. B., Moscoso, M., Cousin, L., … Park, J. Y. (2018). Feasibility of the mobile mindfulnessbased stress reduction for breast cancer (mMBSR(BC)) program for symptom improvement among breast cancer survivors. *Psycho-Oncology, 27*(2), 524–531. https://doi.org/10.1002/pon.449.

Liu, X., Ping, S., & Gao, W. (2019). Changes in undergraduate students' psychological well-being as they experience university life. *International Journal of Environmental Research and Public Health, 16*(16), 2864. https://doi.org/10.3390/ijerph16162864.

Lomas, T., Medina, J. C., Ivtzan, I., Rupprecht, S., & Eiroa-Orosa, F. J. (2019). A systematic review and meta-analysis of the impact of mindfulness-based interventions on the well-being of healthcare professionals. *Mindfulness, 10*(7), 1193–1216.

Long, R. F., Huebner, E. S., Wedell, D. H., & Hills, K. J. (2012). Measuring school-related subjective well-being in adolescents. *American Journal of Orthopsychiatry, 82*(1), 50.

Mahmood, L., Hopthrow, T., & de Moura, G. R. (2016). A moment of mindfulness: Computer-mediated mindfulness practice increases state mindfulness. *PLoS ONE, 11*(4). Retrieved from https://search.ebscohost.com/login.aspx?direct=true&db=psyh&AN=2016-59889-001&site=ehost-live&scope=site.

Mason, A. E., Jhaveri, K., Cohn, M., & Brewer, J. A. (2018). Testing a mobile mindful eating intervention targeting craving-related eating: Feasibility and proof of concept. *Journal of Behavioral Medicine, 41*(2), 160–173. https://doi.org/10.1007/s10865-017-9884-5.

Mikolasek, M., Berg, J., Witt, C. M., & Barth, J. (2018). Effectiveness of mindfulness- and relaxation-based eHealth interventions for patients with medical conditions: A systematic review and synthesis. *International Journal of Behavioral Medicine, 25*(1), 1–16. https://doi.org/10.1007/s12529-017-9679-7.

Mitchell, A. M., Crane, P. A., & Kim, Y. (2008). Perceived stress in survivors of suicide: Psychometric properties of the Perceived Stress Scale. *Research in Nursing & Health, 31*, 576–585.

Noguchi, R., Sekizawa, Y., So, M., Yamaguchi, S., & Shimizu, E. (2017). Effects of five-minute internet-based cognitive behavioral therapy and simplified emotion-focused mindfulness on depressive symptoms: A randomized controlled trial. *BMC Psychiatry, 17*. https://doi.org/10.1186/s12888-017-1248-8.

Özyeşil, Z. (2012). Mindfulness and psychological needs: A cross-cultural comparison. *Elementary Education Online, 11*(1), 151–160.

Pagoto, S., Schneider, K., Jojic, M., DeBiasse, M., & Mann, D. (2013). Evidence-based strategies in weight-loss mobile apps. *American Journal of Preventive Medicine, 45*(5), 576–582.

Palai, P. K., & Kiumar, P. (2016). Relationship among stress, adjustment and homesickness in university students. *International Journal of Innovative Research in Multidisciplinary Field, 2*(6), 100–106.

Pennebaker, J. W., Colder, M., & Sharp, L. K. (1990). Accelerating the coping process. *Journal of Personality and Social Psychology, 58,* 528–537.

Ramasubramanian, S. (2017). Mindfulness, stress coping and everyday resilience among emerging youth in a university setting: A mixed methods approach. *International Journal of Adolescence and Youth, 22*(3), 308–321. https://doi.org/10.1080/02673843.2016.1175361.

Ranney, J. D., & Troop-Gordon, W. (2012). Compter-mediated communication with distant friends: Relations with adjustment during students' first semester in college. *Journal of Educational Psychology, 104*(3), 848.

Rosen, K. D., Paniagua, S. M., Kazanis, W., Jones, S., & Potter, J. S. (2018). Quality of life among women diagnosed with breast cancer: A randomized waitlist controlled trial of commercially available mobile app-delivered mindfulness training. *Psycho-Oncology*. https://doi.org/10.1002/pon.4764.

Schraml, K. P., A, Grossi, G., & Makower, I. (2012). Chronic stress and its consequences on subsequent academic achievement among adolescents. *Journal of Education and Developmental Psychology, 2*(1), 69–79.

Schulenberg, J. E., Sameroff, A. J., & Cicchetti, D. (2004). The transition to adulthood as a critical juncture in the course of psychopathology and mental health. *Development and Psychopathology, 16*(4), 799–806. https://doi.org/10.1017/S0954579404040015.

Segal, Z. V., Williams, M., & Teasdale, J. (2018). *Mindfulness-based cognitive therapy for depression.* New York: The Gulford Press.

Shapiro, S. L. (2009). The integration of mindfulness and psychology. *Journal of Clinical Psychology, 65*(6), 555–560.

Spijkerman, M., Pots, W. T. M., & Bohlmeijer, E. T. (2016). Effectiveness of online mindfulness-based interventions in improving mental health: A review and meta-analysis of randomised controlled trials. *Clinical Psychology Review, 45,* 102–114.

Stjernswärd, S., & Hansson, L. (2017). Outcome of a web-based mindfulness intervention for families living with mental illness—A feasibility study. *Informatics for Health & Social Care, 42*(1), 97–108. https://doi.org/10.1080/17538157.2016.1177533.

Wang, L.-F., & Heppner, P. P. (2002). Assessing the impact of parental expectations and psychological distress on Taiwanese college students. *The Counseling Psychologist, 30*(4), 582–608. https://doi.org/10.1177/0010000203004006.

Wolgast, M., Lundh, L.-G., & Viborg, G. (2011). Cognitive reappraisal and acceptance: An experimental comparison of two emotion regulation strategies. *Behaviour Research and Therapy, 49,* 858–866.

Wong, J. G. W. S., Cheung, E. P. T., Chan, K. K. C., Ma, K. K. M., & Tang, S. W. (2006). Web-based survey of depression, anxiety and stress in first-year tertiary education students in Hong Kong. *Australian and New Zealand Journal of Psychiatry, 40*(9), 777–782. https://doi.org/10.1111/j.1440-1614.2006.01883.x.

Yang, E., Schamber, E., Meyer, R. M. L., & Gold, J. I. (2018). Happier healers: Randomized controlled trial of mobile mindfulness for stress management. *The Journal of Alternative and Complementary Medicine, 24*(5), 505–513. https://doi.org/10.1089/acm.2015.0301.

**Dr. Raymond Chi Fai Chui** is an Associate Professor of the Hong Kong Shue Yan University and serves as Director of the Industrial Attachment Office and Director of Social Work Research and Development Center. His recent publications can be found in academic journals, which include *Journal of College Student Development, Gerontology and Geriatric Medicine, The Open Psychology Journal, Asia Pacific Journal of Tourism Research, International Journal of Tourism Research, Hong Kong Journal of Social Work and Hong Kong Journal of Social Sciences.* His research interests include positive psychology, cyberpsychology, youth well-being and tourism study.

**Dr. Nicolson Yat-Fan Siu** is an Assistant Professor and Associate Director of The Positive Psychology Laboratory of The Department of Counselling and Psychology at The Hong Kong Shue Yan University. His research interests lie in revealing the relationship between the brain and behaviours. He has been focusing on studying frontal lobe functions and devising innovative and evidence-based interventions for improving cognitive and emotional functions. Regarding the community service, Nicolson is currently serving as The Chairperson of The Research Committee for The Hong Kong Association of Therapeutic Horticulture and Program Consultant for The St. James' Settlement Continued Education Center.

**Ms. Wai-Man Tsui** received her Bachelor's degree in Chemistry and Psychology from the University of Hong Kong. She is a Research Assistant in the Department of Social Work of the Hong Kong Shue Yan University. She wishes to pursue further studies with areas of interest being brain imaging and psychological interventions.

**Ms. Chloe Yuet-Tung Wong** is studying her Bachelor's degree in the Department of Counselling and Psychology at the Hong Kong Shue Yan University. Her research focuses on personality and individual difference. Chloe is ambitious to become a Counsellor in the future. She is interested in knowing different approaches and techniques to assist people with emotional disturbance or malfunction.

# Revolutionising Language Teaching and Learning via Digital Media Innovations

**Noble Po Kan Lo**

**Abstract** By analysing reading and writing in a specific context online, we can better understand evolving social and teaching practices. For instance, various online platforms, such as The Open University of Hong Kong's OLE and The Hong Kong Polytechnic University's Moodle, and Facebook, have been gradually incorporated into teaching and learning. "The medium, or process, of our time—electric technology—is reshaping and restructuring patterns of social interdependence and every aspect of our personal life" (McLuhan, 1967, p. 8). The technology-related transformation is embedded in broader social changes, influencing people's language and communicative practices. The domestication of technology (Berker et al., 2005) reveals that people are digitally transformed in their everyday lives. Teenagers are considered as "digital natives" who are specifically adept at using innovative technological devices whilst older people, or "digital immigrants," have to become familiar with new technologies (Prensky, 2001). Nevertheless, it is of great significance not to stereotype a generation of people via this division because technology expands the variety of knowledge and experience in teenagers and the elderly alike (Bennett et al., 2008; Hargittai, 2010). In this global era, research on new media has followed a wider range of how language and literacy practices can transform educational practices. As Barton (2009) notes, "...by examining the changing role of texts we uncover the central tensions of contemporary change: new literacy practices offer exciting possibilities in terms of access to knowledge, creativity and personal power" (p. 39). This paper, therefore, aims to examine how language teaching and learning can be changed through innovative digital media, particularly in relation to educational settings. In doing so, it is found that computer-mediated discourses can be highly effective in promoting literacy via online language learning spaces.

N. P. K. Lo (✉)
Division of Languages and Communication, College of Professional and Continuing Education, The Hong Kong Polytechnic University, 9 Hoi Ting Road, Yau Ma Tei, Kowloon, Hong Kong, China
e-mail: pokan.lo@oxon.org



W. W. K. Ma et al. (eds.), *Learning Environment and Design*, Educational Communications and Technology Yearbook,
https://doi.org/10.1007/978-981-15-8167-0_15

**Keywords** Language teaching and learning · Technology · Literacy · Online platforms · Media

## 1 A New Era of Telecollaboration

Online language learning sites provide innovative opportunities to use the Internet to learn new languages, thus exploiting the advantages offered by the digital medium. Allen and Seaman (2013) note that a third of all students in higher education in the US are taking at least one online course. The opportunities for finding resources online to learn a target language may affect students' perceptions in that they believe that they no longer need to travel abroad to practise the foreign language and learn about culture. For these students, the Internet brings it to their own home (Kern, 2014). However, the interactivity mediated by technology is quite different from face-to-face encounters (Develotte, Kern, & Lamy, 2011). Communication technologies, in fact, transform temporal and spatial relations. Educators also need to consider that the traditional distinction of the forms and functions of speech and writing may overlap in online environments. Similarly, the Internet allows language learners to come into contact with native speakers of different varieties, as well as second language speakers who may not have standard norms of language use. Another issue is that materials available online do not have the desirable cultural authenticity for language learners. For these reasons, Kern (2014) compares the Internet to Plato's *pharmakon:* it presents both promise and challenges for language learners.

With this shift in opportunities to utilise digital learning spaces and communications technologies to revolutionise language learning, as well as the challenges it presents, it is necessary to thoroughly explore the concept of telecollaboration. This paper, therefore, aims to examine how language teaching and learning can be made efficient and effective via digital media innovations. It is important to explore educational settings in this respect in order to facilitate the assessment of the thesis that computer media discourses can promote literacy via online language learning spaces, with multiple platforms being capable of providing telecollaborative spaces that are conducive to providing new and innovative learning opportunities.

## 2 Strengths and Challenges for Digital Media for Language Teaching

There are numerous learning and pedagogical theories that underpin the desire to embrace innovative digital media within education, some of which offer broad theoretical insights into online learning opportunities whereas others are specific to language teaching. Formal education has long held the aim of accumulating objective knowledge, specifically knowledge imparted by a teacher via institutional and structural frameworks (Bower, 2017; Gulati, 2004), but new digital paradigms challenge traditional systems by providing new pedagogical options and frameworks. Community language pedagogy is still preferable where teaching is specifically based on audio-lingual methods (Ellis, 2012). Indeed, e-learning pedagogical techniques and tools have been found to effectively enhance students' learning capabilities. A study by Mehanna (2016) found that effective e-learning practices were broadly dependent on learning frameworks that promoted direct and indirect interaction, including the practice of providing feedback as per face-to-face interaction. In fact, the study contends that e-learning should utilise techniques and methods of teaching online that are effective offline to maximise the success of strategies for teaching languages (Mehanna, 2016). Mehanna's logic makes sense and draws attention to shifts in teaching practices rather than wholesale movements away from what has gone before. Best practice still applies to methods of teaching regardless of the system of delivery. This premise will form the basis of the analysis here.

In addition to the application of a coherent pedagogical theoretical framework, it is necessary to acknowledge the strengths and challenges that utilising digital media for language teaching presents. Firstly, it is important to note that video conferencing is currently used by many language teachers. Video conferencing allows teachers to develop long distance collaborations with two or more classrooms in different countries. Guth and Helm (2010) refer to this as telecollaboration. In these internal partnerships, there is typically an emphasis on culture in language learning and use. Even though some studies have shown the potential benefits of telecollaboration, others, such as Ware (2005), have noted that the intercultural contact given in these telecollaborations does not necessarily mean that students gain any cultural understanding. Video conferencing adds voice, gesture, gaze, and movement in a way that provides communication practice with speakers at a distance. That is, it is the closest approximation to a conversation face to face.

In Kern's (2014) recent study, students in an intermediate-level French section at Berkeley had a weekly video conferencing exchange with students, who were acting as tutors, in Lyon. For this program, MSN Messenger, Skype, and VISU were used. With the first two platforms, the students worked in pairs. VISU allowed students to work individually with their own computer. In this study, student interactions were occasionally recorded. Students also kept journals and were interviewed after the video conferencing sessions finished. Students also completed written evaluations and questionnaires on their online experiences. Most of the students responded positively to the incorporation of video conferencing exchanges, noting an enhanced ability to deal with communicative pressure, a boost in self-confidence, and a higher motivation to study or work abroad. In their interviews, students noted that they regarded these interactions as authentic, engaging, and as a positive addition to classes.

Kern (2014) notes that there are issues with the mediational features of video conferencing that must be taken into account when evaluating the medium. While video conferencing gives the appearance of immediacy, it is filtered by hardware and software. For instance, the webcam is a fixed part of the computer, and it is not easily repositioned. This means that participants have to stay in a position to be visible to their partners. Similarly, if there is a group of two working together, they have to sit close to each other. Hence, Kern (2014) notes that the webcam may introduce ambiguity of interpretation of physical proximity and exaggeration of the effects of physical movement as in Fig. 1.

**Fig. 1** Shifts in position are exaggerated by the webcam

Figure 1 shows how a short-range view may create a sense of immediacy and intimacy. On the other hand, a distance of three feet may appear as one that is distant. Parkinson and Lea (2011) argue that video conferencing can produce less intimacy than other types of communication because participants may look to increase the emotional relevance of the conversation. This appears to be in contrast with other conversations where intimate visual contact tends to result in speaking about less personal topics in order to create social distance between speakers. Another issue with webcams is that they can create a false illusion of contact. For example, a technical issue may mean that a group believes that there is contact, but one cannot hear or see the others. Kern (2014) also notes that the fixed position of the webcam makes it impossible for any real eye contact. When speakers look at each other, it appears as if they are looking down. If one wants to give the impression to look into someone's eyes, they have to look directly to the camera but then cannot see the other party. For instance, Fig. 2 shows a French tutor who looks at the camera while asking a question, but then she is looking at the students when they are answering (Kern, 2014).

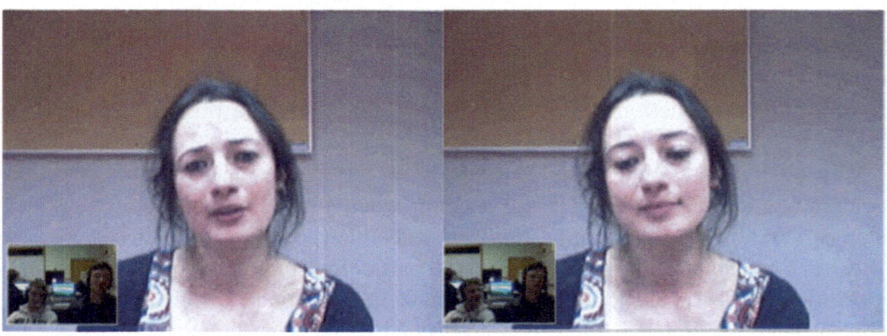

**Fig. 2** Looking at the webcam (left) and looking at interlocutors (right)

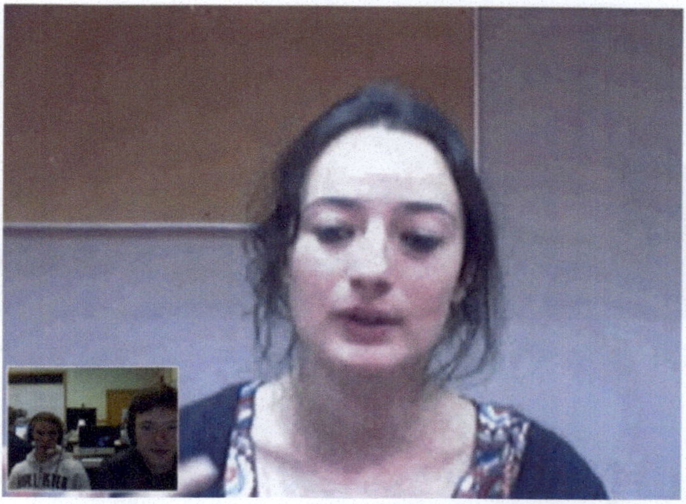

**Fig. 3** A gesture obscured by the inset self-monitor window at bottom left

Further, webcams also mediate gestures. In Kern's (2014) study, not all the gestures were captured by the webcam. These gestures become invisible to online partners, hence they cannot be used to monitor meanings or manage turn-taking. Similarly, a gesture may be blocked by monitor windows as in Fig. 3 (Kern, 2014), and thus video conversing is a skill that develops over time.

As video conferencing limits the vision of gestures and body language, speakers tend to compensate by increasing their facial expressiveness. Grahe and Bernieri (1999) note that this exaggeration is a wish to create liveliness and enhance rapport. An example of such increased facial expressivity is found in Fig. 4 (Kern, 2014).

Kern (2014) also notes that a webcam may also result in speakers projecting images of themselves that they do not wish to. For instance, a student reported straining to listen in a conversation and not looking very friendly before forcing a smile on her face. A further distancing effect can arise from garbled speech when using Skype, as well as the use of headphones and microphone. In Kern's (2014) study, students commented on the self-consciousness they felt as soon as they put on a set of headphones, a feeling that was exacerbated by students seeing themselves on the computer screen.

Despite the shortcomings, technological mediation is beneficial as it offers a way to analyse, discuss, and learn from any possible misunderstandings produced during an interaction. The recording of these interactions gives the students an opportunity to evaluate their own performance (Guth & Helm, 2012). That is, students can perceive details, and pay special attention to the moments when there may be a misunderstanding. In other words, revising these online conversations is comparable to revising an essay: the student can self-assess and think about alternative ways to use language and thereby improve their language use.

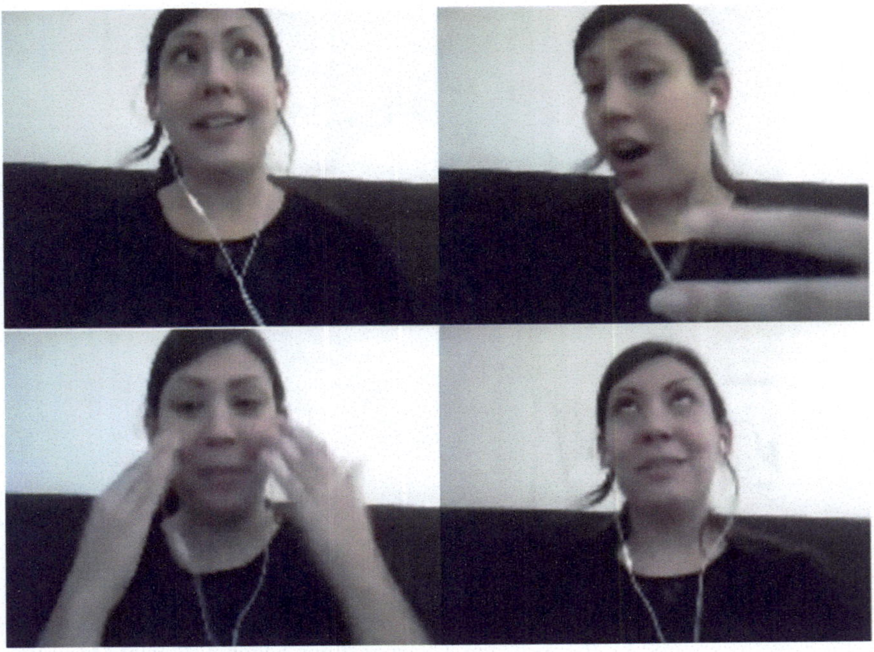

**Fig. 4** Animated facial expressivity online

## 3 An Authentic Learning Environment: Mobile-Assisted Language Learning (MALL)

MALL allows students to learn at any place, any time and, as such, the technology has been incorporated in a variety of devices, including mobile phones and tablets. This allows students from anywhere in the world to download and use apps, including students from Hong Kong. However, it appears that MALL studies do not dominate the attention of second language pedagogy (Burston, 2014, 2015). In fact, there does not seem to be any large-scale implementation to date. There have been some attempts that remain marginal to the curriculum or are restricted to the use of voluntary complementary materials. The first attempt to use MALL was based on the use of PDA word processing programs. These were designed to improve the L1 English writing skills of Canadian secondary school students (Callan, 1994). Only recently, mobile devices have been used to support the reading and writing of L1 Chinese in Taiwan and Singapore (Wong, Song, Chai, & Ying Zhan, 2011). The application of MALL to second languages began later, prompted by the popularity of pocket bilingual dictionaries amongst Japanese students (Weschler & Pitts, 2000). The high rate of mobile phone ownership among Japanese students sparked an interest in the use of text messaging for L2 English vocabulary acquisition (Thornton & Houser, 2002). Burston (2014, 2015) notes that device usage dominates out-of-class applications with over 90%

share of all MALL implementations. It appears that only about 20 MALL implementations have focused on the in-class usage of mobile devices. Thornton and Houser (2003) provided an early application where students could confirm comprehension in L2 lessons. However, most in-class use of mobile devices have been introduced as inexpensive alternatives to a computer lab installation for vocabulary and grammar learning (Begum, 2011), discussion activities (Brown, Castellano, Hughes, & Worth, 2012), reading (Chang & Hsu, 2011), listening practice (Oberg & Daniels, 2012), and video production (Brown, 2012). They have also been used to monitor pronunciation and note-taking in L2 English classrooms (Baleghizadeh & Oladrostam, 2010; Ghorbandordinejad, Aghasafi, Farjadnasab, & Hardani, 2010).

In comparison to desktop use, MALL offers features of portability, social connectivity, context sensitivity, and individuality (Chinnery, 2006). That is, mobile devices give the chance to make learning movable, real time, collaborative, and seamless (Wong & Looi, 2011). This is due to the fact that mobile devices are relatively small and lightweight, meaning they can be easily carried. Further, Wood, Jackson, Hart, Plester, and Wilde (2011) used the portability feature of mobile phones and text messaging so students could use it after school hours to improve their English reading, spelling, and phonological awareness. Similarly, mobile devices allow students to share information, collaborate, and communicate with ease. These characteristics allow students to enhance the efficiency of group learning as well as improve the quality of interaction (Lan, Sung, & Chang, 2007). For instance, Zurita and Nussbaum (2004) developed a learning environment based on wireless interconnected devices to allow children to learn Spanish syllables through collaborative dialogues.

In addition, mobile devices have the necessary functions to offer students more flexibility and accessibility to record and deliver learning experiences. Sandberg, Maris, and De Geus (2011) highlight the portability and context-sensitivity features of mobile devices. These characteristics may help elementary-school learners with their English reading and writing skills at various settings. A further advantage of mobile devices is their individuality. That is, they can be customised and personalised for individual use depending on learning needs, styles, and interests. Hence, a language teacher may design learning materials taking the users' learning behaviour into account, so learning activities designed for mobile devices can be tailored to meet the students' learning needs as well as pace. Therefore, students are not only receiving authentic learning materials, but they are also empowered in their learning. Mobile devices also allow learners to receive individualised feedback, which enhances learning. Teachers may also use mobile devices to monitor and regulate learners' learning process. As an example, Chang, Lan, Chang, and Sung (2010) conducted a study using a mobile-device-assisted Chinese reading system that allowed students to share their thoughts. This allowed students to have supported discussions in a cooperative learning environment.

Steel (2012) reviewed the perspectives of language students when using mobile apps. The student data revealed that students appreciated the flexibility and convenience of using apps. This allowed them to meet their personal learning needs as and when it suited their lifestyles. Students also noted that using apps was efficient as they could spend short periods of time learning as and when they had the chance.

Similarly, students could use the app without a lot of forethought and preparation. As students often carry a mobile device, they are more likely to have immediate access to their apps. This portability extended to workplaces, which enabled students to revise and review their in-class learning in that environment as well. Overall, Steel (2012) reports that students found mobile apps "easy-to-use and understand" and "accessible anywhere anytime" (p. 309). A further advantage mentioned by students is that apps are either free or of low cost. Moreover, they tend to have various resources, including a dictionary, textbook type exercises, flash cards, audio, and so on. The fact that there is a continuous development of new apps was also viewed positively by students. The potential of using mobile apps is increasing as most students own a mobile phone and realise that it allows them to learn and achieve learning tasks quickly, easily, spontaneously, and habitually.

## 4 Application of MALL

Alvarado, Cohelo, and Dougherty (2016) highlight three apps that may be used in an English classroom, and that this part will review Kahoot, EdPuzzle, and AudioBoom. These three platforms have been specifically chosen for this study because they offer tools that are purposely designed for education and therefore give students the opportunity to work on various language skills on their own as well as with others. Further, they are also suitable for all levels of English and may be easily adapted to individual learning styles, thus offering enhanced accessibility. First of all, Kahoot (https://kahoot.it/) is an interactive game that may be used generally for any subject in any school. Kahoot games are a collection of questions on a range of topics. These quizzes may be created by teachers as well as by the students. The advantage of this app is that there is no limit on the number of players that can participate in any given game. Users play the game in real time by using a code provided by the quiz designer. The types of questions that can be created include multiple choice, true or false, and sentence completion. Players can review their scores and their scoreboard place whilst playing the game, a characteristic that Alvarado et al. (2016) argue can be very motivating. The authors mainly use the quizzes to review the students' knowledge before a test or examination. The main use would be then to review concepts, as well as vocabulary and grammar topics. Figure 5 is a screenshot of possible Kahoot activities (Alvarado et al., 2016).

In Alvarado et al. (2016), the Kahoot quiz was played with higher education students in a foundation program mainly to review vocabulary. The questions presented to the students were simple, and the pictures had a key role in the quizzes. The authors suggested that lower level English students would require a higher use of visual aid in order to give them confidence and motivation to win the game. The students could then see which questions had been more difficult for them by observing their mistakes in the results (Fig. 6). The data allowed students to receive information on the areas where they needed to improve. This not only highlights the strengths of the app, but it also acknowledges its challenges. Specifically, it draws attention to the

| **Level:** Beginner (A1–CEFR) | **Topic and skill focus:** Vocabulary–Science |
|---|---|

**Screenshot and link**

Kahoot Quiz Page

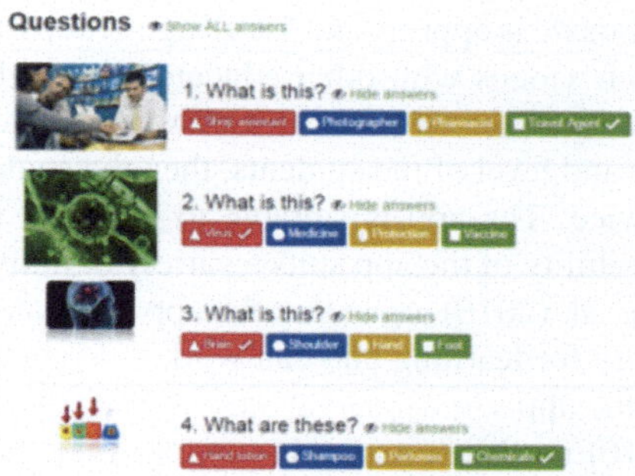


Link to quiz: https://play.kahoot.it/#/k/c4453e8b-0695-4072-8243-dc22eb0e5434

**Fig. 5** Kahoot activities used in the classroom

| STUDENT | CORRECT ANSWERS | INCORRECT ANSWERS | SCORE | Which one is correct? |
|---|---|---|---|---|
| Student A | 11 | 1 | 8869 | They watch films. |
| Student B | 10 | 2 | 8034 | They watch films. |
| Student C | 9 | 3 | 8284 | They watch films. |
| Student D | 9 | 3 | 6983 | They watch films. |
| Student E | 9 | 3 | 6883 | They watch films. |
| Student F | 7 | 5 | 5366 | They play computer |
| Student G | 6 | 6 | 5167 | He watches movies |
| Student H | 6 | 5 | 5117 | They play computer |
| Student I | 6 | 6 | 5067 | They watch films. |
| Student J | 6 | 6 | 5034 | They watch films. |

**Fig. 6** Students' view of their errors (Alvarado et al., 2016, p. 48)

disparate levels that individuals are working at but can undermine self-confidence. Further, there is no direct interaction with teachers face to face, meaning that it is impossible to determine whether students are learning effectively or correctly.

EdPuzzle is an app that teachers may use to turn a video into an engaging and interactive lesson. A teacher may choose a video from any online platform (e.g., YouTube, National Geographic, etc.) and customise it using the tools of the EdPuzzle app. In this way, videos may be edited, cropped, voice-overs can be added, and breaks can be introduced for students to respond to questions. This allows teachers to turn any video into an effective and engaging tool. These videos may be embedded on other sites such as Blackboard and Moodle in the form of quizzes, or they can be used in specific classes. They can even be sent to individual students. Teachers may track the progress of students as EdPuzzle allows for this. Students may access any EdPuzzle assigned to them in their EdPuzzle account from either their home or the classroom.

As with Kahoot, EdPuzzle is appropriate for all age groups and varieties of levels. Teachers may share their videos with other educators through different platforms, including Facebook and YouTube. An educator only needs to choose an appropriate video for the age group and level of the students, then the students may access these videos on a digital device. The app is suitable to integrate reading, writing, and listening skills. The flexibility of the app suits a variety of individual learning styles, and as such, Alvarado et al. (2016) argue that the app motivates the students as they assume the responsibility for learning and can work independently. Figure 7 shows a set of screenshots of the app.

Alvarado et al. (2016) used this particular app so students could reflect on the speaker's performance. Other videos may be used to practise listening skills through the use of authentic materials or teach vocabulary in a fun manner. The videos, as previously noted, may include quizzes that are automatically graded by EdPuzzle, and the results are given directly to the students. However, again, this does not facilitate

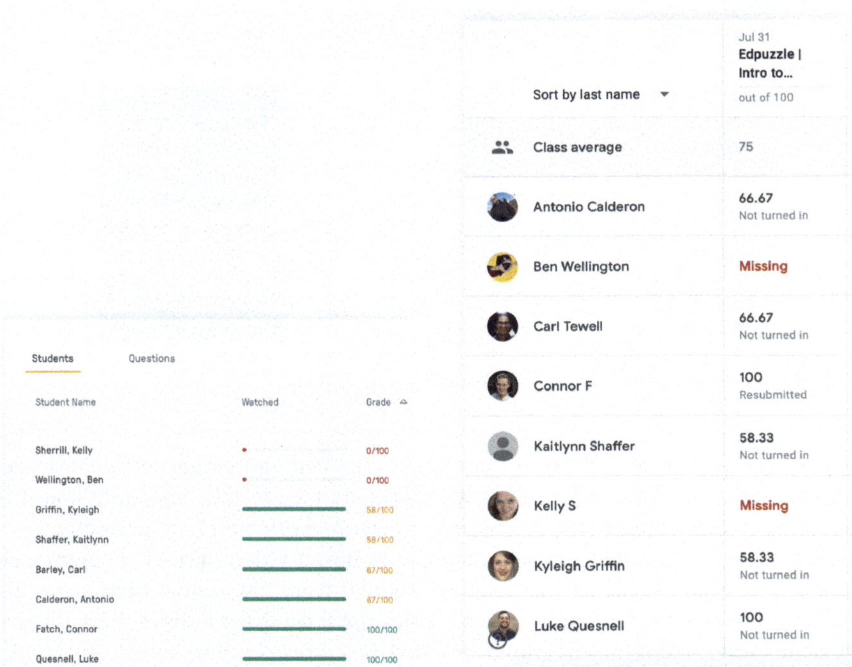


**Fig. 7** Screenshots of EdPuzzle

face-to-face interaction between teachers and students. Further, it is not possible to provide personalised feedback when the app itself marks the quiz. These are both issues that are problematic in the context of developing new language skills.

Alvarado et al. (2016) also mention the app AudioBoom, a voice-based audio podcast and social sharing platform. This app allows users to record, listen to, and publish an unlimited number of podcasts or audio files. It can also be used to purchase audiobooks or follow podcasts from other broadcasters, such as BBC, NFL, and other similar providers. The app also allows users to share and embed content in other websites or social media platforms such as Facebook, Twitter, and Google+. This means that this platform is dynamic, interactive and, according to Alvarado et al. (2016), fun. AudioBoom is also suitable for all age groups, though it is only recommended for intermediate or advanced language levels. The main skills that may be practised in this app are listening and speaking, as well as interpersonal communication skills. Figure 8 shows a screenshot of AudioBoom (Alvarado et al., 2016).

Figure 8 shows an example of a scripted dialogue recorded by two students. Similarly, students may record podcasts of an image they have been shown or recall information on a piece of news. This offers a more practical experience of learning languages but still does not provide a means of learning directly from a teacher in a face-to-face capacity. It is pragmatic but reflects upon the absence of learning tools that pay attention to the special requirements needed to effectively learn languages.

| | |
|---|---|
| **Level:** Elementary (A2–CEFR) | **Topic and skill focus:** Aviation English–Dialogue practice (Speaking) |

**Screenshots and links to samples**

https://audioboom.com/boos/2392298-rbh-and-an

**Fig. 8** AudioBoom activities used in the classroom

AudioBoom, EdPuzzle, and Kahoot are only three examples of the variety of mobile apps in the market that complement students' classroom activities. These apps allow learners to develop language skills using authentic, culturally familiar tools that may be accessed anytime anywhere (Jackson, 2015). Apps like these may be used to incorporate and develop language skills in a practical and fun way. By using these apps, students take responsibility for their own learning and consequently develop learner autonomy and the capacity to work independently.

## 5 Online Gaming as a Learning Environment

Chik (2014) notes that in Hong Kong, Mainland China, and Taiwan, many use commercial off-the-shelf (COTS) games to learn a second language. In fact, research points out that there has been an increasing interest on learning through digital games (Reinders, 2012). On the other hand, schools in East Asia tend to view digital gaming as addictive and non-educational (Gentile et al., 2011). This means that L2 digital gaming tends to take place in out-of-school environments as opposed to the games described in the previous section. However, there are some studies of Hong Kong Chinese gamers which note that self-directed L2 learning was done primarily for pleasure (Chik, 2012). These gamers argued that playing games and navigating through them was a motivation to learn an L2. Most of these gamers also used online communities to support one another. This meant that L2 gamers developed autonomy in L2 learning precisely when playing games in an L2. This made them gain confidence and even shared contributions in the L2 in game walkthroughs, or strategy guides, to gaming communities.

In a further study, Thorne (2008) looked into the game World of Warcraft (WoW), which contains an in-game chat. The author observed exchanges between gamers in an American university and a Ukrainian university. These exchanges exemplified collaborative gameplay and were used by Thorne to illustrate naturally occurring L2 learning episodes. Thorne (2008) noted that in task-based role-playing games, gamers needed to use an L2 to interact in multilingual online game worlds. Thorne's analysis showed that gamers reached a linguistic middle ground by affirming their passion for WoW at first. The gamers then took turns in being learners and teachers for language exchange between Russian and English. The conclusions of Thorne's research show that native speakers of different languages may achieve natural and autonomous learning moments within the multilingual WoW game world. Similarly, Rama et al. (2012) highlight that games like WoW offer safe learning environments. These games even facilitate interactions to develop communicative competence. In their study, the authors showed that gaming expertise could compensate for the lack of L2 ability in cooperative gameplay. As one of the gamers in the study was lending their gaming expertise, they gained language support from the gaming community. Rama et al. (2012) express concerns over the unstructured L2 learning progress of gaming. However, the fact that gaming presents users with authentic L2 interactions counterbalances this, and any disadvantages could be overcome with the use of a dictionary or translation add-ons.

It is obvious that gamers actively organise their L2 gaming and learning practices (Reinhardt, 2019). According to Chik (2014), this organisation is related to learner autonomy. Aside from their own learner autonomy, which in turn increases motivation (Ushioda, 2007), gamers are also likely to be part of a learning community with the game they are playing. Murray and Fujishima (2013) note that this learning community may offer learners the necessary environment to act and interact so as to increase their own autonomy. In particular, Murray and Fujishima (2013) define a learning community as "consist[ing] of individuals who come together to accomplish a specific end or goal" (p. 70). In digital gaming and L2 learning, the concepts of autonomy and community are relevant in two aspects. First of all, gamers regularly take independent decisions on gaming choices; and secondly, the use of game-external websites and other group resources is integral to the gaming experience (Thorne, Fischer, & Lu, 2012). Thus, as digital gaming is a community-based activity, gamers are autonomous learners involved in a wider community.

Chik (2014) explored how gaming out-of-class can change from being incidental learning to a kind of intentional learning. Gamers typically play out-of-class, thus L2 learning is generally informal. Hulstijn (2008) notes that everyday activities may be a way of intentional learning when there is an explicit intention of learning and learners use a set of learning strategies. However, in Chik's (2014) study, some gamers noted that gaming was the primary motive for L2 learning. In this case, these learners rendered L2 learning as incidental. For instance, a gamer noted, "I usually skip looking up new words and continue playing the game, you can't keep stopping to use the dictionary…that's just insane" (Chik, 2014, p. 91). On the other hand, other participants of the study said that they "…jotted down words quickly and then looked them up later." That is, certain participants noted the importance of learning new vocabulary as essential knowledge. In the study, Chik (2014) also noted that there were examples of gamers who specifically used games in order to learn a second language. Chik (2014) added that research on L2 gaming has shown L2 learning can arise both from textual and social interactions within the gaming environments, as well as from the instructional and advisory roles taken up by gamers. Typically, gamers first learn to play by following oral instructions from their immediate social circle. Then they go online to read written instructions from communities with similar interests. In effect, gamers create community pedagogical resources, and act as language advisers, teachers, and translators for those with similar interests.

This collaboration, according to Chik (2014), allows learners the opportunity to take on an instructional role: they can assist other game players in a variety of different media. Some will help other gamers within the same game whereas others will help them on a discussion board or wiki. This instructional role results in a community that Gee (2005) defines as an affinity space. This affinity space exists because of a common shared interest between gamers. There are no set boundaries, and affiliation does not depend on formal qualifications. Anyone can gain acceptance in the space, where all users will be valued because of their knowledge and experience of the game. Other gamers will offer encouraging feedback so players can follow instructions and make sure that they complete tasks in the designed order and with meaningful progression. Many authors have emphasised the great role that games may have on an increased motivation to learn a language. Most note that gaming offers fun and entertainment, as well as establish relationships with other people who play the same games. Wang, Khoo, Liu, and Divaharan (2008) add that gamers are motivated to play games and learn an L2 because they are immersed in a fun virtual fantasy world and the game marks clear achievements.

## 6 Conclusion: Language Learning in Online Spaces as an Extension of Real-Life Dialogue

The primary use of social media for the creation of an academic discourse community is to facilitate a different kind of student–tutor relationship based on reduced social distance. The use of online platforms follows easily from the perception of a real-life community and provides room for an extension of real-life dialogues. After all the case studies, a key characteristic of the online interactions observed was the use of language varieties common to a range of computer-mediated communication. Although it is true that these forms of language sit uneasily with the notion of the "legitimate" academic language (Bourdieu, 1992) expected of undergraduates' coursework, there was no evidence in this study to support the fear that the computer-mediated discourse (CMD) literacies of undergraduates would undermine their ability to produce coursework in acceptable academic English when required. It would be a crude conclusion that students are unable to switch styles according to different situations. On the contrary, it is time CMD as a product of new age technology could be manipulated even better to serve its real strength in the area of language learning in online spaces.

The analysis here points to the need to utilise a hybrid model for language teaching, specifically one that incorporates a range of communicative tools that tap into the individual strengths of the learner and that facilitates face-to-face discourse. This would provide tailored solutions to learning and also facilitate interaction that has a practical appeal in that telecollaboration still establishes a connection between teacher and student. Although the three platforms explored have their individual strengths and are all accessible, the construction of new platforms that are tailored to language teaching would, therefore, be more appropriate and effective as more language courses become accessible via digital media. Ultimately, these recommendations tap into the final conclusions drawn, specifically that computer-mediated discourses can be highly effective in promoting literacy via online language learning spaces.

## References


Allen, I. E., & Seaman, J. (2013). *Changing course: Ten years of tracking online education in the United States: Babson Survey Research Group*. Retrieved March 13, 2019, from https://www.onlinelearningsurvey.com/reports/changingcourse.pdf.

Alvarado, N. C., Coelho, D., & Dougherty, E. (2016). Mobile apps for ELLs: Supporting language learning with engaging digital tools. *Argentinian Journal of Applied Linguistics, 4*(1), 43–58.

Baleghizadeh, S., & Oladrostam, E. (2010). The effect of mobile assisted language learning (MALL) on grammatical accuracy of EFL students. *MEXTESOL Journal, 34*(2), 77–86.

Barton, D. (2009). Understanding textual practices in a changing world. In M. Baynham, & M. Prinsloo (Eds.), *The Future of Literacy Studies* (pp. 38–53). Palgrave Macmillan.

Begum, R. (2011). Prospect for cell phones as instructional tools in the EFL classroom: A case study of Jahangirnagar University, Bangladesh. *English Language Teaching, 4*(1), 105–115.

Bennett, S., Maton, K., & Kervin, L. (2008), The 'digital natives' debate: A critical review of the evidence. *British Journal of Educational Technology, 39*, 775–786. http://dx.doi.org/10.1111/j.1467-8535.2007.00793.x.

Berker, T., Hartmann, M., & Punie, Y. (2005). Introduction. In T. Berker, M. Hartmann & Y. Punie (Eds.), *Domestication of Media and Technology* (pp. 1–16). Berkshire, UK: Open University Press.

Bourdieu, P. (1992). Language and Symbolic Power. Cambridge: Polity Press.

Bower, M. (2017). *Design of technology-enhanced learning: Integrating research and practice*. Bingley: Emerald Group Publishing.

Brown, M. (2012). Tablet computing to cultivate Japanese EFL digital literacy: A study on video production in the classroom. In J. Colpaert, A. Aerts, W. Wu, & Y.-C. Chan (Eds.), *The Medium Matters: Proceedings of the Fifteenth International CALL Conference*, Providence University, Taichung, Taiwan, 25–27 May 2012 (p. 38). Antwerp: University of Antwerp.

Brown, M., Castellano, J., Hughes, E., & Worth, A. (2012). Integration of iPads in a Japanese university's freshman curriculum. In *Proceedings of the JALT CALL Conference 2012*.

Burston, J. (2014). MALL: The pedagogical challenges. *Computer Assisted Language Learning, 27*(4), 344–357. https://doi.org/10.1080/09588221.2014.914539.

Burston, J. (2015). Twenty years of MALL project implementation: A meta-analysis of learning outcomes. *ReCALL, 27*(1), 4–20. https://doi.org/10.1017/S0958344014000159.

Callan, S. (1994). *Can the use of hand-held personal computers assist transition students to produce written work of excellent quality?* Wentworth County Board of Education, Ontario, Canada.

Chang, C.-K., & Hsu, C.-K. (2011). A mobile-assisted synchronously collaborative translation-annotation system for English as a foreign language (EFL) reading comprehension. *Computer Assisted Language Learning, 24*(2), 155–180.

Chang, K. E., Lan, Y. J., Chang, C. M., & Sung, Y. T. (2010). Mobile-device-supported strategy for Chinese reading comprehension. *Innovations in Education and Teaching International, 47*(1), 69–84.

Chik, A. (2012). Digital gameplay for autonomous foreign language learning: Gamers' and language teachers' perspectives. In H. Reinders (Ed.), *Digital games in language learning and teaching* (pp. 95–114). London, UK: Palgrave Macmillan.

Chik, A. (2014). Digital gaming and language learning: Autonomy and community. *Language Learning & Technology, 18*(2), 85–100.

Chinnery, G. M. (2006). Emerging technologies going to the mall: Mobile assisted language learning. *Language Learning & Technology, 10*(1), 9–16.

Develotte, C., Kern, R., & Lamy, M.-N. (2011). *Decrire la conversation en ligne: Le face a face distanciel* [Describing online conversation: Face-to-face at a distance]. Lyon, France: ENS Editions.

Ellis, R. (2012). *Language teaching research & language pedagogy*. Boston: Wiley.

Gee, J. P. (2005). Semiotic social spaces and affinity spaces. In*Beyond communities of practice language power and social context* (pp. 214–232).

Ghorbandordinejad, F., Aghasafi, A., Farjadnasab, A., & Hardani, A. (2010). Mobile handheld recording devices in the general English classroom. In *The First Symposium of Educational Technology*, Oman.

Grahe, J. E., & Bernieri, F. J. (1999). The importance of nonverbal cues in judging rapport. *Journal of Nonverbal Behavior, 23,* 253–269.

Gentile, D. A., Choo, H., Liau, A., Sim, T., Li, D., Fung, D., & Khoo, A. (2011). Pathological video game use among youths: A two-year longitudinal study. *Pediatrics, 127*(2), e319–e329. https://doi.org/10.1542/peds.2010-1353.

Gulati, S. (2004). *Constructivism and emerging online learning pedagogy: A discussion for formal to acknowledge and promote the informal*. Leeds University. Retrieved April 2020, from https://www.leeds.ac.uk/educol/documents/00003562.htm.

Guth, S., & Helm, F. (2010). *Telecollaboration 2. 0: Language, literacies and intercultural learning in the 21st century.* New York: Peter Lang.

Guth, S., & Helm, F. (2012). Developing multiliteracies in ELT through telecollaboration. *ELT Journal, 66,* 42–51.

Hargittai, E. (2010). Digital na(t)ives? Variation in Internet skills and uses among members of the "Net Generation" February 2010. *Sociological Inquiry, 80*(1), 92–113. http://dx.doi.org/10.1111/j.1475-682X.2009.00317.x.

Hulstijn, J. H. (2008). Incidental and intentional learning. In C. J. Doughty & M. H. Long (Eds.), *The handbook of second language acquisition* (pp. 349–381). Oxford, UK: Blackwell Publishing.

Jackson, D. B. (2015). A targeted role for L1 in L2 vocabulary acquisition with mobile learning technology. *TESOL Arabia Perspectives, 23*(1), 6–11.

Kern, R. (2014). Technology as Pharmakon: The promise and perils of the internet for foreign language education. *The Modern Language Journal, 98*(1), 340–357.

Lan, Y. J., Sung, Y. T., & Chang, K. E. (2007). A mobile-device-supported peer-assisted learning system for collaborative early EFL reading. *Language Learning & Technology, 11*(3), 130–151.

McLuhan, M. (1967). *The Medium is the Massage: An Inventory of Effects.* London: Penguin Books.

Mehanna, W. (2016). e-Pedagogy: The Pedagogies of e-Learning. *ALT-J, 12*(3), 279–293.

Murray, G., & Fujishima, N. (2013). Social language learning spaces: Affordances in a community of learners. *Chinese Journal of Applied Linguistics, 36*(1), 141–157.

Oberg, A., & Daniels, P. (2012). Analysis of the effect a student-centred mobile learning instructional method has on language acquisition. *Computer Assisted Language Learning, 25*(2), 1–20.

Parkinson, B., & Lea, M. (2011). Video-linking emotions. In A. Kappas & N. C. Kramer (Eds.), *Face-to-face communication over the Internet: Emotions in a web of culture, language and technology* (pp. 100–126). Cambridge: Cambridge University Press.

Prensky, M. (2001). Digital Natives, Digital Immigrants, Part 1. *On The Horizon, 9,* 3–6. http://dx.doi.org/10.1108/10748120110424816.

Rama P. S., Black R. W., Van Es, E., & Warschauer, M. (2012). Affordances for second language learning in world of Warcraft. *ReCALL Journal, 24,* 322–338. http://dx.doi.org/10.1017/S0958344012000171.

Reinders, H. (Ed.) (2012). *Digital games in language learning and teaching.* New York, NY: Palgrave Macmillan.

Reinhardt, J. (2019). Digital gaming in L2 teaching and learning. In C. Chapelle & S. Sauro (Eds.), *The handbook of technology and second language teaching and learning* (pp. 202–216). London: Wiley-Blackwell.

Sandberg, J., Maris, M., & De Geus, K. (2011). Mobile English learning: An evidence-based study with fifth graders. *Computers & Education, 57*(1), 1334–1347.

Steel, C. (2012). Fitting learning into life: Language students' perspectives on benefits of using mobile apps. In *ascilite* (pp. 875–880).

Thorne, S. L. (2008). Computer-mediated communication. In N. Hornberger & N. Van Duesen-Scholl (Eds.), *Encyclopedia of Language and Education*, 2nd Edition, Volume 4: Second and Foreign Language Education (pp. 325–336). New York: Springer. January 2008. http://dx.doi.org/10.1007/978-0-387-30424-3_108.

Thorne, S. L., Fischer, I., & Lu, X. (2012). The semiotic ecology and linguistic complexity of an online game world. *ReCALL, 24*(3), 279–301.

Thornton, P., & Houser, C. (2002). M-learning in transit. In P. Lewis (Ed.), *The changing face of CALL* (pp. 229–243). Lisse, The Netherlands: Swets & Zeitlinger.

Thornton, P., & Houser, C. (2003). EduCall: Adding interactivity to large lecture classes in Japan via mobile phones. In D. Lassner & C. McNaught (Eds.), *Proceedings of the World Conference on Educational Multimedia, Hypermedia and Telecommunications 2003* (pp. 1871–1874). Chesapeake, VA: Association for the Advancement of Computing in Education.

Ushioda, E. (2007). Motivation, autonomy and sociocultural theory. In P. Benson (Ed.), *Learner autonomy 8: Teacher and learner perspectives* (pp. 5–24). Authentik: Dublin, Ireland.

Wang, C. K. J., Khoo, A., Liu, W. C., & Divaharan, S. (2008). Passion and intrinsic motivation in digital gaming. *CyberPsychology & Behavior, 11*(1), 39–45.

Ware, P. D. (2005). "Missed" communication in online communication: Tensions in a German-American telecollaboration. *Language Learning & Technology, 9*, 64–89.

Weschler, R., & Pitts, C. (2000). An experiment using electronic dictionaries with EFL students. *The Internet TESL Journal, 6*(8).

Wong, L. H., & Looi, C. K. (2011). What seams do we remove in mobile assisted seamless learning? A critical review of the literature. *Computers & Education, 57*(4), 2364–2381.

Wong, L.-H., Song, Y., Chai, C.-S., & Ying Zhan, Y. (2011). Analyzing students' after-school artifact creation processes in a mobile-assisted language learning environment. In T. Hirashima et al. (Eds.), *Proceedings of the 19th International Conference on Computers in Education*. Chiang Mai, Thailand: Asia-Pacific Society for Computers in Education.

Wood, C., Jackson, E., Hart, L., Plester, B., & Wilde, L. (2011). The effect of text messaging on 9- and 10-year-old children's reading, spelling and phonological processing skills. *Journal of Computer Assisted Learning, 27*(1), 28–36.

Zurita, G., & Nussbaum, M. (2004). Computer supported collaborative learning using wirelessly interconnected handheld computers. *Computers & Education, 42*(3), 289–314.

**Noble Po Kan Lo** completed his postgraduate degrees at HKU and the University of Oxford. He also obtained his Cambridge CELTA and Harvard CAEL. He is a Fellow of RSA and a Chartered Linguist of CIOL. He is currently teaching at the College of Professional and Continuing Education of PolyU.

# Gamification, AI, and Innovative Learning Technologies

# The Influence of Online Gaming Experience and Its Social Components on Gamers' Mental Health

**Chi-Keung Chan and Kwan-Lun Cheung**

**Abstract** The present study aims to investigate whether online gaming experience and social components in online gaming can influence gamers' mental health. It is hypothesized that more intensive and frequent online gaming experience is negatively related to gamers' mental health. Besides, this study hypothesizes that the social components (positive) in online gaming can buffer the negative effect of an online game on gamers' mental health. There were 157 participants who completed a self-reported questionnaire, including three key measures—Game Experience Questionnaire (GEQ), Social Presence in Gaming Questionnaire (SPGQ), and Game Addiction Scale (GAS). Results from correlational and regression analyses confirmed that more intensive and frequent online gaming experience was significantly and negatively related to gamers' mental health. Nevertheless, social components in social gaming could not moderate the negative impact of online gaming experience on gamers' mental health. Surprisingly, more social components in online gaming (high SPGQ) strengthened the negative impact for people with high frequency and intensity of online gaming (high GEQ).

**Keywords** Online gaming · Social components · Mental health · Gaming addiction

## 1 Introduction

Online gaming has become a high demanding and fast developing market around the world. In the United States, there were nearly over 130 million Americans (about 40% of the U.S. population) who played commercial online games in 2017 (Carras et al. 2018). According to the report from the Pew Internet and American Life Project (Lenhart, Kahne, Middaugh, Macgill, Evans, & Vitak, 2008), online games have provided a diverse set of experiences and related activities which become part of

C.-K. Chan (✉) · K.-L. Cheung
Department of Counselling and Psychology, Hong Kong Shue Yan University, LG 122, 10 Wai Tsui Crescent, Braemar Hill Road, North Point, Hong Kong, Hong Kong
e-mail: alexchan@hksyu.edu



W. W. K. Ma et al. (eds.), *Learning Environment and Design*, Educational Communications and Technology Yearbook,
https://doi.org/10.1007/978-981-15-8167-0_16

the lives of almost all teens in America. Nearly 97% of teens aged 12–17 play computer, web, portable, or console games. It is obvious to see that online gaming is a mainstream activity, and most teenagers frequently play online games for a few hours daily (Lenhart et al., 2008).

Along with this trend, research related to online games has been substantially increased over the last decade (Griffiths, Kuss, & Pontes, 2016). Although online games have been associated with many positive educational and psychological benefits, such as cognitive, developmental, emotional, motivational, and social (Granic, Lobel, Engels, 2014), however, the vast majority of research have focused on the negative sides of online games rather than the positive sides (Andreassen et al., 2016; Kuss, Louws, & Wiers, 2012). Researchers have been concerning gamers with excessive exposure to online gaming may become 'addicted gamers' and negatively affect their mental health.

Some previous studies have shown that problematic online gaming is associated with depressive mood, loneliness, social anxiety, and negative self-esteem (Rooij, Ferguson, Mheen, & Schoenmakers, 2017). Nevertheless, these studies overemphasized the negative sides of online gaming disorder. These studies argued that feelings of loneliness and lack of social connection are critical risk factors that cause gamers to engaged in problematic (pathological) online gaming behaviors. In recent years, Massive Multiplayer Online Game (MMOG) or Massive Multiplayer Online Role-Playing Game (MMORPG) with immersive social contexts have been getting popular. Indeed, over 70% of gamers play their games with a friend/family member or a group of friends/whole family, either cooperatively or competitively. Thus, gamers are not socially isolated and they can rapidly learn social skills and pro-social behaviors that might generalize to their peer and family relations outside the gaming environment (Gentile et al., 2009; Gentile & Gentile, 2008). Therefore, this study attempts to investigate whether social components in online games can buffer the pathological sides of online gaming.

In short, the purpose of this study is to investigate the relationship between online gaming and gamers' mental health, particularly, whether playing online games more intensively and frequently can increase the tendency of gaming addiction or gaming disorder. If loneliness and social disconnection are the risk factors for problematic gaming behaviors, can the social elements (components) in multiplayer online gaming minimize these social dysfunctions and buffer the negative effect of online gaming on gamers' mental health? Understanding the effect of social elements in a gaming environment not only is beneficial to gamers' mental health, but also provides valuable insights to educators on how to design curriculum and instruction in a gaming environment that can enhance interactive and cooperative game-based learning experience to motivate learners.

### *1.1 Online Gaming*

Online game is simply defined as a game which one plays on an audiovisual apparatus and which can be based on a story (Esposito, 2005). Esposito (2005) mentioned that online game is a game which is not only a fictional activity with rules, time, and space limit, but also a voluntary interactive activity with one or more gamers. Esposito (2005) explained that online games mostly take place in an audiovisual apparatus which is an electronic system with computing capabilities, and input and output devices. An online game seems like a story or a film which has the element of narratives, backstory, cut scene, special event, and discussion with other characters (Esposito, 2005). Online games can be further categorized from simple text-based games to complex 3D graphics and virtual worlds (Uz & Cagiltay, 2015). There are lots of variations in different types of online games, such as first-person shooter games, strategy games, and massive multiplayer online games. According to the report from Digital Education, online (digital) games have been transformed from a single into socially oriented platforms (Uz & Cagiltay, 2015).

Over the last decade, the demand for online gaming had been raising due to the fast development in digital mobile technologies (e.g. smartphone and tablet). According to the statistical report from Newzoo (2017), online gaming using smartphones has been a popular mobile entertainment and has obtained more than 42% from the global game market. These findings obviously indicated that smartphones have been a big catalyst for boosting the online gaming market in recent years (Fernandez, Männikkö, Kääriäinen, Griffiths, & Kuss, 2018).

Within the online gaming in smartphones, there are lots of 'freemium' games which are so-called mobile gaming apps, placed on the IOS's 'App Store' or Android's 'Play Store'. The player can download for free, play for free, and pay for an extra feature without the time and place limitations (Fernandez et al., 2018). Moreover, the freemium provides options for single or multiplayer games. Furthermore, social components of these mobile (freemium) games are major features of current online games because social networking sites (SNS) have successfully integrated and have been used across many gaming platforms (Fernandez et al., 2018). According to the Global Games Market Report (Newzoo, 2017), nearly 78% of European mobile gamers have played freemium games. Thus, online gaming with mobile device represents one of the fastest growing sectors in the mobile application industry (Fernandez et al., 2018).

## *1.2 Online Gaming and Mental Health*

Some previous research claimed a strong association between addictive use of technology and comorbid psychiatric disorders (Andreassen et al. 2016). The addictive use is characterized by being overly concerned about online activities, driven by an uncontrollable motivation to perform the behavior, and devoting so much time and effort to it that it impairs other important life areas (Andreassen & Pallesen, 2014). Many studies further suggested that addictive online gaming is associated with symptoms of psychiatric disorders. These findings showed that addictive online gaming is related to ADHD, obsessive–compulsive disorder, and social anxiety. More seriously, online gaming provides a way for adolescents to escape the realistic life problems and alleviate their feelings of depression, loneliness, social anxiety, and social disengagement (Mercola, 2017; Merelle, Kleiboer, Schotanus, & Rooij, 2017; Rezaeidehaghani, Mohammadi, Mehrabi, & Rezaeidehaghani, 2016). Furthermore, pathological gamers lack social skills in problem-solving and emotional management. Online games provide gamers excitement and relief so that they can escape from daily stressors and problems (Wong & Lam, 2016). The cognitive, emotional, and social functions could be influenced by gaming addition, for instance, pathological gamers have several symptoms including salience, tolerance, withdrawal, mood modification, and losing self-control (Wong & Lam, 2016). Thus, pathological gaming negatively impacts one's cognitive, emotional, physical, psychological, and social well-being (Wong & Lam, 2016).

Although many literature have focused on the negative influence of problematic online gaming on gamers' mental health (associated with gaming addiction), some recent researches have argued that the appropriate uses of online games can be beneficial to users' mental health. These empirical researches have advocated the positive uses of online games in the areas of education, health, therapy, etc. Online game-based health interventions and applications (e.g. serious games for eHealth, a recent healthcare practice supported by electronic processes and communication using online/mobile games) have been innovated and developed (Carras et al., 2018). These health-related virtual games emphasize on highly engaging, realistic, interactive environments and social support networks so as to support the healthy and positive development of users. For example, the computer solitaire game FreeCell may be useful to monitor cognitive status in adults with mild cognitive impairment. Over the past few years, the development of augmented reality (AR) mobile games (e.g. Pokémon GO) could be useful to promote physical activities and exercises that promote positive health. Puzzle games such as Tetris and Bejeweled have been shown to reduce stress, improve attention, and even prevent flashbacks after a traumatic event (Carras et al., 2018).

Furthermore, online games can provide unique benefits beyond what traditional therapies offer (Carras et al., 2018). Literature have shown that online games act as an important tool in psychotherapy, for instance, diagnostic assessment, building rapport, and social skills training. Wulf, Bowman, Rieger, Velez, & Breuer2018 suggested that online gaming can create a sense of nostalgia, which can serve as a psychological resource for the player's self and well-being. Wulf et al. (2018) defined nostalgia as a feeling that can reflect people in their meaningful past events by wallowing in their memories and entering an emotional state that contains both positive emotions (joy, warmth) and negative emotions (sadness, loneliness). Their study discovered that gamers' well-being is boosted by recalling enjoyable online gaming experience. Furthermore, the feeling of nostalgia in online gaming experience has been found to be positively associated with higher fulfillment of competence, higher life satisfaction, and increased optimism (Wulf et al., 2018) as well as acting as an antidote to counter gamers' negative states, sadness, boredom, loneliness, and meaningless or social exclusion. Finally, the nostalgic feeling in online gaming can enhance reflection of inner self, self-acceptance, and positive self-evaluation.

## 1.3 Social Components in Online Gaming

There has been growing empirical support for the social richness in online gaming (IJsselsteijn, de Kort, & Poels, 2013). These empirical findings showed that there are social processes and interpersonal dynamics that have been underrepresented in conceptualizations and theoretical deliberations of game experience and game enjoyment. Sweetser & Wyeth (2005) noted that social interaction is not an element of flow and can often interrupt immersion in games, however, it is clearly a strong element of enjoyment in games. The online gaming environment can also motivate gamers to interact with other gamers. Hence, online games should not only be regarded as individual enjoyment but should also be recognized as a sociocultural context that can enhance and enrich social presence, social interaction, and social dynamics (IJsselsteijn et al., 2013). Definitely, there are key social components of online gaming that can be considered as beneficial social features (Hou, 2011).

According to the report from the Pew Internet and American Life Project, 76% of teens play online games with other people. Among them, 65% of teens play games with family members or friends in the same physical location and more than one-quarter (27%) of them play games with others who are connected to them online from different places (Lenhart et al., 2008). Thus, online gaming provides teens with various types of social interactions in their virtual and real lives. There are different social interactions among gamers in multiplayer online games (especially MMOGs and MMORPGs), for instance, competition and cooperation (Hou, 2011). Social relationships in MMOGs and MMORPGs can be explicit or implicit (Jia, Shen, Bovenkamp, Iosup, Kuipers, & Epema, 2015). Explicit social relationships are formed by the player's own initiative, for example, the player establishes friendships with others by joining a clan (self-organized group of gamers). Implicit social relationships are formed passively among gamers by joining different interactions or dialogues (e.g. team talk and teens' talk) during online gaming (Jia et al., 2015). Thus, these social components in online gaming allow teens to gain benefits to improve peer relationships, learn cooperative skills, and gain social support, which are important for their well-being.

As mentioned before, MMOGs is one of the most cooperative and socially oriented game types which provides a lot of rich interactive experiences for gamers to interact and form relationships during a gaming interaction process. Some of the studies claimed that MMOG gamers who form relationships in gaming environments can be equalized to forming face-to-face relationships (Stiles, 2010). In addition, peer social support is also critical to gamers in online gaming (Beard & Wickham, 2016). MMOGs and MMORPGs can foster support and teamwork among game gamers (Wong & Lam, 2016). Furthermore, teenagers watch and imitate peers to seek relaxation, to meet others, and to cope with upsetting emotions and problems they face during social interactions in online gaming (Wong & Lam, 2016). Therefore, suitable and appropriate use of the social components in online games can potentially be beneficial to gamers' mental health.

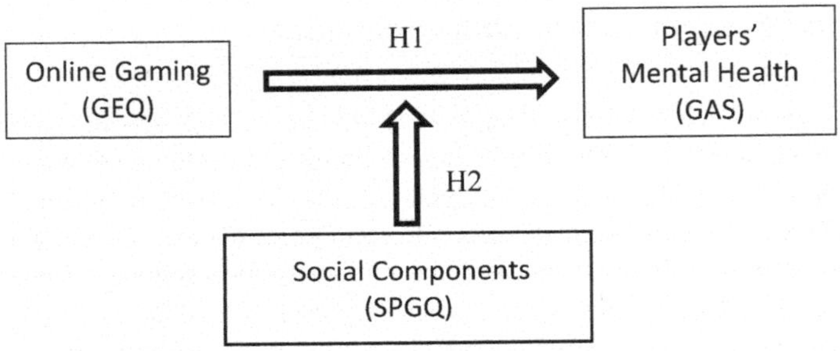


**Fig. 1** Conceptual framework of the relationship among online gaming, social components, and gamers' mental health

### 1.4 The Present Study, Conceptual Framework, and Hypotheses

Based on the above literature review, the conceptual framework of the present study is depicted in Fig. 1 (see next page). The present study attempts to address two research questions. The first research question (RQ1): Is online gaming negatively associated with gamers' mental health (operationalized as addictive gaming)? The frequency and intensity of online gaming are also considered in this study. Based on the pathological literature, it is hypothesized that playing more online games (more intensive and frequent online gaming experience) is significantly and negatively related to gamers' mental health (H1).

The second research question (RQ2): Do the social components (positive social elements) in online gaming act as a moderator to the relationship between online gaming and gamers' mental health (addictive gaming)? In other words, do the social components in online gaming buffer the negative influence of online gaming on gamers' mental health? From a positive technology perspective, it is hypothesized that a higher level of social components (more social interactions) in online gaming can significantly moderate the negative effect of online games on gamers' mental health (H2).

## 2 Method

### 2.1 Participants

In this study, about 160 (n = 157) participants aged 16–30 living in Hong Kong who are fluent in written Chinese were recruited between September and October in 2018 to complete an online survey which measured their online gaming experience, social components in online gaming, and addictive gaming tendency (gamers' mental health). Convenient and snowball sampling methods were employed to recruit participants via the social media (e.g. Instagram and Facebook) or by sending the invitation with a survey link or QR code via popular mobile communication apps (e.g. WhatsApp and Snapchat). Among the participants, 36.3% were males (n = 57) and 63.7% were females (n = 100). For the age range, 31.8% of participants were teenagers/adolescents, 43.9% were emerging adults, and the remaining 24.3% were classified as adults.

## 2.2 Measures

In this study, three measures were adopted to capture gamers' mental health (additive gaming tendency), participants' online gaming experience, and social components in online gaming.

***Gamers' Mental Health***. Game Addiction Scale (GAS; Lemmens, Valkenburg, & Peter, 2009) was used to measure gamers' mental health (additive gaming tendency). GAS measures how the frequency and intensity of online gaming experience affect the gamers' mental health over the last 6 months. GAS is consisted of seven items that assess the symptoms of addictive behavior in online gaming (Griffiths et al., 2016). The respondents anchored their answers for each item on a 5-point rating (never, rarely, sometimes, often, and very often). Therefore, a higher total score in GAS indicates lower degree of gamers' mental health. There is a high reliability for GAS with a Cronbach alpha of 0.89 (Khazaal et al., 2016). Positive correlation was found between the weekly time spent on games (frequency and intensity) and GAS scores. Furthermore, high GAS scores were highly correlated with a number of constructs associated with game addiction, such as lower life satisfaction, lower social competence, and higher loneliness (Khazaal et al., 2016). The reliability of the GAS in this study was very high ($\alpha = 0.91$).

**Online Gaming Experience**. The core module of the Game Experience Questionnaire (GEQ; IJsselsteijn, de Kort, & Poels, 2013) was used to measure the participants' online gaming experience. GEQ assesses the online game experiences on seven components: immersion (6 items), flow (5 items), competence (5 items), positive effect (5 items), negative effect (4 items), tension (3 items), and challenge (5 items) (IJsselsteijn et al., 2013). A total number of 33 items (5-point Likert scale) were included in this study. Higher scores means more exposure to experience in each component. Higher total GEQ scores also reflect higher intensity and frequency in online gaming. The reliability of GEQ in this study was very high ($\alpha = 0.97$). Furthermore, subjects with high intensity and frequency in gaming experience usually represent high-level gamers while subjects with low intensity and frequency in gaming experience usually represent low-level gamers.

**Social Components in Online Gaming**. Social Presence in Gaming Questionnaire (SPGQ; de Kort, de, IJsselsteijn, & Poels, 2007)was used to assess the realization of social interaction during online gaming. SPGQ measures how gamers interact with co-gamers in mediated collaborative gaming environments. SPGQ consists of 3 dimensions:Psychological involvement—empathy (6 items), psychological involvement—negative feelings (5 items), and behavior involvement (6 items). A total of 17 questions (5-point Likert scale) were included in this questionnaire. The reliability of SPGQ was very high ($\alpha = 0.96$). The internal consistency of the three dimensions is moderate to high: Psychological involvement—empathy ($\alpha = 0.86$), psychological involvement—negative feelings ($\alpha = 0.68$), and behavioral engagement ($\alpha = 0.84$). Higher scores indicate higher involvement for each component. After reversing the ratings for items in the psychological involvement—negative feelings, the total SPGQ was computed. A higher total SPGQ score reflects more positive social components (interactions) perceived by gamers in online gaming.

### 2.3 *Procedures*

Prior to taking the online survey, an informed consent page was shown on the first page of the online questionnaire to obtain participants' consent. The consent page stated that there are no harmful and sensitive questions included in the questionnaire, and that the potential risk to the participants was very minimal. Once the participants clicked and gave the consent, the purpose and the procedure of the study were introduced to the participants in detail. Furthermore, all participants had the right to obtain further information about the purpose of this study and had the right to withdraw from the study at any time without any consequences. Although the survey was anonymous, all collected data were kept securely and could only be accessible by the two researchers. All participants completed a survey questionnaire that consisted of GAS, GEQ, and SPGQ, plus some demographic and background information such as sex, age, and years of playing online games. The collected data were retrieved and read into SPSS 23.0. Multiple linear regression (stepwise approach) was employed to test the two hypotheses and address the research questions.

## 3 Results

In this section, descriptive statistics of GAS, GEQ, and SPGQ scales and bivariate relationships among the three scales were reported first. Then, the results from the linear regression analyses were reported. Using the stepwise approach, the total GEQ scores were entered to predict GAS and to test the first hypothesis—whether playing more online games is significantly and negatively related to gamers' mental health (H1). Afterwards, the main effect of SPGQ and the interaction effect of GEQ x SPGQ were entered into the regression model to predict GAS and to test the second hypothesis—a higher level of positive social components in online games can significantly moderate the negative effect of online games on gamers' mental health (H2). If the interaction effect was significant, supplementary analyses were conducted to assist the interpretation of the interaction effect.

**Table 1** Descriptive statistics for GAS, GEQ, and SPGQ and pearson correlation coefficients among GAS, GEQ and SPGQ

| Variables | | | | Mean | SD | Std. error |
|---|---|---|---|---|---|---|
| | GAS | GEQ | SPGQ | | | |
| GAS | – | 0.65** | 0.52** | 14.82 | 5.96 | 0.48 |
| GEQ | | – | 0.74** | 87.71 | 23.22 | 1.85 |
| SPGQ | | | – | 45.85 | 14.02 | 1.12 |

## *3.1 Descriptive Statistics*

Table 1 summarizes the mean and standard deviation for each of the three scales and subscales used in the study. The mean of Game Addiction Scale (GAS) was 14.82 (SD = 5.96), the mean of Game Experience Questionnaire (GEQ) was 87.71 (SD = 23.22), and the mean of the Social Presence in Gaming Questionnaire (SPGQ) was 45.85 (SD = 14.02). Table 1 also shows the bivariate correlations among GAS, GEQ, and SPGQ. All Pearson correlation coefficients were significant and moderate. The bivariate correlational results showed that (1) more intensive and frequent online gaming experience was associated with higher gaming addictive tendency or lower gamers' mental health ($r = 0.65$), (2) more social interactions in online gaming were associated with higher addictive gaming tendency or lower gamers' mental health ($r = 0.53$), and (3) more intensive and frequent online gaming experience was associated with more social interactions in online gaming ($r = 0.74$).

## *3.2 Regression Analysis*

Table 2 summarizes the results of the regression analyses. The first model (Model 1) was a simple linear regression using GEQ (online gaming experience) to predict GAQ (gamers' mental health). The overall regression model was significant, $F(1, 155) =$

**Table 2** Regression analyses of GEQ, SPGQ, and GEQ x SPGQ on GAS

| Model | | Unstandardized coefficients | | Standardized coefficients | *t* | Sig |
|---|---|---|---|---|---|---|
| | | B | Std. error | Beta ($\beta$) | | |
| 1 | (Constant) | 0.180 | 1.421 | | 0.127 | 0.899 |
| | GEQ | 0.167 | 0.016 | 0.650 | 10.654 | <0.001 |
| 2 | (Constant) | 0.499 | 1.369 | | 0.364 | 0.716 |
| | GEQ | 0.159 | 0.022 | 0.620 | 7.157 | <0.001 |
| | SPGQ | 0.010 | 0.037 | 0.023 | 0.266 | 0.791 |
| | GEQ*SPGQ | 0.004 | 0.001 | 0.256 | 4.364 | <0.001 |

113.50, $p < 0.001$, and GEQ alone explained 42.3% of variance in gamers' mental health. In addition, GEQ significantly and strongly predicted GAQ ($\beta = 0.650$, $t = 10.654$, $p < 0.001$). In other words, more intense and frequent online gaming experience was negatively related with gamers' mental health, measured with their gaming addiction tendency. In Model 2, the main effect of social components in online gaming (SPGQ) and the interaction term between GEQ and SPGQ were entered into the regression model.

Results showed that the overall multiple regression model was significant, $F(3, 153) = 48.99$, $p < 0.001$. Furthermore, GEQ, SPGQ, and GEQ x SPGQ explained 48.2% of variance in gamers' mental health. After controlling for SPGQ and GEQ x SPGQ, Table 2 shows that the relationship between online gaming and gamers' mental health was still significant ($\beta = 0.620$, $t = 7.160$, $p < 0.001$). Nevertheless, the relationship between social components in online gaming (SPGQ) and gamers' mental health was insignificant ($\beta = 0.023$, $t = 0.266$, $p = 0.791$). Most importantly, the interaction effect between online gaming experience and the social components in online gaming on gamers' mental health was significant ($\beta = 0.256$, $t = 4.364$, $p < 0.001$). The significant interaction effect implied that the negative relationship between online gaming experience and gamers' mental health depended on the levels of social components in social gaming. The following subsection will further elaborate and interpret this significant interaction effect.

### 3.3 Supplementary Analyses for GEQ x SPGQ Interaction

Supplementary analyses were conducted to examine the significant interaction effect (GEQ × SPGQ). First, the researchers categorized the mixture of intensity of GEQ and levels of SPGQ into four groups using the medium score in each scale—Group 1 (low GAS and low SPGQ), Group 2 (low GAS and high SPGQ), Group 3 (high GAS and low SPGQ), and Group 4 (high GAS and high SPGQ). Figure 2 shows that for gamers with low intensity of online gaming experience (low GEQ), there was significant difference on the levels of gamers' mental health between high SPGQ versus low SPGQ (social components). For gamers with high intensity of online gaming experience (high GEQ), there were significant differences on the levels of gamers' mental health between high SPGQ versus low SPGQ (social components), ($t = 11.765$, $p < 0.001$). Contrary to the anticipation, social components in online gaming strengthen the negative impact of online gaming experience on gamers' mental health for gamers with high GEQ (high intensity and frequency of online gaming experience).

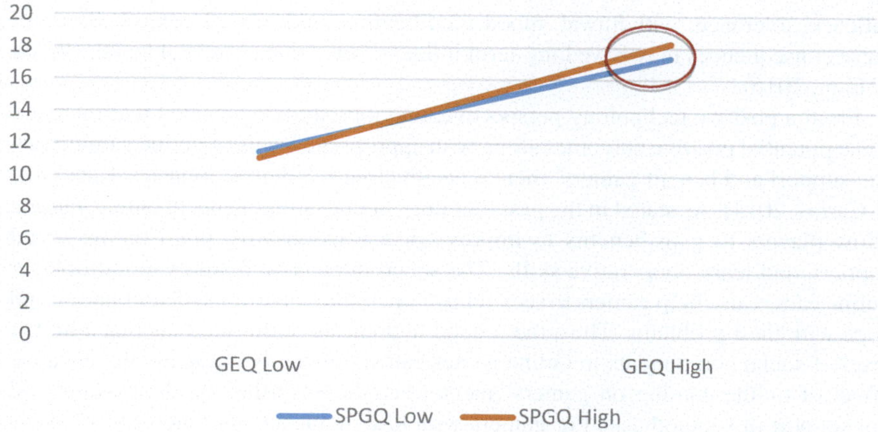


**Fig. 2** Visualization of the Interaction Effect of GEQ (Online gaming experience) and SPGQ (Social Components in Online Gaming) on GAQ (Gamers' Mental Health)

## 4 Discussion

The present study investigates the relations of online gaming experience and social components in online gaming with gamers' mental health (addictive gaming tendency), based on the arguments of the pathological perspective about the problematic online gaming versus the positive technology perspective about social benefits of online interactive gaming.

First, this study hypothesized that more online gaming experience is significantly associated with higher tendency in gaming addiction which is associated with psychological problems and psychiatric disorders. Consistent with previous research findings, results of the present study showed that playing more online games (more intensive and frequent online gaming experience) was significantly and negatively related to gamers' mental health. Thus, hypothesis 1 (H1) was fully supported.

With technological advancement, the development and design of online games have become increasingly attractive and easily accessible (Wong & Lam, 2016). Over the past few years, the evolutionary development of AR mobile games (e.g. Pokemon GO and PUBG which are location-based augmented reality mobile games) make online games more accessible, enjoyable, and immersive (Gortari, 2018). Thus, the intensity and frequency of online gaming experience are expected to have substantially increased for gamers. As a result, gamers become more addicted and find it more difficult to maintain the balance among their studies, work, friends, and family. In fact, the level of online gaming experience indicates problematic use of online games, particularly when the gamers lose their control and disrupt their normal life. The high intensity and frequency of online gaming negatively impact gamers' cognitive, developmental, physical, psychological, and social functions (Wong & Lam, 2016). Eventually, pathological gamers exhibit addiction symptoms including salience, tolerance, withdrawal, mood modification, and loss of control. Some of these consequences may have long-term influence on gamers' mental health (Wong & Lam, 2016).

From a positive technology perspective, previous study advocated that there are some potential positive outcomes along with appropriate levels of online gaming that can support and benefit gamers' mental health (Jones, Scholes, Johnson, Katsikitis, & Carras, 2014). As stated in the past literature, social components in online gaming allow players to gain benefits to improve peer relationships, build online social support, and learn cooperative skills. The social interactive features in multiplayer online games also help gamers to seek relaxation, reduce emotional disturbances, and cope with their problems. Thus, the second hypothesis of this study is that a higher level of social components in online games can significantly moderate the negative effect of online gaming on gamers' mental health. Nevertheless, the findings did not support this hypothesis. For gamers with low frequency and intensity of online gaming, the social components (high vs. low SPGQ) did not make any significant difference on the level of gamers' mental health (addictive gaming tendency). More surprisingly, for gamers with high frequency and intensity of online gaming, higher social components (high SPGQ > low SPGQ) even strengthened the negative impact on gamers' mental health.

These surprising findings could be explained by the differences in gamers' beliefs between the high-level gamers and low-level gamers. The high-level gamers usually have high intensity and frequency of online gaming (Lobel, Engels, Stone, & Granic, 2019); even some beneficial social components in gaming cannot easily moderate the negative impact on their mental health (additive gaming tendency). Besides, the high-level gamers believe that an online game is considered to be a tool to enhance their self-esteem within their peer group rather than just an entertainment tool. When those high-level gamers interact with others, they need to be compelled to meet peer expectations which are not easily satisfied. Due to the desire to influence others, these high-level gamers also want to show the 'best' side to their peers (Wong & Lam, 2016) which resulted in high peer pressure and may cause arguments in social interactions during gaming. Therefore, this may confirm the positive correlation between GAS and GEQ and further explain why high-level gamers with more intensive and frequent online gaming experience the social components in online gaming even strengthened the negative impact on their mental health.

There were a few limitations for this study. First, the sample size was still relatively small and therefore the researchers did not control any demographic variables in this study. Future studies can consider to enlarge the sample size including the potential demographic moderators (age and gender) in the prediction model. Second, the convenient and snowball sampling might affect the validity and generalizability of this study. A larger scale survey study using cluster or stratified sampling can be planned in the future to enhance the validity and generalizability of the study. Third, this study used online survey to collect data. Nevertheless, it is difficult to capture the actual online gaming behaviors and social interactions (positive versus negative) of the participants. Future studies can consider adopting daily time sampling with multiple time points (e.g. 7 time points in a week or 8 time points in a month) in actual gaming platforms to gather more detailed data about the gamers' online gaming experience and social interactions.

## 5 Conclusion

In conclusion, the pathological perspective believed that online gaming is equalized to other addictive behaviors. The findings of this study seemed to confirm this belief that more intensive and frequent online gaming experience was negatively associated with gamers' mental health (higher addictive gaming tendency). Furthermore, the social components in online gaming even strengthened the negative impact on gamers' mental health for gamers with a higher level of online gaming experience. Although the results for positive social components was inconsistent with our anticipation, the present study has highlighted the importance of further investigating how a balanced gaming life with beneficial social components could potentially reduce gamers' additive tendency and improve gamers' mental health and other adaptive and psychological functions (e.g. cognitive, emotional, and social). With a balanced use of gaming, the educational implication of the present study is to incorporate the interactive and cooperative social elements in gaming environments into the instructional design (e.g. using rich interactive gameplay data and social elements in gaming to track students' learning styles, motivation, and behaviors in different types of groupings) so as to fully support and tailor the gamification of learning to address the various educational needs and learning experiences of different types of learners.

## References


Andreassen, C. S., Billieux, J., Griffiths, M. D., Kuss, D. J., Demetrovics, Z., … Pallesen, S. (2016). The relationship between addictive use of social media and online games and symptoms of psychiatric disorders: A large-scale cross-sectional study. *Psychology of Addictive Behaviors, 30*(2), 252–262. https://doi.org/10.1037/adb0000160

Andreassen, C. S., & Pallesen, S. (2014). Social network site addiction—An overview. *Current Pharmaceutical Design, 20*(25), 4053–4061.

Beard, C. L., & Wickham, R. E. (2016). Gaming-contingent self-worth, gaming motivation, and Internet Gaming Disorder. *Computers in Human Behaviors, 61,* 507–515.

Carras, M. C., Rooij, A. J., Spruijt-Metz, D., Kvedar, J., Griffiths, M. D., … Labrique, A. (2018). Commercial online games as therapy: A new research agenda to unlock the potential of a global pastime. *Frontiers in Psychiatry, 8,* 1–7. https://doi.org/10.3389/fpsyt.2017.00300

De Kort, Y. A. W., de, IJsselsteijn, W. A., & Poels, K. (2007). *Digital games as social presence technology: Development of the Social Presence in Gaming Questionnaire* (SPGQ).

Esposito, N. (2005). *A short and simple definition of what an online game.* https://www.utc.fr/~nesposit/publications/esposito2005definition.pdf.

Fernandez, L. O., Männikkö, N., Kääriäinen, M., Griffiths, M. D., & Kuss, D. J. (2018). Mobile gaming and problematic smartphone use: A comparative study between Belgium and Finland. *Journal of Behavioral Addictions, 7*(1), 88–99. https://doi.org/10.1556/2006.6.2017.080

Gentile, D. A., Anderson, C. A., Yukawa, S., Ihori, N., Saleem, M., … Sakamoto, A. (2009). The effects of prosocial video games on prosocial behaviors: International evidence from correlational, longitudinal, and experimental studies. *Personality and Social Psychology Bulletin, 35,* 752–763. https://doi.org/10.1177/0146167209333045

Gentile, D. A., & Gentile, J. R. (2008). Violent video games as exemplary teachers: A conceptual analysis. *Journal of Youth and Adolescence, 9,* 127–141. https://doi.org/10.1007/S10964-007-9206-2

Gortari, A. B. (2018). Empirical study on game transfer phenomena in a location-based augmented reality game. *Telematics and Informatics, 35*(2), 382–396. https://doi.org/10.1016/j.tele.2017.12.015

Granic, I., Lobel, A., & Engels, R. (2014). The benefits of playing video games. *American Psychologist, 69*(1), 66–78.

Griffiths, M. D., Kuss, D. J., & Pontes, H. M. (2016). A brief overview of Internet gaming disorder and its treatment. *Australian Clinical Psychologist, 2*(1), 20108.

Hou, J. (2011). Uses and gratifications of social games: Blending social networking and game play. *First Monday, 16*(7). https://doi.org/10.5210/fm.v16i7.3517.

IJsselsteijn, W. A., de Kort, Y. A. W., & Poels, K. (2013). *The Game Experience Questionnaire.* Eindhoven: Technische Universiteit Eindhoven.

Jia, A. L., Shen, S., Bovenkamp, R. V., Iosup, A., Kuipers, F., & Epema, D. H. (2015). Socializing by gaming. *ACM Transactions on Knowledge Discovery from Data, 10*(2), 1–29. https://doi.org/10.1145/2736698

Jones, C. M., Scholes, L., Johnson, D., Katsikitis, M., & Carras, M. C. (2014). Gaming well: Links between online games and flourishing mental health. *Frontiers in Psychology, 5.* https://doi.org/10.3389/fpsyg.2014.00260

Khazaal, Y., Chatton, A., Rothen, S., Achab, S., Thorens, G., Zullino, D., & Gmel, G. (2016). Psychometric properties of the 7-item game addiction scale among French and German speaking adults. *BMC Psychiatry, 16*(1). Article 132. https://doi.org/10.1186/s12888-016-0836-3.

Kuss, D. J., Louws, J., & Wiers, R. W. (2012). Online gaming addiction? Motives predict addictive play behavior in Massively Multiplayer Online Role-Playing Games (MMORPG). *Cyberpsychology, Behavior, and Social Networking, 15*(9), 480–485. https://doi.org/10.1089/cyber.2012.0034

Lemmens, J. S., Valkenburg, P. M., & Peter, J. (2009). Development and validation of a game addiction scale for adolescents. *Media Psychology, 12*(1), 77–95.

Lenhart, A., Kahne, J., Middaugh, E., Macgill, A. R., Evans, C. & Vitak, J. (2008). *Teens' gaming experiences are diverse and include significant social interaction and civic engagement.* U.S.A.: Pew Internet & American Project.

Lobel, A., Engels, R. C., Stone, L. L., & Granic, I. (2019). Gaining a competitive edge: Longitudinal associations between children's competitive online game playing, conduct problems, peer relations, and prosocial behavior. *Psychology of Popular Media Culture, 8*(1), 76–87. https://doi.org/10.1037/ppm0000159

Mercola. (2017). *How online games affect your brain.* https://articles.mercola.com/sites/articles/archive/2017/07/27/how-online-games-affect-your-brain.aspx.

Merelle, S. Y. M., Kleiboer, A., Schotanus, M., & Rooij, A. J. (2017). Which health-related problems are associated with problematic online-gaming or social media use in adolescents? A large-scale cross-sectional study. *Clinical Neuropsychiatry, 14*(1), 11–19.

Newzoo. (2017). Newzoo global games market report 2017 (light version). https://newzoo.com/insights/trend-reports/newzoo-global-games-market-report-2017-light-version/.

Rezaeidehaghani, A., Mohammadi, M., Mehrabi, T., & Rezaeidehaghani, A. (2016). Association between playing computer games and mental and social health among male adolescents in Iran in 2014. *Iranian Journal of Nursing and Midwifery Research, 21*(2), 153. https://doi.org/10.4103/1735-9066.178236

Rooij, J., Ferguson, C. J., Mheen, D., & Schoenmakers, T. M. (2017). Predicting problematic Internet, game, and social media use from psychosocial well-being and application use. *Clinical Neuropsychiatry, 14*(1), 113–121.

Stiles, S. L. (2010). *Relationship between online role-playing games, personality and interpersonal relationships*. Unpublished master's thesis, Western Carolina University.

Sweetser, P., & Wyeth, P. A. (2005). Gameflow: A model for evaluating player enjoyment in games. *Computers in Entertainment*, *3*(3). https://doi.org/10.1145/1077246.1077253.

Uz, C., & Cagiltay, K. (2015). Social interactions and games. *Digital Education Review, 27,* 1–12.

Wong, I. L., & Lam, M. P. (2016). Gaming behavior and addiction among Hong Kong adolescents. *Asian Journal of Gambling Issues and Public Health, 6*(1). https://doi.org/10.1186/s40405-016-0016-x.

Wulf, T., Bowman, N. D., Rieger, D., Velez, J. A., & Breuer, J. (2018). Online games as time machines: Online game nostalgia and the success of retro gaming. *Media and Communication, 6*(2), 60–68. https://doi.org/10.17645/mac.v6i2.1317

**Dr. Chi-Keung (Alex) Chan** is currently an Associate Professor in the Department of Counselling and Psychology at the Hong Kong Shue Yan University (HKSYU). Alex is also the Director of Positive Psychology Laboratory at HKSYU. Alex was the former Vice-President of HKAECT in 2016 and 2017. His research interests are positive technologies, serious gaming, and prevention of online risky behaviors.

**Kwan-Lun (Alvin) Cheung** received his bachelor degree in Psychology at the Hong Kong Shue Yan University (HKSYU). His research interest is social interaction in gaming process.

# English Vocabulary Building via Language Learning Apps: Voices from Hong Kong University Students and Teachers

**Anna Wing Bo Tso**

**Abstract** Myriads of digital learning resources have been developed to help students who are learning English as a foreign language (EFL). Nevertheless, English vocabulary learning, as an ongoing process (Schmitt et al., *Language Testing 18*:55–88, 2001), means much more than just the convenient access of audiovisual materials. Vocabulary learning strategies cover seven main categories of vocabulary learning strategies, which range from learners' beliefs about vocabulary learning, metacognitive regulation, guessing strategies, dictionary strategies, note-taking strategies, memory strategies (rehearsal), memory strategies (encoding), and activation strategies (Gu and Johnson, *Language Learning 46*:643-679, 1996). To what extent vocabulary learning apps are helping students' vocabulary skills is of paramount concern to EFL teachers and apps designers. In light of this, this paper collected data from questionnaires, vocabulary tests, and interviews with EFL students and teachers at a university in Hong Kong. It reveals and discusses the effectiveness of current English language learning apps for Hong Kong university students.

**Keywords** Digital learning · English vocabulary · Language apps · Learning English as a foreign language

## 1 English Vocabulary Building: Overlooked and Misunderstood

Vocabulary building is central to second language learning. It is commonly agreed among linguists that "lexis is the core or heart of language" (Lewis, 1993, p. 89), and that "without vocabulary *nothing* can be conveyed" (Wilkins, 1972, pp. 111–112). As McCarthy (1992) explains, "without words to express a wider range of meanings, communication in L2 just cannot happen in any meaningful way" (p. 50). Numerous

A. W. B. Tso (✉)
M537, The School of Humanities and Social Science, The Hang Seng University of Hong Kong, Hang Shin Link, Siu Lek Yuen, Shatin, N.T., Hong Kong, Hong Kong SAR, China
e-mail: awbtso@gmail.com



W. W. K. Ma et al. (eds.), *Learning Environment and Design*, Educational Communications and Technology Yearbook, https://doi.org/10.1007/978-981-15-8167-0_17

research studies also reveal that for basic English language use, a vocabulary size of 2,000 is the least requirement (Nation, 2005); for successful English reading comprehension, a vocabulary size of 3,000 is essential (Laufer, 1997); and to function effectively in English, all learners should reach the threshold vocabulary size of 5,000 words (Schmitt, 2000). It is reported that the average English vocabulary size for university students is 10,000 words in Japan, 10,000 words in Holland, and 15,000 words in Russia (Allen, 1983).

Unfortunately, for over 40 years, vocabulary learning research had fallen largely under the radar. From the 1940s to the 1960s, most dominant teaching approaches were influenced by Fries' structural linguistics (1945) and Chomsky's generative transformational linguistics (1957), where learners were falsely expected to have the competence to fill in any lexical items once they learned the English grammatical rules. The role of grammar was overestimated while that of vocabulary was underestimated. Afterwards, from the 1970s, Hymes' (1972) communicative teaching approach took over, and vocabulary was again considered subordinate to functional language use. Oxford (2003) states, "Vocabulary is not explicitly taught in most language classes" (p. 9). Fan (2003) and Tso (2020) further confirm that in the university curriculum of most Asian countries, vocabulary is given little emphasis.

Today, the importance of vocabulary learning and teaching has gained recognition, but most ESL and EFL teachers are still unfamiliar with vocabulary learning strategies, not to mention introducing the strategies to their students. Wu (2005) reflects that in Taiwan, English teachers are still using traditional teacher-centered methods such as the use of dictionaries, verbal and written repetition, and dictation. The traditional methods may not be effective enough for vocabulary learning and building, because unlike syntactic structures and grammatical items, "lexical items…are an open set, constantly being added to (and lost, as archaic words gradually go out of use)" (Ur, 2012, p. 3). Rather than always using static and mechanical means to teach English vocabulary, teachers need to come to the realization that the most natural and effective way of learning English vocabulary is through metacognitive strategies (Schmitt, 1997), where social learning takes place as learners read books and authentic materials, watch films, play online games, and/or interact with native speakers as well as other EFL learners. In other words, vocabulary building is largely a self-directed learning process beyond the classroom.

## *1.1 The Shortfall of Vocabulary Learning Apps*

Numerous recent education studies (Kinshuk & Huang, 2015; Tso & Lau, 2016, 2018; Zhang & Hung, 2015) have shown that apps can be effective learning tools for learners. The most apparent advantage of learning apps is that they offer a combination of flexibility, accessibility, and interactivity (Liu, Tan, & Chu, 2009), which encourages learners' motivation and self-regulated learning, improving learners' academic performance and learning attitude (Kramarski & Gutman, 2006). For English vocabulary learning, a sizable amount of apps and digital games have been designed for EFL learners as well, most of which help to enhance learners' vocabulary retention through audiovisual texts, note-taking and highlight functions, sound effects, animation, games, assessment, and/or instant links to online dictionaries and Wikipedia. Nevertheless, vocabulary building is an ongoing process which involves communication and interactions in the discourse community. "To understanding and use a term or an expression, one needs to be familiar with the system of concepts in the specific domain, examining how it functions and how it is utilized in the particular discourse" (Tso & Chung, 2016, p. 53). Vocabulary learning means way much more than behaviorist retention, repetition and drilling. As Gu and Johnson point out:

> Learners should not…depend on visual repetition and fanciful imagery techniques to remember the words […]. Learning a word includes much more than remembering the orthographic and phonological forms and their corresponding meanings (1996, p. 659).

Rather than seeing vocabulary as isolated items to memorize, Gu and Johnson (1996) suggest "vocabulary in action" (ibid.), which includes the following vocabulary learning strategies:

| | |
|---|---|
| 1. Beliefs about vocabulary learning | (a) Words should be memorized<br>(b) Words should be acquired in context (bottom-up)<br>(c) Words should be studied and put to use (top-down) |
| 2. Guessing strategies | (a) Using background knowledge/wider context<br>(b) Using linguistic cues/immediate context |
| 3. Dictionary strategies | (a) Extended dictionary strategies<br>(b) Looking-up strategies |
| 4. Note-taking strategies | (a) Meaning-oriented note-taking strategies<br>(b) Usage-oriented note-taking strategies |
| 5. Memory strategies: rehearsal | (a) Making vocabulary lists and/or cards of new words<br>(b) Remembering new words by repeating the pronunciation<br>(c) Remembering new words by writing them repeatedly |
| 6. Memory strategies: encoding | (a) Visual encoding<br>(b) Auditory encoding<br>(c) Word structure<br>(d) Semantic encoding |
| 7. Activation strategies | (a) Extensive reading<br>(b) Making sentences with newly learnt words<br>(c) Using newly learnt words in speech and writing |

Research has shown that EFL university students with the best vocabulary skills are those who learn vocabulary "through natural exposure" and "careful studying" (Gu & Johnson, 1996, p. 662). Similarly, Lessard-Clouston (2008) also reveals that proficient learners often employ a wide range of vocabulary strategies instead of relying on one single strategy. For example, successful learners would take part in such vocabulary learning activities as taking notes, forming word lists, creating personal flashcards, consulting dictionaries, listening to recordings of teachers, and having peer discussions. While researchers advocate apps as a promising "platform for vocabulary learning and retention" (Yafei & Osman, 2016, p. 286), it is highly unlikely that vocabulary learning apps will suffice to provide an all-round training for EFL university students. After all, what university students need is mostly academic vocabulary which appears in various academic text types (Coxhead & Nation, 2001). What is at issue is: to what extent are vocabulary learning apps useful for EFL university students who need academic and/or professional vocabulary?

## 2 Research Questions

To find out the answer to the above query, the most practical means is to listen to the voices of the apps users. With the aim of revealing the usefulness of vocabulary learning apps on EFL students studying in the tertiary sector in Hong Kong, this case study poses these research questions:

1. Is there a direct relation between EFL undergraduates' English vocabulary performance and their use of vocabulary learning apps?
2. What is the difference in vocabulary strategies between students who enjoy using vocabulary learning apps and students who prefer conventional vocabulary methods to vocabulary learning apps?
3. What are the university teachers' and undergraduate students' attitudes and opinions on English vocabulary learning apps?

## 3 Research Methodology

To get a fuller picture of the impact of vocabulary learning apps on EFL university students in Hong Kong, this study used The Open University of Hong Kong (OUHK) as the base, collecting data from both EFL undergraduate students and their teachers. The focus of the study is as follows:

(1) EFL students' background and English vocabulary levels;
(2) EFL students' vocabulary building strategies;
(3) EFL students' views of vocabulary learning apps; and
(4) University teachers' views of vocabulary learning apps.

Data were collected and triangulated through (1) a questionnaire survey; (2) a diagnostic test with reference to IELTS general and academic vocabulary; (3) a focus group interview with three EFL undergraduate students; and (4) an interview with three lecturers who used English as a medium of instructions in their class.

On September 27, 2019 (Friday), all student participants were invited to fill in a survey which collected their demographic information, vocabulary learning skills and strategies. From the survey, it was found that the group of student participants consisted of 10 males and 20 females who had a mean age of 22.8 years and a median age of 20 years. The ages ranged from 17 to 33 years. Among the 30 EFL undergraduates, 13 were Putonghua-speaking students from Mainland China, and 17 were local Cantonese-speaking Hong Kong students. Besides the questionnaire, the 30 student participants were asked to work on a 30-minute, multiple-choice (MC) diagnostic test based on the IELTS vocabulary. The first 20 MC questions were to test learners' general vocabulary proficiency, whereas the last 20 MC questions were to measure learners' levels of academic vocabulary for academic reading and writing.

Following up the survey and the diagnostic test, a face-to-face focus group interview was conducted with three full-time students from the disciplines of Humanities, Social Sciences, and Creative Arts. For easy communication, interviews were conducted in the mother tongue of the students. Similarly, in-depth interviews with three English lecturers were conducted to collect EFL university teachers' views on vocabulary learning apps and their comments on EFL students' English vocabulary needs and challenges.

## 4 Research Findings and Discussion

Surprisingly, data from the survey indicate that while most participants use vocabulary learning apps for self-learning, only half of the participants suggested their teachers to use vocabulary apps in class:

| | Yes | No |
|---|---|---|
| 1. Do you use vocabulary learning apps to learn English words? | 80% (24) | 20% (6) |
| 2. Should teachers utilize language apps as part of their pedagogies when they teach you vocabulary in the classroom? | 50% (15%) | 50% (15%) |

This finding is in line with the research results of recent studies on digital learning, which show that digital natives do not agree with using digital devices in English class all the time (Tso, 2019), nor do digital immigrants always find it enjoyable to have digital settings implemented for education purposes (Tso & Lau, 2019). The popularity of digital learning, in particular that of English vocabulary learning apps, may have been overestimated.

### *4.1 Relation Between EFL Undergraduates' English Vocabulary Performance and Their Use of Vocabulary Learning Apps*

More disappointingly, no significant relation or positive correlation can be observed between EFL undergraduates' English vocabulary performance and their use of

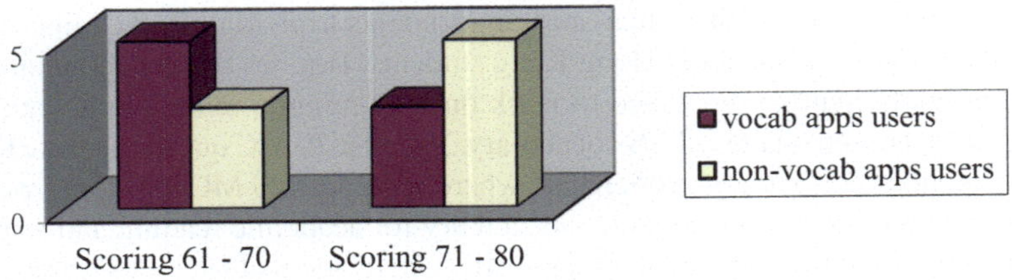


**Fig. 1** A comparison of the test scores obtained by vocabulary apps users and non-users.

vocabulary learning apps either. The diagnostic test results show that the average test score obtained by the 30 participants was 62.75 out of 100, with about 53% of the participants getting a test score between 61 and 80, as well as about 47% of them getting a test score between 40 and 50:

| Test score (Full score = 100) | 40–50 (Bare pass) | 51–60 (Fair) | 61–70 (Good) | 71–80 (Very good) |
|---|---|---|---|---|
| Percentage of students | 16.67% (5) | 30% (9) | 26.67% (8) | 26.67% (8) |

Half of the low achievers (those who obtained a test score of 40–60) were students who reported that they welcomed English vocabulary learning apps for self-learning and classroom learning. Among the high achievers, 5 out of 8 participants who obtained a test score of 71–80 were actually students who preferred not to use vocabulary learning apps in class (Fig. 1).

It appears that there is no direct relation between EFL undergraduates' English vocabulary performance and their use of vocabulary learning apps. As a matter of fact, students who opted not to have apps in class were even found to be performing slightly better than keen supporters of vocabulary learning apps.

### *4.2 Difference in Vocabulary Strategies Between Vocabulary App Learners and Non-vocabulary App Learners*

Students' vocabulary learning strategies and habits may as well reflect the reason why people who preferred not to use apps in class performed better than keen vocabulary apps users. Following Gu and Johnson's vocabulary skills framework (1996), the survey of this study categorizes students' vocabulary acquisition habits into seven learning strategies: (1) beliefs about vocabulary learning; (2) guessing strategies; (3) dictionary strategies; (4) note-taking strategies; (5) memory strategies (rehearsal); (6) memory strategies (encoding); and (7) activation strategies. By expressing through a 6-point Likert scale showing how much they agreed and disagreed (1 equals strongly disagree and 6 equals strongly disagree) with the vocabulary learning habits and strategies, student participants revealed their vocabulary learning attitudes and skills.

| | Average Likert scale point obtained by students who opted not to use vocabulary learning apps for classroom learning | Average Likert scale point obtained by students who welcomed English vocabulary learning apps for both self-learning and classroom learning |
|---|---|---|
| **1. Beliefs about vocabulary learning** | **4.38** | **4.11** |
| 1.1 Words should be memorized | 3.53 | 3.27 |
| 1.2 Words should be acquired in context | 4.53 | 4.20 |
| 1.3 Words should be studied and put to use | 5.09 | 4.85 |
| **2. Guessing Strategies** | **4.52** | **4.49** |
| 2.1 Using background knowledge/wider context | 4.89 | 4.68 |
| 2.2 Using linguistic cues/immediate context | 4.15 | 4.30 |
| **3. Dictionary Strategies** | **4.47** | **4.06** |
| 3.1 Extended dictionary strategies | 4.82 | 4.01 |
| 3.2 Looking-up strategies | 4.11 | 4.10 |
| **4. Note-taking Strategies** | **3.94** | **3.62** |
| 4.1 Meaning-oriented note-taking strategies | 4.27 | 3.83 |
| 4.2 Usage-oriented note-taking strategies | 3.60 | 3.40 |
| **5. Memory Strategies (Rehearsal)** | **4.27** | **3.67** |
| **6. Memory Strategies (Encoding)** | **4.09** | **3.96** |
| 6.1 Visual encoding | 4.17 | 3.90 |
| 6.2 Auditory encoding | 4.4 | 5.40 |
| 6.3 Word structure | 3.67 | 3.27 |
| 6.4 Semantic encoding | 4.10 | 3.27 |
| **7. Activation Strategies** | **4.02** | **3.95** |

It is observed that students who suggested teachers not to use vocabulary apps in classroom teaching turned out to be more all-rounded in terms of vocabulary skills. Compared with students who believed that vocabulary learning apps should be used both in classroom and beyond, the former achieved higher scale points in their beliefs about vocabulary learning. In other words, those who preferred not to rely on apps during classroom learning were more diligent and more willing to put their vocabulary skills to use. Those who opted not to use apps in lessons also showed better memory strategies (encoding). To a certain extent, this indicates that the over-reliance on vocabulary learning apps which offer audiovisual and animation effects cannot significantly improve EFL learners' retention. More significantly, participants who suggested teachers not to use vocabulary learning apps in class also got consistently higher scale points in such conventional vocabulary learning skills as the guessing strategies, dictionary strategies, note-taking strategies, memory strategies (rehearsal), and activation strategies. This reflects that for English vocabulary building, EFL learners need to employ a variety of learning methods. Apps may be helpful in the memory strategies, but one should avoid relying too much on digital technology.

## 4.3 University Teachers' and Undergraduate Students' Attitudes and Opinions on English Vocabulary Learning Apps

From the students' and teachers' interviews, it was noticed that EFL students and teachers were well aware of the limitations of vocabulary learning apps. Students in the focus group interview reflected that as far as they understood, apps that are appealing and addictive are usually not educational. One student admitted that he was once obsessed with a digital game, and he could go on playing the game without rest or sleep for three nights, which he confessed was unhealthy. The optimal duration of his game play had shocked his peers. Students also wondered if language apps would have the same powerful appeal on the player, and even if there was such a language app, there might be ethical issues in game addiction. When being asked whether they believed in edutainment, students said they believed in it only to a certain extent. They agreed that while apps with cute animations and audiovisual effects may seem fun at the beginning, the app designs are mostly formulaic, and the quizzes repetitive. Often, app users get bored with the apps in a few weeks' time. Students also raised the problem that there are too many vocabulary learning apps in the market, all of which claim to be effective for English learners. However, not all of them can provide the right kind of academic vocabulary training for undergraduates. Rather than using apps, the student participants pointed out that it was more useful to learn vocabulary using dictionaries and thesauruses. They also found it enjoyable when lecturers used traditional teaching methods in class.

Similar to the sharing of the student interviewees, the three teachers from various disciplines also mentioned that relying on vocabulary learning apps is not adequate. As the lecturer who teaches senior undergraduates English linguistics remarked, "regarding effectiveness of the vocabulary learning apps, it really depends on its design and how students can utilize those apps". While students who use vocabulary learning apps may have their learning motivation enhanced, the linguistics lecturer still commented that, in general, his students need to make more effort in increasing their vocabulary. Spending more time on building vocabulary, regardless of using apps or traditional methods, is what truly matters:

> Students come across many words in everyday life but they don't really know the correct way to use them or their real meaning. They know the words cognitively or have a basic impression of the words but when using the words, the usage is not correct. The thing is, their foundation has not been very solid since they were young. They don't have an effective and persistent habit to learn vocabulary. Another factor for their poor vocabulary skills is that they don't invest a lot of time in it. Their focus is not on spending time to learn English.

A similar view was shared by another lecturer who teaches translation. Although digital learning is in the trend, she was not confident about the effectiveness of vocabulary learning apps on university students, "I'm only familiar with dictionaries like Longman or Cambridge, or I will use a thesaurus. Regarding vocabulary learning apps, perhaps I'm not that familiar with them, or maybe I'm prejudiced that I think they are more suitable for primary or secondary students but not for tertiary level students. I think if you are at university you need to depend on yourself to learn words through reading books". Instead of recommending her students to use more language learning apps, she advised her students to make full good use of all learning materials, be they in the printed form or digital form:

> I think it sounds like clichés but the basics are to read newspapers and articles, and listen to songs, and watch movies. Perhaps my background is related to language studies. I still think you need to read a lot. Nowadays, it is very easy to get access to news and articles. I teach translation and I know it is easy to find bilingual materials online. For example, one can read *The Economists* and *TIMES*. I know in mainland China there is a website called *Yi Yan Wang* where they have the translated version of English newspapers. When you read such information bilingually, you must be able to learn some new vocabulary.

By and large, although it is commonly believed that apps can increase learners' motivation, both student interviewees and teachers shared the view that learning English vocabulary requires various study skills. Using apps to learn vocabulary may bring no harm, but relying only on the apps will not suffice.

## 5 Conclusion

With a view to informing further large-scale, longitudinal investigations about language learning apps, this pilot study sheds light on the directions which researchers may follow when undergoing full-launch studies on EFL undergraduates' English vocabulary proficiency, learning strategies, and attitudes. While the study sample size is small and more scientific analytical tools are needed to bring on statistical significance, both the initial diagnostic test on EFL learners' vocabulary abilities and the students' and teachers' interviews in this study indicate that vocabulary learning apps may just be adjunct, but not core to EFL learners' vocabulary building. Unlike the learning process in other fields or subjects, vocabulary building includes an authentic English environment, the appropriate discourse community, and a wide range of learning strategies. It is not easy to design an app that can fully satisfy every aspect of the EFL learners' needs. There is no one way of expanding one's vocabulary, but in the long run, those who put their vocabulary into practical everyday use are the winners.

**Acknowledgements** This paper is based on the research work funded by the research and development school project R0144 [2018/19 A&SS] of The Open University of Hong Kong. I acknowledge my research assistant, Eric Chui, who helped prepare the questionnaires, vocabulary tests, and interview questions for my research.

## References

Allen, V. F. (1983). *Techniques in teaching vocabulary*. Oxford: Oxford University Press.

Chomsky, N. (1957). *Syntactic structures*. The Hague: Mouton.

Coxhead, A., & Nation, I. S. P. (2001). The specialized vocabulary of English for academic purposes. In J. Flowerdew & M. Peacock (Eds.), *Research perspectives on English for academic Purposes* (pp. 252–267). Cambridge: Cambridge University Press.

Fan, M. F. (2003). Frequency of use, perceived usefulness and actual usefulness of second language vocabulary strategies: A study of Hong Kong learners. *The Modern Language Journal, 87*, 222–241.

Fries, C. (1945). *Teaching and learning English as a foreign language*. Ann Arbour: University of Michigan Press.

Gu, P. Y., & Johnson, K. (1996). Vocabulary learning strategies and learning outcomes. *Language Learning, 46*(4), 643–679.

Hymes, D. (1972). On communicative competence. In J. P. Pride & J. Holmes (Eds.), *Sociolinguistics: Selected readings*. Harmondsworth, UK: Penguin.

Kinshuk & Huang, R. (Ed.). (2015). *Ubiquitous learning environments and technologies*. New York: Springer.

Kramarski, B., & Gutman, M. (2006). How can self-regulated learning be supported in mathematical e-learning environments? *Journal of Computer Assisted Learning, 22*(1), 24–33.

Laufer, B. (1997). What's in a word that makes it hard or easy: Some intraleixical factors that affect the learning of words. In N. Schmitt & M. McCarthy (Eds.), *Vocabulary: Description, acquisition, and pedagogy* (pp. 140–155). Cambridge: Cambridge University Press.

Lessard-Clouston, M. (2008). Strategies and success in technical vocabulary learning: Students' approaches in one academic context. *Indian Journal of Applied Linguistics, 32*(2), 31–63.

Lewis, M. (1993). *The lexical approach: The state of ELT and a way forward*. Hove, UK: Language Teaching Publications.

Liu, T. Y., Tan, T. H., & Chu, Y. L. (2009). Outdoor natural science learning with an RFID supported immersive ubiquitous learning environment. *Journal of Educational Technology & Society, 12*(4), 161–175.

McCarthy, M. (1992). *Vocabulary* (2nd ed.). Oxford: Oxford University Press.

Miangah, T. M., & Nezarat, A. (2012). Mobile-assisted language learning. *International Journal of Distributed and Parallel Systems, 3*(1), 309–319.

Nation, P. (2005). Teaching and learning vocabulary. In E. Hinkel (Ed.), *Handbook and research in second language teaching and learning* (pp. 581–595). New Jersey: Lawrence Erlbaum.

Oxford, R. (2003). Toward a more systematic model Pf L2 learner autonomy. In P. Palfreyman & R. Smith (Eds.), *Learner autonomy across cultures: Language education perspectives* (pp. 75–91). New York: Palgrave Macmillan.

Schmitt, N. (2000). *Vocabulary in language teaching*. New York: Ernst Klett Sprachen.

Schmitt, N. (1997). Vocabulary learning strategies. In N. Schmitt & M. McCarthy (Eds.), *Vocabulary: Description, acquisition, and pedagogy* (pp. 140–155). Cambridge: Cambridge University Press.

Schmitt, N., Schmitt, D., & Clapham, C. (2001). Developing and exploring the behaviour of two new versions of the vocabulary level test. *Language Testing, 18*(1), 55–88.

Tso, W. B. (2019). Learning English as a foreign language through social media: First-hand perspectives from Hong Kong adolescents. In W. W. K. Ma, W. L. Chan, & C. M. Cheng (Eds.), *Shaping the future of education, communication and technology* (pp. 107–115). Singapore: Springer.

Tso, W. B. (2020). An investigation of the effectiveness of English language apps for EFL vocabulary building. *Journal of Global Literacies, Technologies, and Emerging Pedagogies, 5*(1), 1083–1102.

Tso, W. B., & Chung, J. S. K. (2016). Academic literacy development: University students' perceptions and experiences of English academic writing in Southeast Asia. *Pacific-Asian Education, 28*, 51–61.

Tso, W. B., & Lau, M. Y. (2016). Digital game-based learning: A case study of a digital educational game in Hong Kong. *Journal of Global Literacies, Technologies, and Emerging Pedagogies, 3*(2), 433–447.

Tso, W. B., & Lau, M. Y. (2018). An explorative study on the pedagogical potential of gamification. In K. C. Li, K. S. Yuen, & B. T. M. Wong (Eds.), *Innovations in open and flexible education* (pp. 143–152). Singapore: Springer.

Tso, W. B., & Lau, M. Y. (2019). Visitors' perceptions of a multimodal exhibition: A case study at the Hong Kong Heritage Museum. In A. W. B. Tso (Ed.), *Digital humanities and new ways of teaching* (pp. 177–193). Singapore: Springer.

Ur, P. (2012). *Vocabulary activities*. Cambridge: Cambridge University Press.

Wilkins, D. (1972). *Linguistics in language teaching*. London: Arnold.

Wu, W. S. (2005). Use and helpfulness rankings of vocabulary learning strategies employed by EFL learners in Taiwan. *Journal of Humanities and Social Sciences, 1*(2), 7–13.

Yafe, O. A., & Osman, M. E. (2016). Mobile phone apps: An emerging e-platform for vocabulary learning and retention. *Journal of Applied Linguistics and Language Research, 3*(7), 286–308.

Zhang, X.Y., & Hung, S.C. (2015). Designing mobile apps to enhance the learners' engagement with online learning. In K. S. Yuen & K. C. Li (Eds.), *Proceedings of the second international conference on open and flexible education*. Hong Kong: The Open University of Hong Kong.

**Dr. Anna Wing Bo Tso** is an associate professor of English at The Hang Seng University of Hong Kong. With over 10 years of English teaching experience at various universities in Hong Kong, her research interests include applied linguistics, literature, and language education. In 2020, she won the Eric Hoffer Book Award Second Prize for her edited volume, *Digital Humanities and New Ways of Teaching* (Springer, 2019).

# Algorithms as the New Gatekeepers of Knowledge: Prospects and Challenges Regarding the Use of Artificial Intelligence in Education

**Aynur Sarısakaloğlu**

**Abstract** Many parts of our lives are being fundamentally altered by the ongoing process of digitalisation and the consequent implementation of artificial intelligence technologies. Education is one of the areas of life that is being moulded by these changes. The aim of this study is to provide a constructive discussion regarding the use of artificial intelligence technologies in higher education, as well as the significance of algorithms and their role as the new gatekeepers of education. Through in-depth expert interviews with eight software providers from the USA, this study reveals that while the use of artificial intelligence technologies may increase at university level in the long run, ethical concerns and bridging the digital divide will remain the most significant challenges to deal with. It can be assumed that by using artificial intelligence, individualised learning environments will be created by algorithms that will act as the new gatekeepers of providing knowledge.

**Keywords** Education · Artificial Intelligence · Algorithm · Gatekeeper · Knowledge

## 1 Introduction

The transformation and acquisition of knowledge has changed fundamentally due to the continuously increasing use of digital media in education. The Internet in particular enables the free disposal of knowledge and information that can be accessed at any time from any place as requested. However, applications of artificial intelligence as well as the Internet are now finding their way into the education system. Artificial intelligence already has promising applications in many areas of everyday life such as voice recognition, text creation, and image recognition, and it will continue to play an important role in the private and professional spheres, especially in the transfer of

A. Sarısakaloğlu (✉)
Faculty of Cultural and Social Sciences, Department of Communication, Turkish-German University, Istanbul, Turkey
e-mail: aynur.sarisakaloglu@tau.edu.tr



W. W. K. Ma et al. (eds.), *Learning Environment and Design*, Educational Communications and Technology Yearbook,
https://doi.org/10.1007/978-981-15-8167-0_18

knowledge. The increasing use of artificial intelligence technologies in the field of education is leading to profound changes in teaching and learning. Algorithm-based applications offer new leverage points for automated analyses of learning processes, which increase the quality of teaching and learning by designing individual teaching materials according to the learner's performance level. Consequently, new forms of teaching and learning are emerging. In the near future, virtual classrooms will assume the place of physical classrooms, aided by technologies such as computer-mediated communications, multimedia, intelligent tutoring systems, and intelligent agents that are able to create virtual learning environments tailored to the individual. These new directions in education will redefine lecturers' and students' roles. The use of artificial intelligence technologies in the transfer of knowledge has therefore become the focus of debates on how new technologies influence the educational landscape. In particular, questions on topics such as possibilities and challenges regarding the integration of artificial intelligence, acquisition of digital skills, adequate technological infrastructure, and ethical factors are moving into the forefront of discussions regarding sustainable development.

Studies on the chances and risks of algorithm-based learning systems are generally based on the perspectives of students and teachers in primary and secondary education (Baker, Smith, & Anissa, 2019; Bryant, Heitz, Sanghvi, & Wagle, 2020). Fewer studies examine students' behavioural data from applied artificial intelligence usage in higher education to identify little-engaged students on the basis of their academic achievements or to analyse the effectiveness of applications such as intelligent tutoring systems, automated grading of assignments, and adaptive systems (Alfarsi, Omar, & Alsinani, 2017; Hussain, Zhu, Zhang, & Abidi, 2018; McNamara, Crossley, Roscoe, Allen, & Dai, 2015; Steenbergen-Hu & Cooper 2014). However, further research is needed regarding the possible applications of new technologies in higher education from the software providers' viewpoint while they develop the artificial intelligence applications expected to be increasingly used and of fundamental importance in the coming years.

The use of artificial intelligence in transforming knowledge is particularly widespread in the USA and in China. According to the 'Global Artificial Intelligence Market in Education Sector' report, the share of artificial intelligence in US education will rise to 47.5% by 2021 (Marr, 2018). China is aiming to become the global leader of artificial intelligence development by 2030 (Jing, 2018). In European countries, artificial intelligence has also already arrived in the education sector (European Commission, 2018c). However, the extent to which artificial intelligence will be used for knowledge transfer depends, among other things, on the attitudes of lecturers towards new technologies, which in turn depends on their assessment of the opportunities and challenges of artificial intelligence in education.

Against this backdrop, the purpose of this study is to discover what artificial intelligence can offer in education so as to determine the way knowledge will be transferred in the future, to improve learning outcomes, and to ensure educational quality. To this end, it is important to discuss capabilities of artificial intelligence technologies for the transfer of knowledge. As a final step, the challenges that new technologies relying on artificial intelligence will bring about in the field of pedagogy should also be examined.

## 2 Theoretical Framework

### *2.1 Artificial Intelligence and Education*

Artificial intelligence is a very broad term, and does not have a widely accepted definition because of its interdisciplinary nature (Organisation for Economic Cooperation and Development, 2019). Researchers from different disciplines such as Anthropology, Philosophy, Communication Science, Computer Science, and Neurology all employ their own terminology and contribute to the development of artificial intelligence in different ways. Generally, every definition assumes a computer system that usually has a connection to human intelligence and is capable of performing tasks such as speech recognition and production, visual recognition, and knowledge transfer (Bellman, 1978; Kurzweil, Richter, Kurzweil, & Schneider, 1990; Russell & Norvig, 2010; Winston, 1992). Artificial intelligence is understood to be 'systems that display intelligent behaviour by analysing their environment and taking action—with some degree of autonomy—to achieve specific goals' (European Commission, 2018a: 1). According to this, we can talk of artificial intelligence when 'machines use language, form abstractions and concepts, solve kinds of problems now reserved for humans, and improve themselves' (McCarthy, Minsky, Rochester, & Shannon, 1955: 2). When it comes to the implementation of artificial intelligence in the field of education, the main objective is to 'make computationally precise and explicit forms of educational, psychological and social knowledge which are often left implicit' (Self, 1998: 350). So, the main focus is on the valorisation of students' personal information, the progress of studies, the previously learned content, and similar themes in order to understand how learning happens. Based on big data—which consists of structured and unstructured data, such as personal data of the students—an intelligent tutorial system could, for example, recognise when students are overwhelmed by learning materials, and as a solution adjust the speed or content depending on the learning process, as well as generate precisely calculated notes and corresponding learning content in seconds, so that each individual student successfully completes his/her university education (Manyika, Chui, Brown, Bughin, Dobbs, Roxburgh, & Byers, 2011). In this case, intelligent systems are aware of the students' knowledge level and learning progress. The use of artificial intelligence technologies could fundamentally change how students study. Algorithms as a set of rules designed for solving a problem or accomplishing a specific task would therefore take control of the education of individual students by providing them with individually designed learning programs based on personal data (Rogers, 1987).

### *2.2 Gatekeeping in Education*

The term gatekeeper primarily refers to the ability of news agencies and journalists to influence the flow of news and to determine which topics are reported on (White,

1950). These are individuals who play a central role in the selection of topics and the interpretation and presentation of content (Lewin, 1947; White, 1950). However, contents are not only selected or rejected, but also edited with regard to certain influencing factors and prepared accordingly. Due to the 'computational turn', the gatekeeping approach is evolving and undergoing changes, especially through the use of algorithms as actors (Berry, 2011; Bro & Wallberg, 2015).

In education, gatekeepers resume influential positions and possess the ability to control educational thresholds. Examples of this are university admission interviews or examinations, which are conducted by so-called gatekeepers. Thus, lecturers at universities also fulfil the function of gatekeepers by determining the content of their courses. Algorithms play a special role due to advancing digitalisation and the use of artificial intelligence technologies in education. The function of gatekeeping points to the fact that algorithms in education allow or deny access to teaching and learning content for students. Algorithms decide how the learning process unfolds and can control the flow of information according to the user. Therefore, algorithms as designers of courses act as gatekeepers with regard to the determination of teaching and learning content. The determination of the content is done automatically according to exact rules based on personal data. With the help of certain filter functions, students are provided with access to content corresponding to their existing knowledge and interests.

## 3 Methodology of the Study

The overall aim of this study is to investigate the potential of artificial intelligence technologies in education. For this, in-depth expert interviews were conducted with eight software providers from the USA. The interviewees were selected randomly on the basis of their specialisation in producing artificial intelligence in education. Software providers were interviewed due to their important position as stakeholders who program the algorithms and who thus have significant power to influence how artificial intelligence technology is used in education. The protocol for the expert interviews was semi-structured and consisted specifically of questions regarding the future of teaching and learning methods, and the opportunities and challenges awaiting lecturers and students.

The interviews were conducted via Skype between September 2019 and January 2020, and lasted between 46 and 67 minutes. As this is a qualitative study, the transcripts of the software providers were analysed utilising MAXQDA software. A thematic analysis was carried out to identify common patterns of meaning referring to specific themes within the interview transcripts (Gibbs, 2007). The analysis was conducted according to Braun and Clarke's (2006) six phases of thematic analysis, as follows: making the data of the transcripts familiar, coding the data, generating themes across the interviews, reviewing identified themes, defining and naming reviewed themes, and writing the report.

# 4 Results

## *4.1 Prospects of Artificial Intelligence Technologies in Transferring Knowledge*

In the first step, interviewees were asked to determine the capabilities of artificial intelligence technologies to find out how knowledge will be transferred in the future. The prospects of artificial intelligence technologies in education can be categorised by utilising a word cloud that depicts the following as the most frequently used terms and phrases: robots as teaching and learning aids, creation of individual learning profiles through personal tutors, individualised learning environments through 'Learning Analytics', support for collaborative learning environments, replacing routine tasks by automation (Fig. 1).

### 4.1.1 Robots as Teaching and Learning Aids

The first scenario, which was mentioned most frequently in the interviews, is the use of humanoid robots as teaching and learning aids. However, the accelerated process of digitalisation and the new possibilities of teaching and learning it evokes do not mean that human lecturers will be replaced in the future. The interviewed experts (6 out of 8) are of the opinion that, new possibilities notwithstanding, lecturers should continue to advise and accompany students in their learning process in order to achieve efficient learning success.

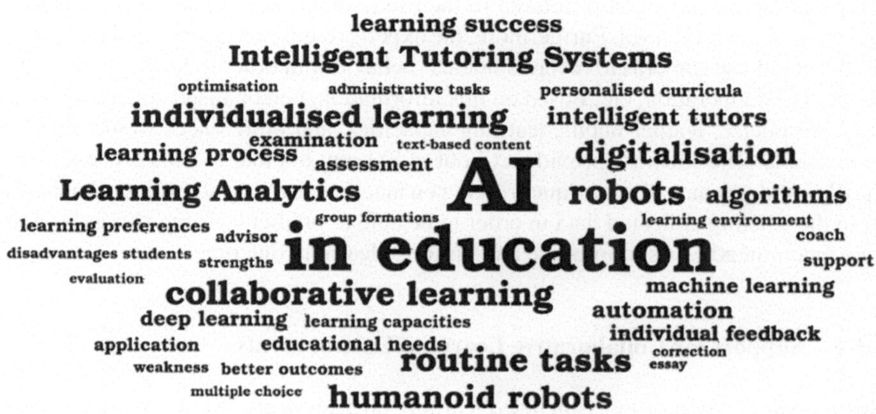


**Fig. 1** Prospects of artificial intelligence in education

### 4.1.2 Creation of Individual Learning Profiles Through Personal Tutors

As we know, face-to-face human tutoring is an effective approach to teaching and learning. According to all interviewed experts, the implementation of artificial intelligence can provide an intelligent tutor for every learner. With the help of machine learning techniques such as self-training algorithms, 'Intelligent Tutoring Systems' can simulate one-to-one human tutoring by giving individual feedback to students through collecting and evaluating data regarding their behaviour. Individual learning profiles of the students can be created in this way. Moreover, they emphasise the inclusiveness of 'Intelligent Tutoring Systems', as these can support students with specific educational needs and provide additional help for disadvantaged students in order to improve their level of learning performance and outcomes.

### 4.1.3 Individualised Learning Environments Through 'Learning Analytics'

Another possible area for the application of artificial intelligence relates to the optimisation of the learning environment. All of the interviewed experts predict that individualised learning environments will be made possible in the near future by using the tool 'Learning Analytics'. 'Learning Analytics' would make it possible to track and analyse the learning behaviour of students through found patterns in large data sets. By using big data and the collected data of the students' learning behaviour, it would be possible to respond to the learning needs and individual existing knowledge of the learners, and thus circumvent failures. With the help of algorithms, it would be possible to respond to students' interests, strengths, and weaknesses, as well as to offer them tasks and support tailored to their individual needs and their bachelor's, master's, or doctoral level. Furthermore, the experts (6 out of 8) mentioned that algorithms could use cameras to record students' behaviour in order to draw conclusions about their cooperation, etc. Based on this information, personalised curricula based on performance, learner habits, learning capacities, and preferences would also be conceivable. The software providers (7 out of 8) point out that for this process, they mostly use data analysis techniques relying on machine learning algorithms that learn from the existing collected data in order to be able to predict learning processes and give recommendations to improve and optimise learning outcomes.

### 4.1.4 Support for Collaborative Learning Environments

Beyond the creation of individualised learning environments, all of the experts also addressed the possible support of algorithm-based tools in collaborative learning when the students are physically not in the same place. Artificial intelligence technologies can, for example, provide support in forming groups best suited for a particular group assignment by analysing the students' behaviour, learning habits, cognitive level, and interests, and then recommending group formations to put a team together with students on the same level.

### 4.1.5 Replacing Routine Tasks Through Automation

Another area for the application of artificial intelligence in the educational sector mentioned by the experts (6 out of 8) was the possible use of automated corrections and evaluations of exercises, examinations, etc. In this way, lecturers could partly be released from administrative and routine tasks. Furthermore, they also stated that assessments would be more objective, especially for text-based content. Deep learning algorithms would make it possible for automated programs to not only assess multiple choice tests but also correct essays.

## *4.2 Challenges Regarding the Implantation of Artificial Intelligence Technologies in Education*

As with any use of new technology, challenges need to be considered and evaluated. For this purpose, the experts were asked to describe the possible challenges they perceived regarding the use of artificial intelligence in education and make recommendations to overcome these difficulties. The responses of the experts provide a valuable insight into the challenges regarding the implementation of artificial intelligence technologies in higher education. The findings can be summarised under four different headings as follows (Table 1).

### 4.2.1 Proficiency Enhancement of Lecturers and Students in Dealing with New Technologies

Results of the study show that proficiency in dealing with new technologies—especially with regard to the use of applications of artificial intelligence—is a fundamental prerequisite of media-pedagogical procedures by lecturers. According to all interviewed experts, lecturers must develop new digital skills and an understanding of what artificial intelligence technologies can offer for their students. Furthermore, they will need training opportunities to develop their proficiency in this area in order to be able to educate and work in a digitalised society. However, the students also need to understand how they can enlist the aid of new technologies in learning, so as to make their education more flexible, personalised, and engaging, as well as to thrive in a digitalised society.

**Table 1** Challenges of artificial intelligence in education

| Challenges | Problem definition | Recommendation |
|---|---|---|
| Proficiency enhancement of lecturers and students in dealing with new technologies | – Inadequate digital skills in using new technologies | – Competency enhancement<br>– Training opportunities |
| The changing professional profile of lecturers | – Implementation of innovative teaching concepts<br>– Expanding the repertoire of possibilities in education<br>– Learner-centred learning instead of teacher-centred education<br>– AI-based learning assistants | – Advisor and coach for the students<br>– Use of remaining time for other teaching activities |
| Granting the adequate technological infrastructure to ensure equality | – Digital divide<br>– Inequity and exclusion of individuals with no access to AI applications | – Ensuring required technological infrastructure |
| Ethical principles as the greatest challenge in the use of intelligent learning systems | – Transparency of data processing<br>– Reliability of collected data<br>– Traceability of the generated data<br>– Bias in the algorithms | – Ensuring compliance with ethical principles<br>– Assurance of data protection<br>– Equal treatment for all students |

### 4.2.2 The Changing Professional Profile of Lecturers

As mentioned above, the endorsement of intelligent learning systems can lead to the implementation of innovative teaching concepts and expand the repertoire of possibilities in university education, as well as enable learner-centred learning processes. Furthermore, digital learning assistants or individualised learning support can be used. However, these new forms of teaching design will also lead to changes in the professional profile of lecturers. All of the experts mentioned that the future job profile and role of the lecturer will continue to fundamentally evolve and change. While some lecturers will consider the use of intelligent learning systems as being supplementary, others will think they might be replaced by them. Nevertheless, lecturers will be primarily responsible for student motivation and the interpersonal aspects of the seminar. Other studies also confirm that lecturers should ensure students' motivation (Sparks, 2019). The most important tasks lecturers will perform will be accompanying and advising students, and thus assuming the task of coaching. This will free up a lot of time which could be used effectively for other course activities.

#### 4.2.3 Granting the Adequate Technological Infrastructure to Ensure Equality

Four of the interviewed experts also mentioned that it is vitally important for the required technological infrastructure to be in place, in order to avoid inequality and the exclusion of individuals with no access to methods offered by new technology; otherwise, this will lead to the emergence of a new digital divide.

#### 4.2.4 Ethical Principles as the Greatest Challenge in the Use of Intelligent Learning Systems

With regard to the challenges in the interaction of artificial intelligence and humans, the experts referred, in particular, to the following ethical issues needing to be taken into account: 'By whom and how are algorithms programmed?', 'What criteria are used to define teaching content as relevant or not relevant?', 'Can algorithms really react to learning interests?', etc.

The determination of ethical principles is therefore considered necessary by all experts. Moral principles should ensure that regulations are in place regarding the use of intelligent learning systems. All of the interviewed engineers stressed that applications of artificial intelligence should have transparent and reliable working processes. The traceability of results and the possible bias contained in algorithms are among the most important challenges. Furthermore, it was also articulated that intelligent learning systems have to be programmed in a way that treats all students equally, regardless of gender, age, physical disability, etc., and that equal access for all students should be provided. Moreover, the confidentiality of the collected learning data should also be guaranteed, and data protection regulations should be observed. Last but not least, the experts also pointed out that there could be difficulties in controlling learning relationships between lecturers and students.

## 5 Discussion

Artificial intelligence technologies hold an immense potential to improve teaching and learning processes in higher education. So far, the use of artificial intelligence in education has been limited, but it can be assumed that it will increase rapidly (Tuomi, 2018). According to the results of the study, artificial intelligence applications have the potential to fundamentally change the process of knowledge transfer. The study highlights the future importance of implementing new artificial intelligence technologies in education. In the future, artificial intelligence will make it possible to create flexible learning environments for individual students in order to respond individually to the needs and existing knowledge of students, and to adapt these according to their learning rate and capacity, so that students with strong learning abilities can be adequately supported and their long-term learning experiences can be secured. However, the objective is not to give students complete control over their learning through computers, but to create the best possible learning environment by taking into account their individual learning needs and requirements. The main aim is to promote the independence of students, as well as to develop their individual competence. As an example, university graduates could assess their course with the help of a questionnaire to ensure greater student satisfaction; this data could then be collected to evaluate their satisfaction or dissatisfaction with the studies they undertook, and subsequently be analysed in order to trace decision-making paths and formulate measures.

Nevertheless, it has to be emphasised that human mediators of knowledge should not be replaced by intelligent learning systems, but rather be supplemented or supported by them. The role of educational institutions is not called into question by the use of artificial intelligence applications, neither from a pedagogical nor a social standpoint. In this context, the professional profile and the role of lecturers should not be rendered meaningless. New technologies should merely substitute and/or support lecturers in routine tasks such as correcting exercises and exams, or other administrative duties able to be performed by programmed algorithms. Thus, artificial intelligence applications could act as supporting and advisory systems.

The increasing use of artificial intelligence in teaching and learning processes also raises ethical questions, particularly with regard to the collection and use of personal data (Ananny, 2015; Mittelstadt, Allo, Taddeo, Wachter, & Floridi, 2016). Responsibility, transparency, and the inspectability of generated data are the most significant criteria to be ascertained (Bostrom & Yudkowsky, 2013). With regard to the ethical concerns raised in this study, it should be noted that institutions alongside lecturers and students should be involved in the process of developing an artificial intelligence-based learning system, so as to ensure responsible use and transparency to meet their needs and reduce uncertainty about the use of new technologies. In order to benefit from the advantages of the possible applications of artificial intelligence, it is particularly necessary to ensure the protection of personal data, to consider ethical issues, and to impart the necessary skills for the handling of artificial intelligence technologies.

Moreover, it is of utmost importance that artificial intelligence technologies should provide equal access to education so that students with disabilities or those who live in remote communities without adequate technological infrastructure or facilities can also make use of these new methods of learning. This can only be possible if lecturers and students acquire the digital literacy skills required to use artificial intelligence technologies, such as knowledge of the fundamentals of hardware and software, how to interpret generated data, the ability to communicate and collaborate through digital technologies, and the creation of digital content (United Nations Educational, Scientific and Cultural Organization, 2018).

Overall, as artificial intelligence transforms the education system in the near future, it can be discerned that algorithms have the potential to become the new gatekeepers that will determine the learning environment of the students. Furthermore, it has to be noted that 'strong' artificial intelligence technologies have the potential to fully support teaching and learning processes, as well as to develop efficient algorithms that will function as gatekeepers or non-human lecturers that define students' learning environment.

## 6 Conclusion

The new possibilities of artificial intelligence play a significant role in education, as in many other areas of our lives. In order to utilise the potential of intelligent systems and to improve learning outcomes, however, it is necessary to ensure the best possible framework conditions and to take measures for the acquisition of digital skills. After all, the use of artificial intelligence technologies can only succeed if lecturers and students can handle these new technologies. To this end, an adaptation of the education system and a new curriculum for an artificial-intelligence-supported education would be of great importance (European Commission, 2018b). Learning how to harness the possibilities of artificial intelligence should involve the whole of society and begin at primary school level by teaching young pupils what artificial intelligence is and how algorithms work in order to prevent digital illiteracy and social division between people with these skills and those without. A high level of readiness to act and measures are essential to promote the skills of lecturers and students. This is also stated in the 'Policy and investment recommendations for trustworthy Artificial Intelligence' formulated by the European Union, which aims to promote the digital competence of all persons (European Commission, 2019).

Although the way knowledge is generated will change in the sense that traditional forms and methods of teaching and learning will be supplemented or partly replaced by the use of new artificial intelligence technologies, it is not to be assumed that human lecturers will be completely replaced or a 'Surveillance University' will be established. Nevertheless, algorithms could become the new gatekeepers in education that decide what, how, and when students learn; even so, we have to take into consideration that the future of implementing artificial intelligence in education is irrevocably correlated with the future development of artificial intelligence technologies.

## 7 Limitations

It should be noted at this point that the results of the study only show possible areas of application of artificial intelligence in education. However, it lays the foundation to analyse future prospects and challenges, and addresses issues that the education sector should take into consideration. Against this background, further research is particularly needed with regard to the use of artificial intelligence technologies in education, which covers not only possibilities and challenges, but also their influence and impact on lecturers and students. Additionally, further research is also required in the field of intelligent systems development to solve technical and ethical problems in teaching and learning processes. In order to improve the use of these intelligent systems, the cooperation of experts with other stakeholders and researchers in the education sector is of great importance here.

## References


Alfarsi, G. M. S., Omar, K. A. M., & Alsinani, M. J. (2017). A rule-based system for advising undergraduate students. *Journal of Theoretical and Applied Information Technology, 95*(11), 2453–2465.

Ananny, M. (2015). Toward an ethics of algorithms: Convening, observation, probability, and timeliness. *Science, Technology & Human Values, 41*(1), 93–117.

Baker, T., Smith, L., & Anissa, N. (2019). *Educ-AI-tion rebooted? Exploring the future of artificial intelligence in schools and colleges*. Retrieved May 12, 2020, from https://media.nesta.org.uk/documents/Future_of_AI_and_education_v5_WEB.pdf.

Bellman, R. (1978). *An introduction to artificial intelligence: Can computers think?* San Francisco, CA: Boyd & Fraser Publishing Company.

Berry, D. M. (2011). The computational turn: Thinking about the digital humanities. *Culture Machine, 12*, 1–22.

Bostrom, N., & Yudkowsky, E. (2013). *The ethics of artificial intelligence*. Cambridge: University Press.

Braun, V., & Clarke, V. (2006). Using thematic analysis in psychology. *Qualitative Research in Psychology, 3*(2), 77–101.

Bro, P., & Wallberg, F. (2015). Gatekeeping in a digital era: Principles, practices and technological platforms. *Journalism Practice, 9*(1), 92–105. https://doi.org/10.1080/17512786.2014.928468

Bryant, J., Heitz, C., Sanghvi, S., & Wagle, D. (2020). *How artificial intelligence will impact K-12 teachers*. Retrieved May 12, 2020, from https://www.mckinsey.com/industries/social-sector/our-insights/how-artificial-intelligence-will-impact-k-12-teachers.

European Commission (2018a). *Artificial intelligence for Europe*. Retrieved May 12, 2020, from https://eur-lex.europa.eu/legal-content/EN/TXT/PDF/?uri=CELEX:52018DC0237&from=EN.

European Commission (2018b). *Coordinated plan on artificial intelligence*. Retrieved May 12, 2020, from https://ec.europa.eu/newsroom/dae/document.cfm?doc_id=56018.

European Commission (2018c). *The impact of artificial intelligence on learning, teaching and education*. Retrieved May 12, 2020, from https://publications.jrc.ec.europa.eu/repository/bitstream/JRC113226/jrc113226_jrcb4_the_impact_of_artificial_intelligence_on_learning_final_2.pdf.

European Commission (2019). *Policy and investment recommendations for trustworthy AI. High-level expert group on artificial intelligence*. Retrieved May 12, 2020, from https://ec.europa.eu/newsroom/dae/document.cfm?doc_id=60343.

Gibbs, G. R. (2007). Thematic coding and categorizing. In G. R. Gibbs (Ed.), *Qualitative research kit: Analyzing qualitative data* (pp. 38–55). London: Sage. https://doi.org/10.4135/9781849208574

Hussain, M., Zhu, W., Zhang, W., & Abidi, S. M. R. (2018). Student engagement predictions in an e-learning system and their impact on student course assessment scores. *Computational Intelligence and Neuroscience*, 1–21. https://doi.org/10.1155/2018/6347186

Jing, M. (2018). China wants to bring artificial intelligence to its classrooms to boost its education system. *Science & Research, South China Morning Post*. Retrieved May 12, 2020, from https://www.scmp.com/tech/science-research/article/2115271/china-wants-bring-artificial-intelligence-its-classrooms-boost.

Kurzweil, R., Richter, R., Kurzweil, R., & Schneider, M. L. (1990). *The age of intelligent machines* (Vol. 579). Cambridge, MA: MIT Press.

Lewin, K. (1947). Channels of group life, social planning and action research. *Human Relations, 1,* 143–153.
Manyika, J., Chui, M., Brown, B., Bughin, J., Dobbs, R., Roxburgh, C., & Byers, A. H. (2011). Big data: The next frontier for innovation, competition, and productivity. *McKinsey Global Institute.* Retrieved May 12, 2020, from https://www.mckinsey.com/Insights/MGI/Research/Technology_and_Innovation/Big_data_The_next_frontier_for_innovation.
Marr, B. (2018). *How is AI used in education—Real world examples of today and a peek into the future.* March 9, 2020. Retrieved May 12, 2020, from https://www.forbes.com/sites/bernardmarr/2018/07/25/how-is-ai-used-in-education-real-world-examples-of-today-and-a-peek-into-the-future/#36e55e39586e.
McCarthy, J., Minsky, M., Rochester, N., & Shannon, C. (1955). *A proposal for the Dartmouth summer research project on artificial intelligence.* Retrieved May 12, 2020, from https://jmc.stanford.edu/articles/dartmouth/dartmouth.pdf.
McNamara, D. S., Crossley, S. A., Roscoe, R. D., Allen, L. K., & Dai, J. (2015). A hierarchical classification approach to automated essay scoring. *Assessing Writing, 23,* 35–59. https://doi.org/10.1016/j.asw.2014.09.002
Mittelstadt, B. D., Allo, P., Taddeo, M., Wachter, S., & Floridi, L. (2016). The ethics of algorithms: Mapping the debate. *Big Data & Society, 3*(2). https://doi.org/10.1177/2053951716679679
Rogers, H. (1987). *Theory of recursive functions and effective computability.* Cambridge: MIT Press.
Russell, S. J., & Norvig, P. (2010). *Artificial intelligence: A modern approach.* Upper Saddle River, NJ: Pearson Education Inc.
Self, J. (1998). The defining characteristics of intelligent tutoring systems research: ITSs care, precisely. *International Journal of Artificial Intelligence in Education, 10,* 350–364.
Sparks, S. D. (2019). *Teachers shape students' motivation. Where do they learn how to do it?* Retrieved May 12, 2020, from https://blogs.edweek.org/edweek/inside-school-research/2019/05/teachers_can_shape_students_motivation.html.
Steenbergen-Hu, S., & Cooper, H. (2014). A meta-analysis of the effectiveness of intelligent tutoring systems on college students' academic learning. *Journal of Educational Psychology, 106*(2), 331–347. https://doi.org/10.1037/a0034752
Tuomi, I. (2018). The impact of artificial intelligence on learning, teaching, and education. Policies for the future. In: M. Cabrera, R. Vuorikari, & Y. Punie, (Eds.), *JRC Science for policy report.* Luxembourg: Publications Office of the European Union.
United Nations Educational, Scientific and Cultural Organization (2018). *A global framework for reference on digital literacy skills for indicator 4.4.2.* Information Paper No. 51. Montreal, Quebec. Retrieved May 12, 2020, from http://uis.unesco.org/sites/default/files/documents/ip51-global-framework-reference-digital-literacy-skills-2018-en.pdf.
White, D. M. (1950). The gatekeeper: A case study in the selection of news. *Journalism Quarterly, 27,* 383–390. https://doi.org/10.1177/107769905002700403
Winston, P. H. (1992). *Artificial intelligence.* Reading, MA: Addison-Wesley.
Organisation for Economic Cooperation and Development. (2019). Artificial intelligence in society. Paris: OECD Publishing. https://doi.org/10.1787/eedfee77-en

**Dr. Aynur Sarısakaloğlu** is Assistant Professor at the Faculty of Cultural and Social Sciences, Department of Communication at the Turkish-German University in Istanbul. She earned her Bachelor's, Master's, and Ph.D. degrees in Communication Science from the Faculty of Cultural and Social Sciences at the Paris Lodron University of Salzburg. She worked as a Research Assistant at the Paris Lodron University of Salzburg and was a Visiting Scholar at the University of San Francisco. Preferring an interdisciplinary approach to research and teaching, her main interests are artificial intelligence in communications, journalism research, media and communication technologies, new media and intercultural communication.

# AI in Music Education: The Impact of Using Artificial Intelligence (AI) Application to Practise Scales and Arpeggios in a Virtual Learning Environment

**Jason Chi Wai Chen**

**Abstract** This pilot study investigates the impact of using mobile devices to facilitate the practising of scales and arpeggios. The AI application was designed by an engineer/musician in Hong Kong. This AI application enables students to practise, record themselves, view their mistakes and prepare for examinations based on the Associated Board of the Royal Schools of Music's scales and arpeggios syllabus for five instruments (violin, flute, clarinet, trumpet and saxophone). This study (a) examines the progress of learners who use the application to practise scales and arpeggios; (b) observes how elementary learners perceive mobile practising in a virtual learning environment and (c) proposes a blended learning model that uses mobile practising with the support of AI.

**Keywords** Mobile practising · Scales and arpeggios · Artificial intelligence · Scalebook · Virtual learning environment

## 1 Introduction

The purpose of this study is to evaluate the impact of using AI in practising music. Learning progress was observed through the database of the application to see whether the frequency of user logins to the program and achievements in scale learning were correlated. Issues such as self-regulated practice, self-efficacy and musical memory were addressed through a survey of the users. The blended learning of scales and arpeggios was studied through teacher interviews. This study asks how this AI application can solve issues in classroom and studio pedagogy such as time limitations and the monitoring of practice. For consistency, the term AI application is adopted in this study. This AI application was designed and trademarked in 2017 and has already been launched for users in the UK, the US, Australia and Hong Kong. The designer has agreed to support the process of data collection through online

J. C. W. Chen (✉)
Flat D, 21/F, Block 1, Highland Park, 11 Lai Kong Street, Kwai Chung, N.T., Hong Kong
e-mail: cwchen@eduhk.hk



W. W. K. Ma et al. (eds.), *Learning Environment and Design*,
Educational Communications and Technology Yearbook,
https://doi.org/10.1007/978-981-15-8167-0_19

surveys and file analysis. The author declares that no conflict of interest is involved in this study.

The following research questions have guided the present study:

1. Is AI application an effective tool for practising scales and arpeggios?
2. To what extent does AI application enhance mobile practising?
3. How can mobile practising be implemented in studio teaching?

## 2 Literature

### 2.1 *Learning Through Mobile Devices*

Today, we are surrounded by rapidly developing technology. Rossing, Miller, Cecil and Stamper (2012) claimed that changes in technology continue to alter possibilities for learning and create new challenges for pedagogy. One new trend is learning through mobile devices, like tablets, at school and at home. A research team from Indianapolis defined mobile learning as the efficient and effective use of wireless and digital devices and technologies to enhance learners' individual outcomes during participation in learning activities.

Mobile devices act as a platform for students to learn and study at any time and in any place. Melhuish and Falloon (2010) stated that in many ways mobile technologies have the capacity to stimulate a redefinition of what constitutes a learning space, away from the constraints of a fixed place and time, towards a conceptualisation based on connecting people with each other and information, through virtual collaborative spaces and communities which are highly fluid, and not bounded by time or location. Chen (2015) studied the use of e-Auralbook to facilitate a mobile learning environment for teachers and students in both school and studio settings. This study extends this line of research by studying how mobile practising can be used by instrument teachers and learners to master scales and arpeggios.

The Horizon Report (Johnson, Levine, Smith, & Stone, 2010) reported that the iPad tablet is suitable for use in fieldwork due to its portability and usefulness for document and e-book transportation, real-time observation recording and reference evaluations.

### 2.2 *Scale Learning*

Scales are an essential element of basic technique. Franke (2013) explained that the primary purpose of scale and arpeggio practice is to develop and enforce the ability to stay within a given key or harmony. Scale learning is one of the requirements for playing an instrument, especially an orchestral instrument.

## 2.3 Artificial Intelligence (AI)

According to Warwick (2012), the field of Artificial Intelligence (AI) really came into existence with the birth of computers in and around the 1940s and 1950s. In the early period of AI's development, scientists focused on getting computers to do things that, if a human did them, would be regarded as intelligent. Essentially, this involved trying to get computers to copy humans in some or all aspects of their behaviour, an approach referred to as classical AI. The AI application as a form of classical AI, acts as a home teacher that simulates private instrument lessons. The AI application provides instant feedback after recording students' performance. Johnson (2014) suggested that real-time feedback has great potential for enhancing learning complex motor skills by enabling people to correct their mistakes as they go.

In addition, the AI application displays the two-dimensional score of students playing with an automatic music transcription (AMT) technique. Automatic music transcription is the process of converting an acoustic musical signal into some form of musical notation. Benetos, Dixon, Giannoulis, Kirchhoff, and Klapuri (2013) defined AMT as the process of converting an audio recording into a piano-roll notation—a two-dimensional representation of musical notes across time—and defined as the process of converting a recording into common music notation.

According to Conati, Porayska-Pomsta, and Mavrikis (2018) advocated that AI in education needs interpretable machine learning as lessons from open learner modelling. They investigated how to leverage AI techniques formally known as Intelligent Tutoring System (ITS) in the 1970s to create educational technologies that are personalised to the needs of the individual learners with the goal to approximate the well-known benefits to one-to-one instruction. The idea is to build intelligent interactive tutors that can model, predict and monitor relevant learner behaviours, abilities, and mental states in a variety of educational activities and provide personalised help and feedback accordingly. In this study, the AI application functions as an ITS with AI to analyse, improve, predict and monitor the learning process of practising scales with a Personal Tutoring System (PTS).

## 2.4 Virtual Learning Environment

Based on the sources from the AI application *Scalebook*, it was designed to use mobile devices to enhance scale and arpeggio practice for elementary learners of orchestral instruments. The AI application is based on the Grade 1–Grade 5 scales and arpeggios syllabus of the Associated Board of the Royal Schools of Music (ABRSM). The AI application provides a platform for elementary orchestral players to practise scales and arpeggios and acts as a tool that allows teachers to monitor and improve students' practice after each private lesson, as shown in Fig. 1. The AI application includes basic functions such as tuners and metronomes. It comes in three languages: English, Cantonese and Mandarin.

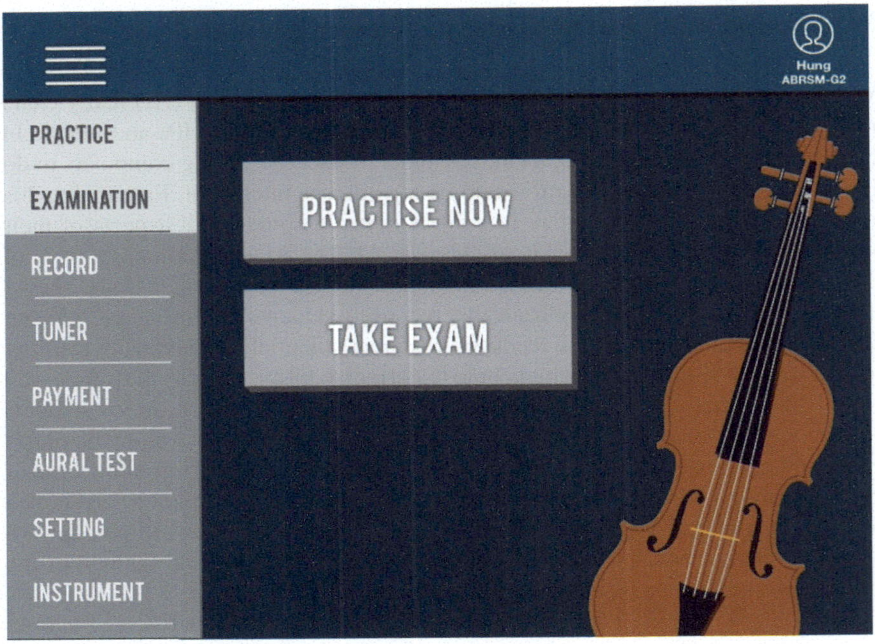


**Fig. 1** Core functions in the AI music application

The AI application provides two platforms: *practice mode* and *examination mode*.

*Practice mode* is designed for students to practise scales and arpeggios at home. It contains two ways to select the scales or arpeggios: *recommended* and *personalised*. The score is displayed on the screen with a key signature. Students can slide their fingers along the notes in the score to view the suggested fingerings and positions in the selected scale or arpeggio. The tempo can be adjusted by pressing the arrows next to the tempo marking, as shown in Fig. 2. Students are asked to play the scales and arpeggios following the beats of the metronome. The AI application analyses the student's practice session and indicates errors in intonation, rhythm and sliding notes with comments and red circles. Students may press the red circles on the score to explore their specific errors. They may also click the demo button to listen to a demonstration of correct playing, or press the playback button to hear their own playing.

*Examination mode* is a platform for rehearsing before attempting the real ABRSM Grade 1–Grade 5 scales and arpeggios examination, shown in Fig. 3. Scales and arpeggios from the same grade are randomly selected to test the student. A standard marking sheet is displayed with comments in Fig. 4. Stars are used to indicate the accuracy of the student's playing. The marking sheet is stored as a record in the AI application.

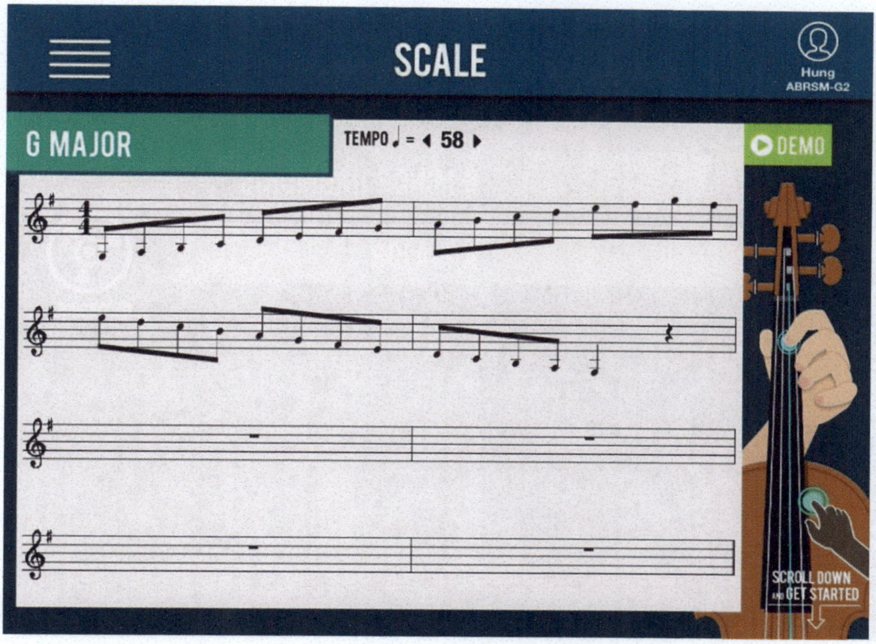


**Fig. 2** Scores with the fingering board in practice mode

# 3 Methodology

## *3.1 Data Sources*

In this study, data were collected from three sources: (1) an online survey, (2) file analysis and (3) individual interviews with studio teachers.

### 3.1.1 Online Survey

Table 1 shows the online survey conducted to learn about the interests and motivations of the AI music application users. In this pilot study, 24 users (n = 24) from around the world were invited to practise scales and arpeggios on five orchestral instruments (violin, flute, clarinet, saxophone and trumpet) over 8 weeks. After 8 weeks of practice, an evaluation questionnaire with seven questions on a 4-point Likert scale (1 = strongly disagree, 2 = disagree, 3 = agree, 4 = strongly agree) was sent to the application users to study how they used it and their perceptions of their self-regulated learning, musical memory and self-efficacy.

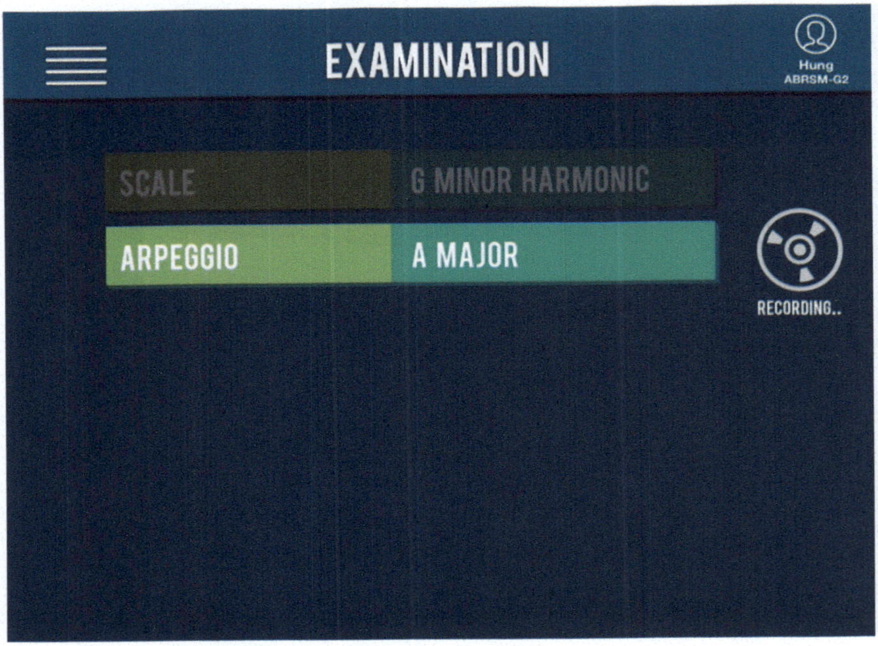


**Fig. 3** Exam mode page

### 3.1.2 File Analysis

An analysis of data files was retrieved from the AI application database. The sample size (n = 24) was 24 valid frequent users. Data were retrieved for beginners (ABRSM Grades 1–5) only. All participants received unlimited access to the AI application. They were asked to practise scales and arpeggios with the AI application for 8 weeks. Specifically, they practised a scale or an arpeggio from the ABRSM exam syllabus (Grade 1–Grade 5) using the AI application at least three times a week. All of their practice records were saved in cloud storage. After 8 weeks of practice, data were retrieved to verify their practice habits. Students' mean scores were calculated based on the starting and ending mean scores and how these scores impact students' music practice.

### 3.1.3 Individual Interviews

Semi-structured individual interviews with studio teachers were conducted. Five instrument teachers were interviewed with the following questions to understand how they used the application with daily instrument lessons and what they thought of it as a mobile practising device.

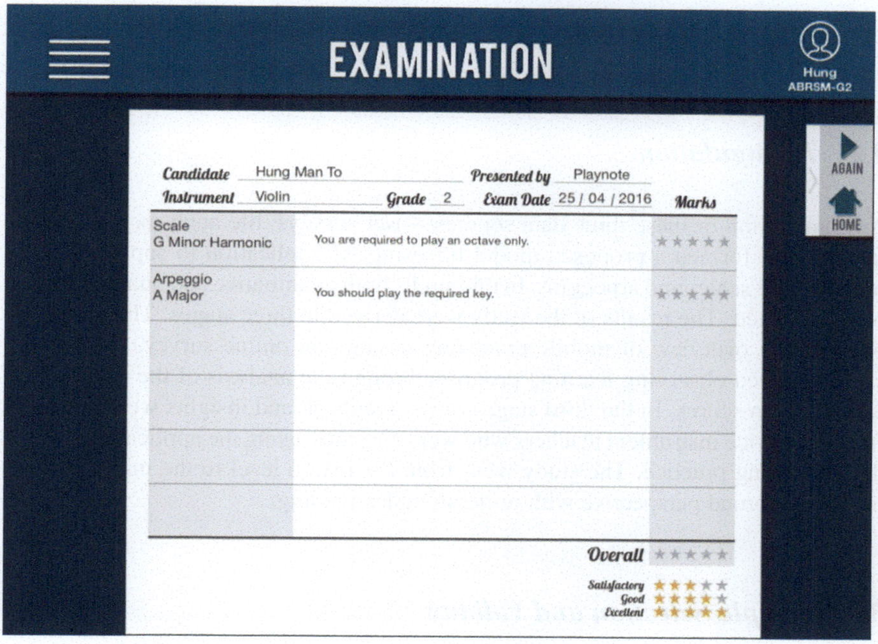


**Fig. 4** Standard mark sheet in examination mode

**Table 1** Online survey questions

| |
|---|
| Q1 Using the AI application to practise scales can enhance my learning |
| Q2 Using the AI application to practise scales can enhance my performance |
| Q3 Using the AI application to practise scales can regulate my practising habits |
| Q4 The AI application can enhance my musical memory (e.g., notes, sharps and flats) |
| Q5 The AI application can enhance my confidence in playing scales and arpeggios |
| Q6 The AI application can improve my motor skills (e.g., fingerings and muscle memory) |
| Q7 Any suggestions/comments (open-ended) |

1. How did you find the motivation of the students after they started to use the AI music application?
2. Did you find the application enhanced your students' musical skills?
3. Do you think the 'blended learning' mode (face-to-face plus mobile learning) is effective in teaching scales and arpeggios?

4. Would you continue to use this blended learning approach in the future? Why or why not?

## 3.2 Triangulation

A triangulation of these three data sources—user surveys, file analysis and teacher interviews—formed a proposed model for using AI application to support mobile practising of scales and arpeggios. In this study, both quantitative and qualitative data were collected. The results of the study were derived in three stages. The first stage provided an overview of mobile practising through an online survey. The second stage involved observing learning progress through file analysis of the starting and ending mean scores. In the third stage, views, feedback and insights were collected from in-service instrument teachers who were currently using the application in their daily teaching practice. The study went from the macro level to the micro level to integrate a broad perspective with in-depth understanding.

## 3.3 Sample Selection and Validity

Only frequent users who had fulfilled the basic practice requirements were counted as valid users. The relationship between the frequency of use of the application over the period of 2 months and the practice scores were analysed.

## 3.4 Data Analysis

The data analysis focused on the differences between the starting and ending mean scores of the valid frequent users. The starting and ending means were calculated by finding the average of the records between 1 December 2018 and 31 January 2019. The following cases were considered invalid records:

1. The recording was not all done by the same person;
2. Two or more people were practising at the same time;
3. No practices were offered in the record;
4. The user was doing something other than answering the questions; and
5. The recording was disturbed by the environment.

The remaining records were considered valid records for this study's purposes.

For the first and second data sources, all volunteers had to tick 'agree' to participate in this pilot study for 2 months through an online agreement. For the third data source, the individual interviews, consent forms were signed by the human subjects when the qualitative data were collected. The researcher declares that there is no conflict of interest in this research between the application company and the researcher. No commercial agreements or benefits were derived from this study.

## 4 Results

### *4.1 Online Survey*

#### 4.1.1 Geographical Distribution of the Respondents

Twenty-four respondents replied to this survey. The majority came from the UK (38%) and Hong Kong (34%), followed by Australia (8%), Spain (8%), Singapore (4%), Malaysia (4%) and Malta (4%). This geographical distribution maintains the reliability of this research as a globalised study. Figure 5 shows the geographical distribution of the respondents.

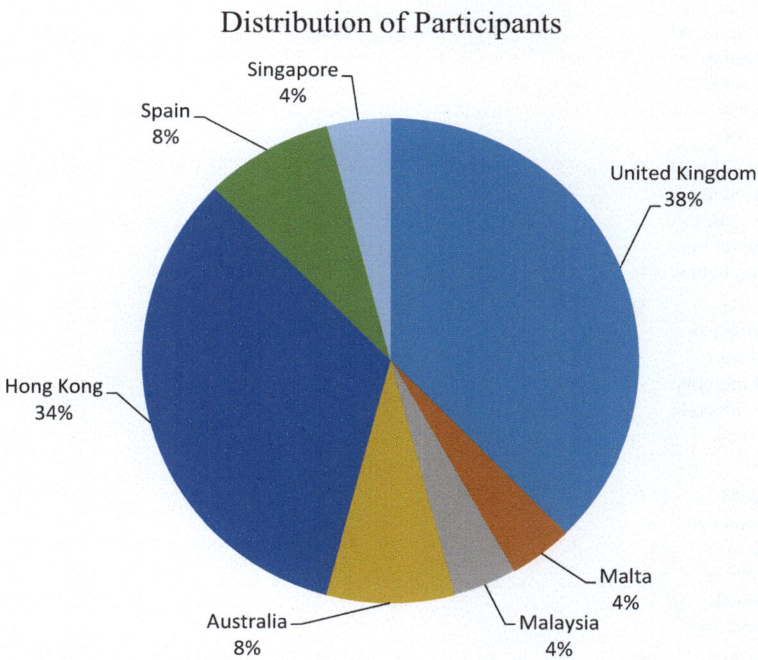


**Fig. 5** Geographical distribution of the respondents

### 4.1.2 Survey Results

In Table 2, the mean scores of Q1 and Q2 are 3.38 and 3.46, respectively. One hundred per cent of the users agreed or strongly agreed that the AI application could enhance their learning, while 95.83% of the users agreed or strongly agreed that the application could enhance their performance. In Q3 and Q4, 16.67% of the users disagreed that the AI application could regulate their practising habits and 91.66% of the users agreed that the AI application could enhance their musical memory (e.g., of accidentals or musical patterns). In Q5 and Q6, 91.66% of the users agreed that

**Table 2** Results of the online survey

| Question | Strongly disagree (%) | Disagree (%) | Agree (%) | Strongly agree (%) | Mean score |
|---|---|---|---|---|---|
| Q1. Using the AI application to practise scales can enhance my learning | 0 | 0 | 62.5 | 37.5 | 3.38 |
| Q2. Using the AI application to practise scales can enhance my performance | 0 | 4.17 | 45.83 | 50 | 3.46 |
| Q3. Using the AI application to practise scales can regulate my practising habits | 0 | 16.67 | 62.5 | 20.83 | 3.04 |
| Q4. The AI application can enhance my musical memory (e.g., accidentals and musical patterns) | 0 | 8.34 | 70.83 | 20.83 | 3.13 |
| Q5. The AI application can enhance my confidence in playing scales and arpeggios | 0 | 8.34 | 45.83 | 45.83 | 3.38 |
| Q6. The AI application can improve my motor skills (e.g., fingerings and muscle memory) | 0 | 4.17 | 62.5 | 33.33 | 3.29 |

the AI application could enhance their confidence in playing scales and arpeggios, and 95.83% of the users agreed that the AI application could improve their motor skills (e.g., fingerings and muscle memory).

## 4.2 *File Analysis (Overall Performance)*

In Fig. 6, the overall performance of the frequent valid users playing five instruments is shown by grade. After 8 weeks of practice with the AI application, a substantial improvement was recorded at the beginner level (Grade 1) with a mean score difference of 13.2. From Grades 2 to Grade 5, good progress was shown, with a mean score difference ranging from 2.5 to 5.56. This trend suggests that the AI application is an effective tool for learning scales and arpeggios at the beginner level (Grade 1) in particular and is also useful in Grades 2 through 5.

In Fig. 7, the change in the mean scores for practice records after 8 weeks of practice is shown. During the 8 weeks of practice, each user was advised to practise each scale or arpeggio at least three times per week. The ABRSM syllabus lists

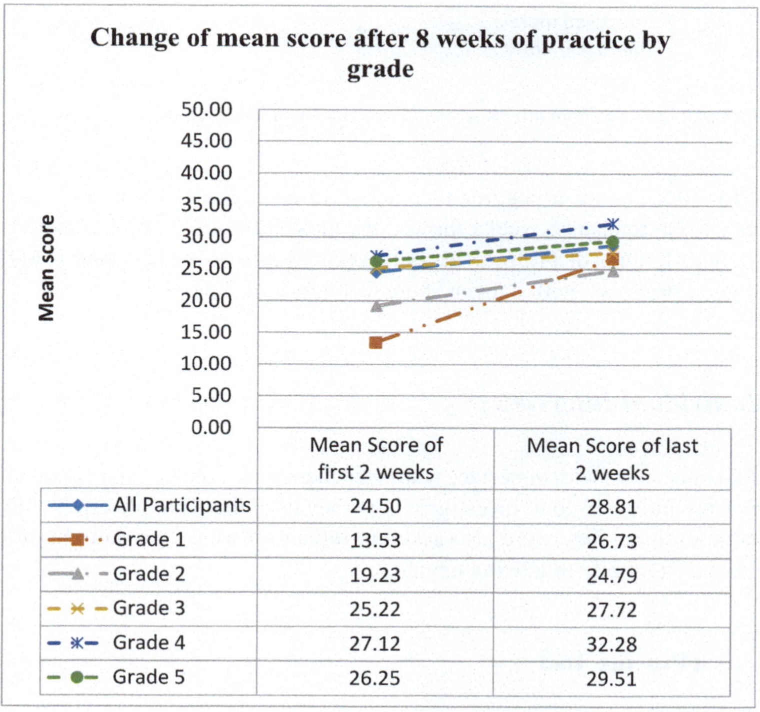


| | Mean Score of first 2 weeks | Mean Score of last 2 weeks |
|---|---|---|
| All Participants | 24.50 | 28.81 |
| Grade 1 | 13.53 | 26.73 |
| Grade 2 | 19.23 | 24.79 |
| Grade 3 | 25.22 | 27.72 |
| Grade 4 | 27.12 | 32.28 |
| Grade 5 | 26.25 | 29.51 |

**Fig. 6** Change in mean scores after 8 weeks of practice by grade

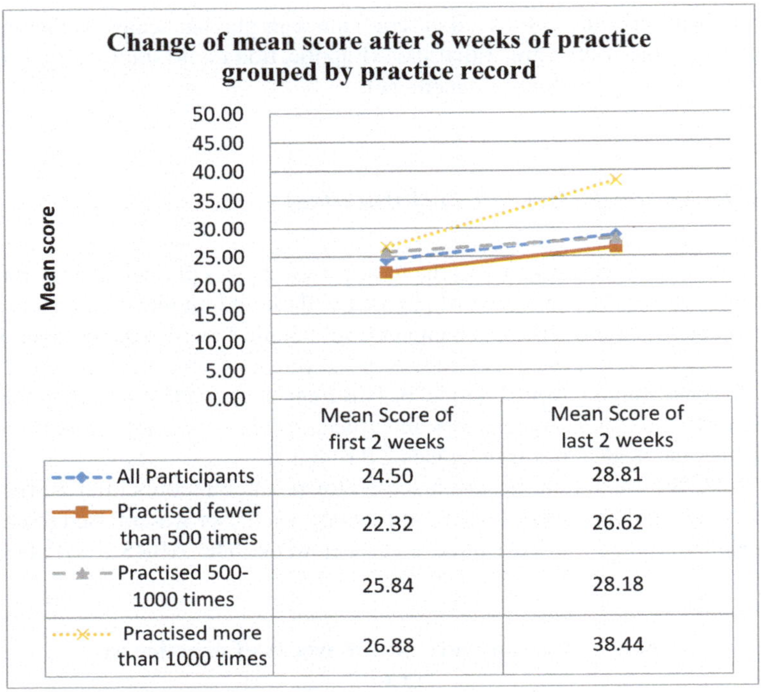


| | Mean Score of first 2 weeks | Mean Score of last 2 weeks |
|---|---|---|
| All Participants | 24.50 | 28.81 |
| Practised fewer than 500 times | 22.32 | 26.62 |
| Practised 500-1000 times | 25.84 | 28.18 |
| Practised more than 1000 times | 26.88 | 38.44 |

**Fig. 7** Changes in mean score after 8 weeks of practice by practice record

around 15–20 scales or arpeggios to practise in each grade. Users who practised more than 1,000 times in 8 weeks showed the most substantial improvement, with a mean score difference of 11.56. Users with a practice record of less than 1,000 times showed good progress, with a mean score difference of 2.34–4.3.

## *4.3 Individual Interviews*

Individual interviews with in-service studio teachers were conducted. Five instrument teachers were interviewed to investigate how they used the AI application with daily instrument lessons. They were also asked to reflect on what they thought of the AI application as a mobile practising device.

### 4.3.1 As a Practice Tool

Most of the teachers agreed that the AI application could enhance students' learning. They reported that it functioned as a 'home teacher,' helping students remember what they learned and what kinds of mistakes they made during lessons. The teachers could use the AI application to monitor their students because it records every instance of scale practice.

> It will be easier for teachers to monitor students' practice record and how they perform. (Teacher C, interview, 1 April 2017)

### 4.3.2 Visualising Fingering Patterns

A chart that shows the correct fingering can help students improve their visual memory of fingering. Most of the teachers mentioned that students could easily forget the fingering after the lesson, and so would not be able to practise the scale at home.

> While there is no teacher there, they can still remember the fingering. (Teacher B, interview, 1 April 2017)

> When the students use the AI application, they pay more attention to the screen and the fingering. (Teacher C, interview, 1 April 2017)

### 4.3.3 Demo Playback and AI Feedback

All of the teachers agreed that the demo and playback function was the most effective function in this application. This function can develop students' ear training.

> After hearing the demo playing, they imitate, practise and perform better. (Teacher A, interview, 1 April 2017)

> Most students do not know what they have played wrong; this function can help them clarify their problem, showing where they have played wrongly so that they can correct it immediately. Also, they can also find out the mistakes by themselves without teachers to train up their self-learning skills and listening skills in order to stimulate their musical sense. (Teacher A, interview, 1 April 2017)

> AI feedback can increase learning efficiency and point out the mistake so that students can correct it immediately. (Teacher C, interview, 1 April 2017)

> Most of the students do not know they are practicing incorrectly, but this AI function can clearly show what mistakes they have made in order to help them correct them at home. (Teacher E, interview, 1 April 2017)

### 4.3.4 Intonation

Intonation is the hardest part of practising scales and arpeggios, as students have no idea what it is or how it can be improved.

> After receiving comments and advice from the AI application, students know where they perform well and where they are not in tune. This can enhance their learning effectiveness. Their intonations have improved since practising with this application. (Teacher C, interview, 1 April 2017)

## 4.4 Response to Research Questions

1. Is the AI application an effective tool for practising scales and arpeggios?

In Table 2, the mean scores of Q1 and Q2 are 3.38 and 3.46, respectively. One hundred per cent of users agreed or strongly agreed that the application could enhance their learning, and 95.83% of users agreed or strongly agreed that the application could enhance their performance. However, in Q3, 16.67% of users disagreed that the AI application could regulate their practising habits. In an interview, Teacher C stated that *'It will be easier for teachers to monitor students' practice record and how they perform.'* This means that even though the teacher may monitor students' practice, the students do not necessarily become self-regulated learners.

From student perspective, 91.66% of users agreed that the AI application could enhance their musical memory (e.g., of accidentals and musical patterns) and 95.83% of users agreed that the AI application could improve their motor skills (e.g., fingerings and muscle memory). Teacher C claimed that *'AI feedback can increase learning efficiency and point out a mistake so that students can correct it immediately.'* The AI application can enhance students' musical memory and motor skills by pointing out intonation mistakes immediately during their daily practice, which enhances the effectiveness of each practice session. However, the open-ended responses on the survey commented that there are delays in the audio signal which could frustrate the learner and discourage further practice with the AI application.

2. To what extent does the AI application enhance mobile practising?

This study has two major findings. The first is that a substantial result is recorded at the beginner level (Grade 1), with a mean score difference of 13.2 for all five instruments. The second is that users who practised more than 1,000 times in 8 weeks had the most substantial result, with a mean score difference of 11.56.

3. How can mobile practising be implemented in studio teaching?

This AI application serves as a post-lesson teacher. Teacher E claimed that *'most of the students do not know they practice incorrectly, but this AI function can clearly show where mistakes they have made in order to help them correct them at home.'* Therefore, blended learning can be adopted in studio teaching in the future. AI makes it possible to incorporate mobile practising as a complement to private studio teaching.

## 5 Discussions and Conclusions

### *5.1 Engage Learners in the New AI Learning Environment*

The use of AI has changed our daily lives, including music education. Students can engage and listen to demonstration recordings and see the fingering on the screens of their mobile devices. AI can provide comments and immediate feedback to improve students' practice at home or elsewhere. The pre-exam mode can provide marks and comments to help students prepare for their ABRSM instrument examinations, which can save time for instructors. Therefore, instrument instructors can spend more time teaching musical styles and instrumental techniques.

### *5.2 The Changing Role of Learners and Educators in Mobile Practising*

The traditional way of practising scales and arpeggios mainly relies on the instrument teacher to demonstrate the fingering and articulation during private lessons or instrument classes. Students may forget the instructions and then practise in their own way at home. Sometimes students practise something incorrectly or contrary to the instruction of the teachers. The teachers, however, must wait for the next lesson to check whether students have practised well. There is no doubt that scales and arpeggios are the foundation of classical music. Effective practising is therefore crucial. Mobile learning and AI can improve the mode of practice. In this pilot study, the feedback of the users and instrument teachers on the use of recorded data was positive.

### *5.3 The Future of AI as Blended Learning in Music Education*

The use of AI in blended learning can prevent errors in practising. This can save teaching time and allow the teacher to monitor students' progress in practising. However, teachers have commented that when users have too many records in the application, delays are caused by the process of rendering the audio signals from the application's server. If this technical issue causes users to lose patience, they may choose to practise in the traditional way. Therefore, the usefulness of blended learning relies heavily on the technical capabilities of the application.

## 5.4 Limitations

As this is a pilot study with 24 valid users (n = 24) on five instruments, the present findings cannot be generalised to all users. Furthermore, this AI application is available only on iPhones and iPads due to the technical requirements of the audio engine. This restriction further limited the recruitment of volunteer users.

**Acknowledgements** Thank you to Mr. Eric Yung for his tremendous support in this research and to my research assistant, Tsang Long Sing, for help during the process of data collection.

**Funding Acknowledgement** This research received no specific grant from any funding agency in the public, commercial, or not-for-profit sectors.

## References

Benetos, E., Dixon, S., Giannoulis, D., Kirchhoff, H., & Klapuri, A. (2013). Automatic music transcription: Challenges and future directions. *Journal of Intelligent Information Systems, 41*(3), 1–28.

Conati, C., Porayska-Pomsta, K., & Marvrikis, M. (2018, June). *AI in education needs interpretable machine learning: Lessons from open learner modelling*. Paper presented at the 2018 ICML Workshop on Human Interpretability in Machine Learning, Stockholm, Sweden. https://arxiv.org/pdf/1807.00154.pdf.

Chen, J. (2015). Mobile learning: Using application Auralbook to learn aural skills. *International Journal of Music Education, 33,* 244–259.

Franke, J. (2013). Violin scales: Strategies for improving intonation. *American String Teacher*, May, *63*(2), 74–75.

Johnson, L., Levine, A., Smith, R., & Stone, S. (2010). *2010 Horizon report*. Austin, TX: New Media Consortium.

Johnson, M. (2014). *In touch with the wild: Exploring real-time feedback for learning to play the violin*. PhD thesis. London, United Kingdom: University College London.

Melhuish, K., & Falloon, G. (2010). Looking to the future: M-learning with the iPad. *Computers in New Zealand Schools: Learning, Leading, Technology, 22*(3), 1–16.

Rossing, J., Miller, W., Cecil, A., & Stamper, S. (2012). iLearning: The future of higher education? Student perceptions on learning with mobile tablets. *Journal of the Scholarship of Teaching and Learning, 12*(2), 1–26.

Warwick, K. (2012). Classical AI. In *Artificial intelligence: The basics*. New York: Routledge, 31.

**Dr. Jason Chi Wai Chen** is an Assistant Professor at the Education University of Hong Kong. He has received a General Research Fund (GRF) to study computer-mediated composition pedagogy. His articles have been published by the top-ranked journals—*International Journal of Music Education, Research Studies in Music Education*, and *Music Education Research*.

# Index




W. W. K. Ma et al. (eds.), *Learning Environment and Design*, Educational Communications and Technology Yearbook,
https://doi.org/10.1007/978-981-15-8167-0

The manufacturer's authorised representative in the EU is Springer Nature Customer Service Centre GmbH, Europaplatz 3, 69115 Heidelberg, Germany. If you have any concerns regarding our products, please contact ProductSafety@springernature.com

Printed and bound by CPI Group (UK) Ltd, Croydon, CR0 4YY
30/09/2026
02225506-0006